AF771883

ACTA PHYSICA AUSTRIACA / SUPPLEMENTUM I

WEAK INTERACTIONS AND HIGHER SYMMETRIES

PROCEEDINGS OF THE
III. INTERNATIONALE HOCHSCHULWOCHEN
FÜR KERNPHYSIK 1964 DER KARL-FRANZENS-UNIVERSITÄT
GRAZ, AT SCHLADMING (STEIERMARK, AUSTRIA)
24th FEBRUARY—7th MARCH 1964

SPONSORED BY
BUNDESMINISTERIUM FÜR UNTERRICHT
STEIERMÄRKISCHE LANDESREGIERUNG AND
THE INTERNATIONAL ATOMIC ENERGY AGENCY

EDITED BY
PAUL URBAN
GRAZ

WITH 94 FIGURES

1964

SPRINGER-VERLAG · WIEN · NEW YORK

Organizing Committee:

Chairman: Prof. Dr. PAUL URBAN
Vorstand des Institutes für Theoretische Physik,
Universität Graz

Committee Members: Dr. P. KOCEVAR, Graz
Dr. H. LATAL, Graz

Secretary: F. KOLLER

ISBN 978-3-211-80746-0 ISBN 978-3-7091-5563-9 (eBook)
DOI 10.1007/978-3-7091-5563-9

Alle Rechte, insbesondere das der Übersetzung
in fremde Sprachen, vorbehalten

Ohne schriftliche Genehmigung des Verlages
ist es auch nicht gestattet, dieses Buch oder Teile daraus
auf photomechanischem Wege (Photokopie, Mikrokopie)
oder sonstwie zu vervielfältigen

Preface

Weak interactions and higher symmetries are nowadays of special importance for elementary particles theory. Lately both theoretical and experimental physicists became more and more interested in the subject. Because of the complicated subject and the scarce available literature proper introductions in the subject are tiresome. The mathematical background such as higher Algebra and Grouptheory etc. cannot be applied immediately in all cases.

The third Schladming University Courses on the above subject belong to this category. The present first supplement volume of "Acta Physica Austriaca" contains all lectures (with literature references) giving not only a review of the fundamentals but also discussing recent research work. I sincerely hope that the publication will find interested readers.

In the last years it became customary in physics and mathematics to arrange summer schools on subjects which allow the experienced as well as the young scientist to get quickly acquainted with special modern fields. Discussions and private contacts make it possible for participants to touch quickly the peripheries of the subject in question, and new research work will be stimulated which otherwise could not be expected.

Graz, October, 1964. **P. Urban**

Contents

Ladies and Gentlemen!

It is a great honour for me to welcome you at the beginning of our conference. The large number of government representatives, and authorities shows that the object and intentions of academic research have become the deep concern of society at large. So I welcome respectfully the representative of the Minister of Education thanking him, at the same time, for the subvention without which these Hochschulwochen could not have taken place. Unfortunately Prof. BUDINI who is the representative of the director-general of the International Atomic Energy Agency, Vienna, and director of the International School of Theoretical Physics at Trieste cabled yesterday that he became sick and cannot attend our conference. We have to thank the organisation he represents for cooperation and financial support. I hope Prof. BUDINI's condition is not serious and that he will soon be restored to good health.

I also welcome the representatives of the Landesregierung of Steiermark, especially the deputies of the Landeshauptmann, Prof. KOREN and Dr. BLANC, who also administered to our enterprise. To our thanks we add the promise to give the conference Styrian atmosphere and to point out the beauties of our country to all of our colleagues from far and near.

I am deeply indebted, in particular, to the authorities of Schladming, our hosts. To express my gratitude for their cooperation let me name two men, Director LAURICH, Mayor of Schladming, and Mr. KASERER, Chairman of the Fremdenverkehrsverein. I am thankful to them, because they have complied with our wants and wishes in a most agreeable way. Moreover I have to mention the director of the Handelskammer, Dr. TÖPFNER, who supported the organisation considerably by giving financial aid to lecturers and students. In addition I thank the representatives of press and radio for their coming. Scientific progress leads, as a matter of fact, to a general progress of all mankind and must therefore never be of scientific interest only. Last not least, I welcome all our lecturers and students. We are glad to have quite a few distinguished physicists from abroad and from several European universities with us. I think we were very fortunate that we could find competent theorists and experimentalists who are willing to give us courses and tell us about recent developments in their special fields. All of us look foreward to the discussions from which we hope to learn and gain deeper insight in our problems. Let me say a few words in general about this third conference which we now begin. The main topic will be properties and behaviour of weak interactions. It was not until recently that all we knew about weak interactions was what we could learn at lower energies. This has now changed. During the last two years two important neutrino experiments were carried out, one at Brookhaven and one at CERN and both these experiments confirmed beyond doubt that there are now two different neutrinos. The CERN experiment has since been continued and this week we will have the opportunity to hear the last results and what the situation from an experimental point of view now is.

The theorists, of course, haven't been idle either and there are accomplishments to talk about. Higher symmetries and the theory of conserved vector current proved to be two particularly fruitful approaches. During the conference competent theorists will inform us of what they have been able to do and which problems they consider as solved and where open questions remain.

All this is only part of this conference, the high-brow part, so to speak. However, our participants aren't just experts but also all our graduate students. We are interested in the most recent discoveries in experiment and theory but we are not less concerned about education without which all progress would be useless. A moment ago I mentioned higher symmetries. To deal with problems of this kind leads into group theory of which we all know that it plays and has ever played an important role in more fundamental investigations in physics. Group theory is not taught in our routine courses during the two academic semesters we have each year. We are therefore very grateful to find an opportunity to fill this gap. We have courses on Lie Algebras and applications to special groups, the SU(3) in particular, in our program and I expect that our students will welcome this second part of our school. Group theory then leads us back to questions of which symmetries play a role in the weak interactions.

What would we like to accomplish? We have two weeks of work and, as I hope, pleasure ahead of us. We will exchange ideas, discuss problems and if we are lucky, we might even solve a few of them. I myself would be very happy if we succeeded to clarify the role of higher symmetries in weak interactions. Perhaps the new results of the CERN experiment will help us and provide unambiguous clues to which of the many suggested theories is the most promising.

So let me wish you all the best. Let us do our best, which is all we can and the only way to success and progress.

We won't be in business with our conference until this evening, for an obvious reason: I am not the only speaker this morning. However, experienced conference participants undoubtedly know how to survive this unavoidable part of every meeting.

P. Urban

Lectures on Lie' Algebras for Physicists

By

H. Joos, DESY, Hamburg

With 2 Figures

I. Introduction

The organizers of this Schladming-Meeting on "SU_3-Symmetry and Weak Interactions" asked me to give some lectures on the mathematics of Lie groups, just similar to those given by Salam at the "Seminar on Theoretical Physics" in Triest 1962. As A. Salam gave an ideal collection of those theorems, which a theoretician working on symmetry problems should know, I had a hard time in thinking how I could present the same stuff in a not identical manner.

These are some of my points I want to emphasize:

(a) I shall start with a discussion of Lie algebras, and I shall mention the Lie groups only occasionally. This is the point of view of a physicist who is mainly interested in systems of commutation relations of observables. I want to mention Gell-Mann's approach to the symmetry group of weak interactions. In our discussion, therefore, formal arguments will be dominant, which correspond to the taste of the majority of physicists.

(b) The main problem of such lectures is always to find the right balance between information on the useful results of the mathematical theory on the one hand and of giving an idea of the proofs on the other. Using freely a physicists knowledge of linear algebra I shall try as much as possible to give an idea of the proofs.

(c) Particularly in the first lecture I shall restrict myself not to Lie algebras which are used in the discussion of symmetry problems. I have two reasons for that: (1) The other types of algebras are important in other branches of theoretical physics. (2) I want to give a mathematical reason, why one has to restrict oneself in symmetry problems on compact real semisimple Lie algebras.

II. Basic notations

1. Definition of a Lie algebra

We call a vector space over the real (complex) numbers a real (complex) Lie algebra, if for the elements of $\mathfrak{L}$ a product with the properties of a commutator is defined:

$$a, b, c \in \mathfrak{L};\ \alpha, \beta \in \mathfrak{R}\,(\mathfrak{C}),\ [a, b] \in \mathfrak{L}$$

$$[\alpha a + \beta b, c] = \alpha\,[a, c] + \beta\,[b, c] \tag{1}$$

$$[a, b] = -[b, a]$$

$$[a, [b, c]] + [b, [c, a]] + [c, [a, b]] = 0. \qquad \text{(Jacoby identity)}$$

We may introduce a base $\{x_i\}$ in $\mathfrak{L}$

$$a = \sum_{i=1}^{n} \alpha^i\, x_i \equiv \alpha^i\, x_i.$$

$n = \dim \mathfrak{L}$ = dimension of the vectorspace of $\mathfrak{L}$.
Then it follows immediately from (1) that

$$[a, b] = \alpha^i\, \beta^k\, \Gamma^l_{ik}\, x_l$$

with

$$[x_i, x_k] = \Gamma^l_{ik}\, x_l \qquad \Gamma^l_{ik} = -\Gamma^l_{ki}. \tag{2}$$

Therefore the product is defined by the product of the base elements. The $\Gamma^l{}_{ik}$ are called structure constants. In a transformed base

$$\bar{x}_i = T_i^{\,k}\, x_k$$

the structure constants are

$$\bar{\Gamma}^l_{ik} = T_i^{i'}\, T_k^{k'}\, \Gamma^{l'}_{i'k'}\, T^{-1}_{l'} \tag{3}$$

i. e. $i, k, l, \ldots$ are covariant and contravariant labels.

2. Examples

From elementary quantum mechanics the following examples are familiar:

(A) The C. R. of the 3 components of linear momentum:

$$[P_1, P_2] = [P_2, P_3] = [P_3, P_1] = 0.$$

A Lie algebra is called *abelian* if all commutators are zero.

(B) The canonical C. R. for one degree of freedom:

$$[P, X] = Q,\ [Q, P] = 0,\ [Q, X] = 0.$$

(Q is for irreducible representations a multiple of the unit operator: $Q = \hbar \cdot 1$.)

(C) The C. R. of the components of linear momenta in a plane P_1, P_2 and the angular momentum perpenticular to this plane M:

$$[P_1, P_2] = 0,\ [M, P_1] = P_2,\ [M, P_2] = -P_1.$$

(D) The C. R. of angular momenta:

$$[M_1, M_2] = M_3,\ [M_2, M_3] = M_1,\ [M_3, M_1] = M_2.$$

(E) The plane center of mass coordinates and the angular momenta in relativistic quantum mechanics (the infinitesimal 3-dim. Lorentz transformations)

$$[M, N_1] = +N_2,\ [M, N_2] = -N_1,\ [N_1, N_2] = -M.$$

(F) The creation operator a^+ with a particle number operator n define a two dimensional Lie algebra

$$[n, a^+] = a^+$$

from which we make a 3-dimensional Lie algebra by adding a "c-number" Q:

$$[Q, n] = [Q, a^+] = 0.$$

All this examples are 3-dimensional (and this are essential all types of 3-dimensional Lie algebra's [3]).

There are the higher dimensional Lie algebra's defined by C. R. of energy, momentum, angular momentum and center of mass coordinates in relativistic as well as non-relativistic quantum mechanics.

3. Sub-algebras

A linear sub-space $\mathfrak{M} \subset \mathfrak{L}$ is called a *sub-algebra* if: $a, b \in \mathfrak{M} \Rightarrow [a, b] \in \mathfrak{M}$. It is an *invariant sub-algebra* or *ideal* if: $a \in \mathfrak{M}, b \in \mathfrak{L} \Rightarrow [a,b] \in \mathfrak{M}$. The *center* $\mathfrak{Z}$ of $\mathfrak{L}$ is the set of elements z: $[z, a] = 0$ for all elements a of $\mathfrak{L}$. $\mathfrak{Z}$ is an abelian invariant sub-algebra. A Lie group is *simple*, if it is non-abelian and has no invariant sub-algebra $\neq \{0\}$, $\mathfrak{L}$; it is *semi-simple* if it has no abelian invariant sub-space $\neq \{0\}$. In our examples are:

$\{\alpha Q\}$ the center of the algebras (B) and (F),

$\{\alpha P_1 + \beta P_2\}$ is an abelian ideal of (C),

$\{\alpha n + \beta a^+\}$ is a non-abelian invariant subgroup of (F).

The algebras (D) and (E) are simple and semi-simple.

We may consider in any Lie algebra the sub-space spanned by the set of commutators $\overline{\{[a, b]\}} \equiv [\mathfrak{L}, \mathfrak{L}] = \mathfrak{L}'$; $a \in \mathfrak{L}, b \in \mathfrak{L}$. $\mathfrak{L}'$ is an ideal of $\mathfrak{L}$, the commutator sub-algebra $\mathfrak{L}'$. $\mathfrak{L}' = 0$ means $\mathfrak{L}$ is abelian; for a non-abelian simple algebra is $\mathfrak{L}' = \mathfrak{L}$. We may construct the commutator sub-algebra of the commutator sub-algebra $\mathfrak{L}'' = [\mathfrak{L}', \mathfrak{L}']$ and so on $\mathfrak{L}^{(k)} = [\mathfrak{L}^{(k-1)}, \mathfrak{L}^{(k-1)}]$ and we get the *derived series*: $\mathfrak{L} \supseteq \mathfrak{L}' \supseteq \mathfrak{L}'' \supseteq \ldots$ $\mathfrak{L}$ is called *solvable* if $\mathfrak{L}^{(k)} = 0$, for a finite k. Abelian algebras are solvable, simple algebras are not. Our examples (B), (C) and (F) are solvable.

The sets of iterated commutators $\mathfrak{L}^n = \overbrace{[\mathfrak{L}, [\mathfrak{L} \ldots}^{n-1} [\mathfrak{L}, \mathfrak{L}]]$ (e. g. $\mathfrak{L}^2 = \mathfrak{L}'$) form the *lower central series*: $\mathfrak{L} \supset \mathfrak{L}^2 \supset \mathfrak{L}^3 \supset \ldots$ If $\mathfrak{L}^n = 0$ for some n, $\mathfrak{L}$ is called *nil-potent*. $\mathfrak{L}^k$ are ideals in $\mathfrak{L}$. As $\mathfrak{L}^n \supset \mathfrak{L}^{(n)}$ nilpotent algebras are solvable. Our example (B) is nil-potent, (C) not.

The qualifications simple, semi-simple resp. nilpotent and solvable reflect the most important characteristics of Lie algebras. Nilpotent and solvable algebras are in some sense generalizations of abelian algebras.

From two Lie algebras $\mathfrak{L}'$ and $\mathfrak{L}''$ we define the direct sum, $\mathfrak{L}' \oplus \mathfrak{L}''$ as the set of elements $a' + a'' = a$, where $a' \in \mathfrak{L}'$ and $a'' \in \mathfrak{L}''$. The commutators in $\mathfrak{L}' \oplus \mathfrak{L}''$ are defined by $[a' + a'', b' + b''] = [a', b'] + [a'', b'']$. It is easily verified that $\mathfrak{L} = \mathfrak{L}' \oplus \mathfrak{L}''$ is a Lie algebra. $\mathfrak{L}' + 0''$ and $0' + \mathfrak{L}''$ are ideals in $\mathfrak{L}$.

There is another concept similar to the direct sum, the semidirect sum. $\mathfrak{L}$ is called a semi-direct sum of $\mathfrak{L}'$ and $\mathfrak{L}''$: $\mathfrak{L} = \mathfrak{L}' \oplus \mathfrak{L}''$ if every $a \in \mathfrak{L}$ can be written as $a = a' + a''$ where $a' \in \mathfrak{L}', a'' \in \mathfrak{L}''$, and $\mathfrak{L}''$ is an ideal in $\mathfrak{L}$, but $\mathfrak{L}'$ is only a sub-algebra, $\mathfrak{L}' \cap \mathfrak{L}'' = \{0\}$.

In our list of examples (F) is the direct sum of the ideals $\{\alpha Q\} \oplus \{\beta n + \gamma a^+\}$. The algebra (D) is $\{\alpha M\} \oplus \{\beta P_1 + \gamma P_2\}$. Other examples are the infinitesimal inhomogeneous Lorentz group: $\{\alpha^{\mu\nu} M_{\mu\nu}\} \oplus \{\beta^\nu P_\nu\}$, the infinitesimal Euklidean group $\{\alpha^r M_r\} \oplus \{\beta^s P_s\}$ etc.

4. Homomorphism, isomorphism and representations

$\mathfrak{L}', \mathfrak{L}''$ are real (complex) Lie algebras, $T\, \mathfrak{L}' \subset \mathfrak{L}''$ a mapping of $\mathfrak{L}'$ in $\mathfrak{L}''$ which satisfies:

$$T(\alpha a' + \beta b') = \alpha T a' + \beta T b'; \quad T[a', b'] = [Ta', Tb'], \tag{4}$$

then T is called a homomorphism of $\mathfrak{L}'$ in $\mathfrak{L}''$. It is easily verified that the elements in $\mathfrak{L}'$ for which $Ta' = 0$, form an ideal in $\mathfrak{L}'$.

If T is a $1-1$ mapping, it is an isomorphism. We consider two Lie algebras which are isomorphic as the same abstract Lie algebras.

The examples (D) and (E) show, that two Lie algebras which are given by the C. R. of a base may be isomorphic as complex Lie algebras, but not as real Lie algebras.

A homomorphic mapping of a Lie algebra on a set of linear transformations of a vectorspace $\mathfrak{M}$: $a \to u\ u(a)$ is called a representation. We assume that the notations "reducible, irreducible, decomposible, faithful, etc." are well known.

III. Adjoint representation

1. Definition

The Lie algebra $\mathfrak{L}$ as a vectorspace is the representation space of the adjoint representation $v(a)$

$$[a, x_i] = x_k\, v_i^{\ k}(a) \quad \text{or general} \quad [a, b] \equiv b v(a)$$

i. e.

$$\begin{gathered} b = \beta^i x_i,\ b' = [a, b] = \beta'^{\,i} x_i,\ \beta'^{\,i} = v_k^{\ i}(a)\, \beta^k \\ [c, b\, v(a)] = [c, b]\, v(a) = b v(c)\, v(a). \end{gathered} \tag{1}$$

The Jacobi identity gives

$$\begin{aligned} \big[[a, b], x_i\big] &= \big[a, [b, x_i]\big] - \big[b, [a, x_i]\big] = \\ &= x_k \big(v(a)\, v(b) - v(b)\, v(a)\big)_i^{\ k}. \end{aligned} \tag{2}$$

As we have further

$$[\alpha a + \beta b, x_k] = x_i \big(\alpha v(a) + \beta v(b)\big)^i{}_k$$

the mapping $a \to v(a)$ is a homomorphism.

From the difinition of the structure constants we see $v(x_l) = \Gamma^k{}_{li}$. They must satisfy

$$\Gamma^k{}_{li}\,\Gamma^i{}_{l'n} - \Gamma^k{}_{l'i}\,\Gamma^i{}_{ln} = \Gamma^m{}_{ll'}\,\Gamma^k{}_{mn}$$

or

$$\Gamma^k{}_{lm}\,\Gamma^i{}_{kn} + \Gamma^k{}_{mn}\,\Gamma^i{}_{kl} + \Gamma^k{}_{nl}\,\Gamma^i{}_{km} = 0 \qquad (3)$$

corresponding to (2). Structure constants which satisfy (3) and $\Gamma^l{}_{ik} = -\Gamma^l{}_{ki}$ define a Lie algebra.

An element $a \in \mathfrak{L}$ is mapped on $v(a) = 0$ if and only if $a \in \mathfrak{Z}$, the center of $\mathfrak{L}$. The adjoint representation $ad(\mathfrak{L})$ is a faithful representation if $\mathfrak{Z} = \{0\}$.

2. *Examples*

(B): $x_1 = Q,\ x_2 = P,\ x_3 = X$

$$v(Q) = 0,\ v(P) = \begin{pmatrix} 0 & 0 & 1 \\ 0 & 0 & 0 \\ 0 & 0 & 0 \end{pmatrix}, \quad v(X) = \begin{pmatrix} 0 & -1 & 0 \\ 0 & 0 & 0 \\ 0 & 0 & 0 \end{pmatrix}$$

(C): $x_1 = P_1,\ x_2 = P_2,\ x_3 = M$

$$v(P_1) = \begin{pmatrix} 0 & 0 & 0 \\ 0 & 0 & -1 \\ 0 & 0 & 0 \end{pmatrix},\ v(P_2) = \begin{pmatrix} 0 & 0 & 1 \\ 0 & 0 & 0 \\ 0 & 0 & 0 \end{pmatrix},\ v(M) = \begin{pmatrix} 0 & -1 & 0 \\ 1 & 0 & 0 \\ 0 & 0 & 0 \end{pmatrix}$$

(D): $x_1 = M_1,\ x_2 = M_2,\ x_3 = M_3$

$$v(M_1) = \begin{pmatrix} 0 & 0 & 0 \\ 0 & 0 & -1 \\ 0 & 1 & 0 \end{pmatrix},\ v(M_2) = \begin{pmatrix} 0 & 0 & 1 \\ 0 & 0 & 0 \\ -1 & 0 & 0 \end{pmatrix},\ v(M_3) = \begin{pmatrix} 0 & -1 & 0 \\ 1 & 0 & 0 \\ 0 & 0 & 0 \end{pmatrix}$$

(E): $x_1 = N_1,\ x_2 = N_2,\ x_3 = M$

$$v(N_1) = \begin{pmatrix} 0 & 0 & 0 \\ 0 & 0 & -1 \\ 0 & -1 & 0 \end{pmatrix},\ v(N_2) = \begin{pmatrix} 0 & 0 & +1 \\ 0 & 0 & 0 \\ +1 & 0 & 0 \end{pmatrix},\ v(M) = \begin{pmatrix} 0 & -1 & 0 \\ 1 & 0 & 0 \\ 0 & 0 & 0 \end{pmatrix}$$

(F): $x_1 = Q,\ x_2 = a^+,\ x_3 = n$

$$v(Q) = 0,\ v(a^+) = \begin{pmatrix} 0 & 0 & 0 \\ 0 & 0 & -1 \\ 0 & 0 & 0 \end{pmatrix}, \quad v(n) = \begin{pmatrix} 0 & 0 & 0 \\ 0 & 1 & 0 \\ 0 & 0 & 0 \end{pmatrix}.$$

3. *A product formula*

As the adjoint representation is so intrinsically connected with a Lie algebra, it is a main tool to analyze the structure of a Lie algebra, particularly one without center. The following trivial formula shows another aspect of the connection between Lie algebra and adjoint representation:

$$[a, [b, [\ldots [c, x_k] \ldots] = x_i\,(v(a)\,v(b) \ldots v(c))_k{}^i. \qquad (4)$$

In a nilpotent algebra the products of k matrices of the adjoint representation vanish.

4. Exponentials

Now we make some remarks which give a clue to an algebraic approach to the group generated by the Lie algebra. Considering the exponentials of the elements of a Lie algebra, as a formal power series: $\exp a = \sum_{n=0}^{\infty} \frac{a^n}{n!}$, the theorem by Campbell and Hausdorff states:

$$(\exp a)(\exp b) = \exp z$$

where z has a formal development in a series of iterated commutators:

$$z = a + b + \tfrac{1}{2}[a, b] + \tfrac{1}{12}\left[[a, b]\, b\right] + \tfrac{1}{12}\left[[a, b]\, a\right] + \dots \tag{5}$$

Provided the convergence of this development to an element of the Lie algebra, this theorem shows, that the exponentials of a Lie algebra contain the germ of a group structure.

$$e^0 = 1,\ (e^a)^{-1} = e^{-a},\ e^a e^b = e^c,\ a, b, c \in \mathfrak{L}. \tag{6}$$

We do not want to discuss the convergence problem and the topological question connected with it: we only add the remark that from Lie's formula:

$$e^a b e^{-a} = b + [a, b] + \tfrac{1}{2}\left[a, [a, b]\right] + \dots$$

and the product formula (4) follows:

$$e^a b e^{-a} = b e^{v(a)}. \tag{7}$$

IV. On the general structure of Lie algebras

The question how we can classify all possible Lie algebras appears quite natural. We shall elaborate this problem with different emphasis, guided by the view points of a physicist who is interested in symmetry problems. One of the main tools for the discussion of this problem is:

1. The invariant Killing form

We define with help of the adjoint representation a bilinear form in the vectorspace of the Lie algebra:

$$\begin{gathered} a, b, c, \in \mathfrak{L},\ \alpha, \beta \in \mathfrak{R}(\mathfrak{L}) \\ (a, b) = \text{Trace}\ (v(a)\, v(b)). \end{gathered} \tag{1}$$

The properties

$$(a, b) = (b, a),\ (c, \alpha a + \beta b) = \alpha (c, a) + \beta (c, b)$$

follow immediately from the properties of the trace

$$\text{Trace}\ (v) = \sum_i v_{ii}.$$

With respect to a base we calculate from $a = \alpha^i x_i$, $b = \beta^i x_i$

$$(a, b) = \alpha^i \beta^k g_{ik} \tag{2}$$

with

$$g_{ik} = \text{Trace}\ (v(x_i)\, v(x_k)) = \Gamma^l{}_{im} \Gamma^m{}_{kl}.$$

This inner product is invariant

$$(a, [c, b]) + ([c, a], b) = 0. \tag{3}$$

This is shown by:

$$(a, [c, b]) + ([c, a], b) = \text{Trace}\,(v(a)\,[v(c), v(b)] + [v(c), v(a)]\,v(b)) =$$
$$= \text{Trace}\,([v(c), v(a)\,v(b)]) = 0.$$

We remark that we can define with help of any representation $a \to u_i^k(a)$ a bilinear inner product $(a, b) = \text{Trace}\,(u(a)\,u(b))$ which is invariant. The (g_{ik}) of our examples are:

$$\text{(A)}\quad (0) \qquad \text{(B)}\quad (0) \qquad \text{(C)}\quad \begin{pmatrix} 0 & 0 & 0 \\ 0 & 0 & 0 \\ 0 & 0 & -2 \end{pmatrix}$$
$$\text{(D)}\quad -\begin{pmatrix} 2 & 0 & 0 \\ 0 & 2 & 0 \\ 0 & 0 & 2 \end{pmatrix} \qquad \text{(E)}\quad \begin{pmatrix} 2 & 0 & 0 \\ 0 & 2 & 0 \\ 0 & 0 & -2 \end{pmatrix} \qquad \text{(F)}\quad \begin{pmatrix} 0 & 0 & 0 \\ 0 & 1 & 0 \\ 0 & 0 & 0 \end{pmatrix}. \tag{2'}$$

2. Compact Lie algebras

Physicists working in symmetry problems are interested in real Lie algebras $\mathfrak{L}$ which possess finite dimensional faithful representations by hermitian matrices:

$$\begin{gathered} a, b \in \mathfrak{L},\ a \to u(a),\ u^+(a) = u(a) \\ [u(a),\ u(b)] = i\,u([a, b]). \end{gathered} \tag{4}$$

[The definition (4) differs from (II,4) by a factor i on the r.h.s., as it is usual in quantum theory to consider hermitian instead of antihermitian matrices.]

By the definition

$$\langle a, b\rangle = \text{Trace}\,(u(a)\,u(b)) \tag{5}$$

we introduce in $\mathfrak{L}$ a positive definite invariant bilinear form. We state:

(a) A Lie algebra with a faithful finite dimensional representation is a direct sum of simple algebras and 1-dim. abelian Lie algebras. (Let $\mathfrak{J}$ be an ideal of $\mathfrak{L}$, then it follows from the invariance of $\langle a, b\rangle$, that the vectorspace orthogonal to $\mathfrak{J}$ is also an ideal $\mathfrak{J}^\perp$: $\mathfrak{L} = \mathfrak{J} \oplus \mathfrak{J}^\perp$

$$a \in \mathfrak{J}, b \in \mathfrak{J}^\perp \text{ then } [a, b] \in \mathfrak{J} \text{ and } [\alpha, b] \in \mathfrak{J}^\perp$$

therefore $[a, b] = 0$.)

(b) The Killing form of such a Lie algebra with center $\mathfrak{Z} = \{0\}$ is negative definite. (Consider:

$$\langle a, b\,v(c)\rangle = \langle a, [c, b]\rangle = -\langle [c, a], b\rangle = -\langle a\,v(c), b\rangle. \tag{6}$$

Therefore in a base $\{x_i\}$, in which the form is diagonal

$$\langle x_i, x_k\rangle = \delta_{ik}$$

the matrices $v^i{}_k(a)$ are antisymmetric, and therefore

$(a, a) = \text{Trace}\,(v\,(a)\,v\,(a)) = -\,\text{Trace}\,(v\,(a)\,v^T\,(a)) < 0$
for $v\,(a) \neq 0$.)

A Lie algebra with negative definite Killing form is called compact. (It can be shown that the groups infinitesimally generated by a compact $\mathfrak{L}$ are compact.) According to (2′) only the angular momentum operators of example (D) have finite dimensional hermitian representations (the trivial abelian case excluded).

(c) In a base in which the positive definite Killing form is diagonal, the structure constants as adjoint matrices of the base elements are antisymmetric [according to (6)].

$$\Gamma^l_{ik} = -\Gamma^l_{ki} = -\Gamma^k_{il}\,. \tag{7}$$

3. Theorems on the general structure of Lie algebras

Because of the statement (b) above we shall study only the structure of simple real compact Lie algebras in some detail. In order to give a review on the general structure of real and complex Lie algebras we mention the following theorems without proofs:

(a) Levi's Theorem: A Lie algebra is isomorphic to a semidirect sum of a solvable Lie algebra $\mathfrak{S}$ and a semisimple algebra $\mathfrak{B}$

$$\mathfrak{L} = \mathfrak{B} \oplus \mathfrak{S}.$$

(b) Lie's Theorem: The adjoint representation of a complex solvable Lie algebra can be taken in simultaneous triangular form [look at examples (B), (C) and (F)].

(c) A semisimple Lie algebra is a direct sum of simple Lie algebras. A Lie algebra is semisimple, if and only if the Killing form is nondegenerate. (Cartan's criterium.)

(d) A complex simple Lie algebra is the "complexification" of a simple real compact Lie algebra.

The discussion of the Lie algebras of the nonrelativistic or relativistic integrals of motion illustrate in essence the content of these theorems.

V. The real simple compact Lie algebras

In this section we treat the real Lie algebras with a negative definite Killing form.

1. "Complexification"

As shown in (IV, 6) the matrices of the adjoint representation are antisymmetric with respect to the Killing form. Therefore $\mathfrak{v}\,(a) = +i\,v\,(a)$ is hermitian which has the familiar consequences for the eigenvalue problem of the $\mathfrak{v}\,(a)$, i. e. $\mathfrak{v}\,(a)$ has real eigenvalues [the eigenvalues of the $v\,(a)$ are pure imaginary] and a complete system of eigenvectors. As the components of the eigenvectors will be in general complex, we enlarge the real Lie algebra $\mathfrak{L}$ to the complex Lie algebra $\bar{\mathfrak{L}}$:

$a = \alpha^i\,x_i,\ a^* = \alpha^{i*}\,x_i,\ \alpha^i \in \mathfrak{C},\ \{x_i\}$ a base of the real $\mathfrak{L}$,

$$[a, b]^* = [a^*, b^*].$$

We call an element real if $a^* = a$. The linear transformations $\mathfrak{v}(a)$ of real elements in the adjoint representation

$$[a, b] = -i\, b\, \mathfrak{v}(a); \quad [\mathfrak{v}(a), \mathfrak{v}(b)] = i\, \mathfrak{v}([a, b])$$

are hermitian with respect to the hermitian form

$$\langle a, b \rangle = -(a^*, b).$$

2. Regular element, rank, Cartan subgroup

If b is an eigenvector of $\mathfrak{v}(a)$ with the eigenvalue 0

$$b\, \mathfrak{v}(a) = 0 = i\, [a, b]$$

then b commutes with a. Therefore every $\mathfrak{v}(a)$ has at least one eigenvalue 0. The multiplicity of the eigenvalue 0 of $\mathfrak{v}(a)$ is a function $n(a)$. The minimum value $l = \min_{a \in \mathfrak{L}} n(a)$ is called the rank of $\mathfrak{L}$, an element for which $n(a) = l$ is a regular element. The set of elements which commute with a regular element a is an abelian subalgebra, called a Cartan subalgebra $\mathfrak{H}_a$. We may choose in $\mathfrak{H}_a$ a real base: $h_1, \ldots, h_l$ which is orthonormal: $\langle h_i, h_k \rangle = \delta_{ik}$. The $\mathfrak{v}(h_i)$ are then a maximal set of mutually commuting matrices in the adjoint representation.

3. The eigenvalue problem of the algebra, roots

Now we start the usual procedure; we solve the simultaneous eigenvalue problem

$$e_\alpha\, \mathfrak{v} = \vec{r}(\alpha)\, e_\alpha = i \left[\vec{h}, e_\alpha\right] \tag{1}$$

where we consider $(h_1, \ldots, h_l)$ as a vector in $\mathfrak{H}$. α labels the different eigenvectors. The simultaneous eigenvalues $\vec{r}(\alpha)$ are called *roots*, e_α are eigenvectors which belong to roots different from 0. They form an orthogonal base (with respect to the hermitian Killing form) of the subspace $\mathfrak{T} \perp \mathfrak{H}_a$; $\mathfrak{H}_a$ is the eigenspace with simultaneous eigenvalues 0. Therefore we have $n - l$ eigenvectors e_α. They form together with $\vec{h}$ an orthonormal base of $\mathfrak{L}$:

$$\langle e_\alpha, e_\beta \rangle = \delta_{\alpha\beta}, \ \langle e_\alpha, h_i \rangle = 0, \ \langle h_i, h_k \rangle = \delta_{ik}. \tag{2}$$

The $v(h_i)$ are diagonal in this base with the eigenvalues $-i\vec{r}(\alpha)$ therefore

$$\langle h_i, h_k \rangle = \sum_\alpha r_i(\alpha)\, r_k(\alpha) = \delta_{ik}. \tag{3}$$

4. Properties of the roots and eigenvectors

We make some simple remarks on the roots $\vec{r}(\alpha)$ and the eigenvectors e_α.

(a) As $\vec{h}$ is real, $e_\alpha{}^*$ is an eigenvector with the root $-\vec{r}(\alpha)$. We write $e_\alpha{}^* = e_{-\alpha}$, $\vec{r}(-\alpha) = -\vec{r}(\alpha)$.

(b) From the Jacobi identity follows:

$$\left[\vec{h}, [e_\alpha, e_\beta]\right] = \left[[\vec{h}, e_\alpha]\, e_\beta\right] - \left[[\vec{h}, e_\beta]\, e_\alpha\right] = -i\left(\vec{r}(\alpha) + \vec{r}(\beta)\right) [e_\alpha, e_\beta],$$

therefore

(c) $[e_\alpha, e_{-\alpha}] = \beta^i h_i \in \mathfrak{H}$ because it commutes with $\vec{h}$

$$\beta^i = \langle h_i, [e_\alpha, e_{-\alpha}]\rangle = -i\, r_i(\alpha).$$

(d) $[e_\alpha, e_\beta] = 0$ if $\vec{r}(\alpha) + \vec{r}(\beta) \neq 0$ and not a root.

(e) Further information on the set of roots Σ we get by an iteration of the calculation (b), which shows that if the elements

$$[e_\beta, e_\alpha],\ \big[[e_\beta, e_\alpha]\, e_\alpha\big], \ldots$$

respectively

$$[e_\beta, e_{-\alpha}],\ \big[[e_\beta, e_{-\alpha}]\, e_{-\alpha}\big], \ldots$$

are different from 0, they are eigenvectors with the roots

$$\vec{r}(\alpha) + \vec{r}(\beta),\ \vec{r}(\beta) + 2\vec{r}(\alpha), \ldots,\ \vec{r}(\beta) + m\vec{r}(\alpha)$$

respectively

$$\vec{r}(\alpha) - \vec{r}(\beta),\ \vec{r}(\beta) - 2\vec{r}(\alpha), \ldots,\ \vec{r}(\beta) - n\vec{r}(\alpha).$$

The sequences must terminate, because there are only dim $\mathfrak{L} - l$ eigenvectors.

(f) If $[e_\gamma, e_{-\alpha}] = 0$ and $u^t = [\ldots\overbrace{[e_\gamma, e_\alpha]\ldots e_\alpha]}^{t}$, we can show by induction:

$$[e_{-\alpha}, u^t] = \left(t\,\big(\vec{r}(\gamma)\vec{r}(\alpha)\big) + \frac{t\,(t-1)}{2}\,\vec{r}^{\,2}(\alpha)\right) u^{t-1}$$

$$(\vec{r}, \vec{r}\,') = \sum_i r_i\, r_i'.$$

(g) There is a smallest $t = g$, so that $u^{t+1} = 0$. From (f) it then follows that if $[e_\gamma, e_{-\alpha}] = 0$ we have

$$-\frac{2\vec{r}(\gamma)\,\vec{r}(\alpha)}{\vec{r}(\alpha)\,\vec{r}(\alpha)} = g = \text{integer larger than zero.} \tag{4}$$

(h) The roots are all different. (From the hypothesis $\vec{r}(\alpha) = \vec{r}(\gamma)$ it follows $[e_\gamma, e_{-\alpha}] = \beta^i h_i$ with $\beta^i = \langle h_i, [e_\gamma, e_{-\alpha}]\rangle = -i r_i(\gamma)\,\langle e_{-\gamma}, e_{-\alpha}\rangle = 0$. Therefore (4) gives a contradiction.)

(i) If $\vec{r}(\alpha)$ and $j\vec{r}(\alpha)$ are roots, it follows that $j = \pm 1$. [We put in (g) $\vec{r}(\gamma) = -k\vec{r}(\alpha)$ for the largest k, for which $-k\vec{r}(\alpha)$ is a root.]

(j) If $\vec{r}(\alpha) + \vec{r}(\beta) \in \Sigma$ it follows $[e_\alpha, e_\beta] \neq 0$. [Using the notation of (c) and (f), we put $\vec{r}(\gamma) = \vec{r}(\beta) - n\vec{r}(\alpha)$ then there are all u^t, $t = 1, 2, \ldots$ $n + m$ different from zero.]

(k) $\vec{r}(\alpha), \vec{r}(\beta) \in \Sigma$; n, (m) is the largest integer for which all vectors $\vec{r}(\beta) \overset{+}{(-)}\, j\vec{r}(\alpha)$, $j = 0, 1, \ldots n, (m)$ are elements of Σ, then it holds

$$n-m=\frac{2\left(\vec{r}(\alpha)\vec{r}(\beta)\right)}{\left(\vec{r}(\alpha)\vec{r}(\alpha)\right)}.$$

$$(2)\quad \left[e_{-\alpha},\left[e_\alpha, e_\beta\right]\right]=\frac{-m(n+1)}{2}\left(\vec{r}(\alpha)\vec{r}(\beta)\right)e_\beta. \tag{1}$$

5. Canonical form

From the properties of section (4) it follows immediately that we can bring the commutation relations in the following form (canonical form):

$$[h_i, e_\alpha]=-i\, r_i(\alpha)\, e_\alpha,\quad [e_\alpha, e_{-\alpha}]=-i\,\Sigma_i\, r_i(\alpha)\, h_i,$$
$$[h_i, h_j]=0,$$
$$[e_\alpha, e_\beta]=\begin{cases}-i N_{\alpha\beta}\, e_\gamma & \text{if } \vec{r}(\alpha)+\vec{r}(\beta)=\vec{r}(\gamma).\\ 0 & \text{otherwise}\end{cases} \tag{5}$$

6. Primitive roots

(a) The subspace of the l-dimensional space which is spanned by the l-dimensional roots, is l-dimensional. (If there would exist a vector $\vec{t}$, $\left(\vec{t},\vec{r}(\alpha)\right)=0$ for all α, then $\left(\vec{t},\vec{h}\right)$ would commute with all h_i, e_α and therefore generate an invariant subspace which contradicts to the assumed simplicity of $\mathfrak{L}$.) Therefore the roots form an l-dimensional "star". Compare Fig. 1.

(b) In order to simplify further the description of $\mathfrak{L}$ we look for a base of the roots. We define a root as positive, if the first non-vanishing component is positive. $\vec{r}(\alpha)$ is called *primitive*, if it is positive and not the sum of two positive roots.

If $\vec{r}(\alpha), \vec{r}(\beta)$ are primitive roots, then $\vec{r}(\alpha)-\vec{r}(\beta)$ is not a root. [Assuming $\vec{r}(\alpha)-\vec{r}(\beta)$ is a root, then either $\vec{r}(\alpha)-\vec{r}(\beta)$ or $\vec{r}(\beta)-\vec{r}(\alpha)$ is positive and therefore $\vec{r}(\alpha)=\vec{r}(\beta)+\left(\vec{r}(\alpha)-\vec{r}(\beta)\right)$ or $\vec{r}(\beta)=\vec{r}(\alpha)+$ $+\left(\vec{r}(\beta)-\vec{r}(\alpha)\right)$. This contradicts the assumed primitivity of $\vec{r}(\alpha)$ and $\vec{r}(\beta)$.]

From the properties quoted above under (4, i) and (4, k) follows that there is a base of l primitive roots $\left\{\vec{s}(k)\right\}$:

$$\vec{r}(\alpha)=\varepsilon\left(n_1\vec{s}(1)+\ldots+n_l\vec{s}(l)\right)$$

$\varepsilon=\pm 1$ for $\vec{r}(\alpha)$ positive or negative

$$n_i \text{ integers} \geq 0. \tag{6}$$

7. Geometric properties of the system of primitive roots

Let $\vec{s}(k)$, $k=1,\ldots, l$ be the system of the primitive roots of the Lie algebra $\mathfrak{L}$. Then from (4, k) it follows

$$-m=\frac{2\left(\vec{s}(k)\vec{s}(i)\right)}{\left(\vec{s}(i)\vec{s}(i)\right)},\; k\neq i$$

m integer >0. $\left(n=0\right.$ because of the primitivity of $\left.\vec{s}(i),\ \vec{s}(k)\right)$.

$$\text{From } \frac{\vec{s}(k)\vec{s}(i)}{\vec{s}(i)\vec{s}(i)} = -\frac{p}{2}; \quad \frac{\vec{s}(k)\vec{s}(i)}{\vec{s}(k)\vec{s}(k)} = -\frac{q}{2}$$

$$\Theta_{ik} = \sphericalangle(\vec{s}(i), \vec{s}(k)), \quad \pi/2 < \Theta < \pi$$

follows

$$\cos^2 \Theta_{ik} = \frac{p\,q}{4} \qquad \text{with } p, q \text{ integers} > 0.$$

Therefore

$$pq = 0,\ 1,\ 2,\ 3$$

$$\Theta = 90°,\ 120°,\ 135°,\ 150°$$

$$\frac{q}{p} = \frac{\vec{s}(i)\vec{s}(i)}{\vec{s}(k)\vec{s}(k)} = x,\ 1,\ 2,\ 3; \quad q \geq p.$$

8. Cartan matrices, generating elements and defining relations

In order to show that a Lie algebra is completely determined by the geometry of the primitive roots, we consider some of the commutation relations of the h_i and the eigenvector e_α and their conjugates which belong to the primitive roots $\vec{s}(k)$: e_k, e_{-k}. From (5) we see

$$[h_i, h_k] = 0 \quad [h_i, e_{\pm k}] = \mp i\, s_i(k)\, e_{\pm k}$$

$$[e_k, e_{-k}] = -i\left(\vec{s}(k)\vec{h}\right), \quad [e_k, e_{-i}] = 0$$

for $i \neq k$, because $\vec{s}(k) - \vec{s}(i)$ is not a root.
We now define

$$g_k = \frac{2\vec{s}(k)\vec{h}}{\vec{s}(k)\vec{s}(k)}, \qquad f_k^{\pm} = \left[\frac{2}{\vec{s}(k)\vec{s}(k)}\right]^{1/2} e_{\pm}$$

and get

$$[g_i, g_k] = 0 \quad [f_i^+, f_k^-] = -i\delta_{ik}\, g_k,$$
$$[f_k^+, g_i] = +i A_{ik}\, f_k^+, \quad [f_k^-, g_i] = -i A_{ik}\, f_k^-. \tag{7}$$

The matrix $A_{ik} = \dfrac{2\left(\vec{s}(i)\vec{s}(k)\right)}{\vec{s}(i)\vec{s}(i)}$ is called *Cartan matrix*. The elements of (A_{ik}) have values $A_{ii} = 2$, $A_{ik} = 0, -1, -2, -3$ for $i \neq k$ according to the geometric properties of the primitive roots.

The f_i^+, f_i^-, g_i, i fixed, generate a sub-algebra isomorphic to the algebra generated by the "angular momenta" (example D).

The relations (7) define the Lie algebra completely in the following sense. We may express the base elements $h_i, e_\alpha, e_{-\alpha}$ of the canonical base (5) of $\mathfrak{L}$ as linear combinations of g_i and iterated commutators of $f_i^{\pm}$. For the h_i this is trivial, as the expression $g_k = \dfrac{2\vec{s}(k)\vec{h}}{\vec{s}(k)\vec{s}(k)}, k = 1, \ldots, l$ can be inverted. An e_α belongs to a root $\vec{r}(\alpha)$, which can be expressed by the primitive roots of $\mathfrak{L}$:

$$\vec{r}(\alpha) = \left(n_1\vec{s}(1) + \ldots + n_l\vec{s}(l)\right). \tag{8}$$

The $N = \sum_{i=1} |n_i|$ is called the level of the root. We can construct the roots $\vec{r}(\alpha)$ and the corresponding eigenvectors e_α by induction on the level. The level one roots are the $\pm \vec{s}$ (k) and the corresponding eigenvectors, the $f_k^\pm = c\, e_k^\pm$. Now suppose we already know the roots of level N and the corresponding eigenvectors $e_{\pm\alpha} = c' [f_{i_N}^\pm, [\ldots [f_{i_2}^\pm, f_{i_1}^\pm]]$. Then the positive roots of level $N+1$ have the form $\vec{r}(\alpha) + \vec{s}(k)$, $\vec{r}(\alpha)$ positive of level N. We may assume that one $n_i > 0$, $i \neq k$ in the expression of $\vec{r}(\alpha)$ (because of (4, i)). Then the linears forms $\vec{r}(\alpha) - \vec{s}(k)$, $\vec{r}(\alpha) - 2\vec{s}(k) \ldots$ are roots which are positive of level less than N. So the number m of remark (4, k) is known and we can determine the number n for which all $\vec{r}(\alpha)$, $\vec{r}(\alpha) + \vec{s}(k)$, $\ldots$, $\vec{r}(\alpha) + n\vec{s}(k)$ are roots with help of (4, k)

$$n = m - \frac{2\,(\vec{r}(\alpha), \vec{s}(k))}{\vec{s}(k)\,\vec{s}(k)} = m - \sum_{i-1}^{l} n_i A_{ik}$$

which allows us to decide with help of the Cartan matrix if $\vec{r}(\alpha) + \vec{s}(k)$ is a root. In this case $[e_\alpha, f_k^+]$ is an eigenvector. In the simple case of the Cartan matrix $(A_{ik}) = \begin{pmatrix} 2 & -1 \\ -1 & 2 \end{pmatrix}$ which belongs to the root diagram of Fig. 1. (The algebra of the SU_3.) The roots are

$$\pm \vec{s}(1), \quad \pm \vec{s}(2), \quad \pm \left(\vec{s}(1) + \vec{s}(2)\right)$$

$$\vec{s}(1) = \left(\frac{1}{2\sqrt{3}}, \frac{1}{2}\right) \qquad \vec{s}(2) = \left(\frac{1}{2\sqrt{3}}, -\frac{1}{2}\right)$$

and the corresponding eigenvectors are

$$\frac{1}{\sqrt{6}} f_1^\pm, \quad \frac{1}{\sqrt{6}} f_2^\pm, \quad e_{\pm 3} = \frac{1}{\sqrt{6}} [f_1^\pm, f_2^+].$$

9. *Classification of the simple compact Lie algebras*

As we have seen in the last section, a Lie algebra is determined by the geometric properties of its primitive roots, i. e. its Cartan matrix. For a given dimension of the root-space, i. e. for a given rank of the algebra, there are only a certain number of configurations geometrically possible. In order to review this possibilities, we introduce a graphical representation of these configurations. (Dynkin diagram.)

(α) The primitive roots are marked by a circle.

(β) If the angle between two roots is 120°, 135°, or 150° the circles are joint by 1, 2, resp. 3 lines. They are not joint if the angle is 90°.

(γ) The roots with the largest length are marked black. By the Dynkin diagram is the system of the primitive roots and therefore the Lie algebra completely given. In Fig. 2 we have given all possible diagrams.

VI. The representations of real compact Lie algebras

Let us discuss now the representation of a real compact simple Lie algebra $\mathfrak{L}$ by hermitian matrices in a finite dimensional Hilbert space.

1. Complete reducibility

The representations $D(\mathfrak{L})$ are completely reducible because of the hermitian character. The non-trivial representations are faithful, because $\mathfrak{L}$ is assumed simple.

2. Commutation relations

We may introduce in $\mathfrak{L}$ a canonical base $\{h_i, e_\alpha, e_{-\alpha}\}$, $i=1, \ldots, l$ $\alpha=1, \ldots, \frac{n-l}{2}$. By the representation these are mapped on H_i, E_α, $E_{-\alpha}$, $E_\alpha{}^+ = E_{-\alpha}$, which satisfy commutation relations corresponding to (5, 5):

$$[H_i, H_k] = 0, \; [H_i, E_{\pm\alpha}] = \pm\, r_i(\alpha)\, E_{\pm\alpha} \qquad [E_\alpha, E_{-\alpha}] = \Sigma_i\, r_i(\alpha)\, H_i$$

$$[E_\alpha, E_\beta] = \begin{cases} N_{\alpha\beta}\, E_\gamma \text{ if } \vec{r}(\alpha)+\vec{r}(\beta)=\vec{r}(\gamma) \\ 0 \quad \text{if } \vec{r}(\alpha)+\vec{r}(\beta) \text{ is not a root} \end{cases} \tag{1}$$

or we regard $\mathfrak{L}$ given by the generating g_i, $f_i{}^+$, $f_i{}^-$; $i=1, \ldots, l$ then these elements are represented by G_i, $F_i{}^+$, $F_i{}^-$ and satisfy according to (V, (7)) the C. R.

$$\begin{aligned} &[G_i, G_k]=0, \quad [F_i{}^+, F_k{}^-]=\delta_{ik}\, G_k \\ &[F_i{}^+, G_k]= -A_{ki}\, F_i{}^+, \; [F_i{}^-, G_k]= +A_{ki}\, F_i{}^-. \end{aligned} \tag{2}$$

(A_{ik}) is the Cartan matrix of $\mathfrak{L}$.

The connection between the (H_i, E_α) and the G_k, $F_i{}^\pm$ is corresponding to (V, (7)):

$$\begin{aligned} G_k &= \frac{\left(2\, \vec{s}(k)\, \vec{H}\right)}{\left(\vec{s}(k)\, \vec{s}(k)\right)} \\ E_{\pm\alpha} &= [F_{i_n}{}^\pm, [\ldots [F_{i_2}{}^\pm, F_{i_1}{}^\pm] \ldots] (-i)^n \cdot C \\ \text{if } e_{\pm\alpha} &= [f_{i_n}{}^\pm, [\ldots [f_{i_2}{}^\pm, f_{i_1}{}^\pm] \ldots] \cdot C. \end{aligned} \tag{3}$$

3. Weights

In order to discuss the representations of $\mathfrak{L}$ we introduce in $\mathfrak{M}$ a base of eigenstates where the H_i are simultaneous diagonal

$$\vec{H}\, \big|\vec{\lambda}\big) = \vec{\lambda}\, \big|\vec{\lambda}\big).$$

The simultaneous eigenvalues $\vec{\lambda}=(\lambda_1, \ldots, \lambda_l)$ are called weights (the roots are the weights of the adjoint representation). As familiar from quantum mechanics of angular momentum, the E_α acts as "shift"-operator:

$E_\alpha \big(\vec{\lambda}\big)$ is an eigenvector with the eigenvalue $\vec{\lambda} + \vec{r}(\alpha)$

$$\left[\vec{H}\, E_\alpha \mid \lambda\right) = E_\alpha \vec{H} \mid \vec{\lambda}) + \vec{r}\,(\alpha)\, E_\alpha \mid \vec{\lambda}) \text{ according (1)} \\ = \left(\vec{\lambda} + \vec{r}\,(\alpha)\right) E_\alpha \mid \vec{\lambda})\Big]. \tag{4}$$

(4) We simplify the discussion of the representations considerably if we start from the defining relations (6, 2). From (6, 3) we see

$$G_k \mid \vec{\lambda}) = \frac{2\,\vec{s}\,(k)\,\vec{\lambda}}{\vec{s}\,(k)\,\vec{s}\,(k)} \mid \vec{\lambda}) \equiv m_k \mid \vec{m}\rangle. \tag{5}$$

Now we use for orientation of our further procedure the fact, that G_k, F_k^+, F_k^- have the C. R. of $(2M_3,\ M_1 \pm i M_2)$; $\vec{M}$ is an angular momentum operator. From this follows our first conclusion:

(a) The eigenvalues m_k of G_k are integers. We introduce now an order in the set $\mathfrak{W}$ of weights $\vec{m}$, by calling $\vec{m} > \vec{m}'$ if the first non-vanishing component of $\vec{m} - \vec{m}'$ is larger than zero. As $\mathfrak{M}$ is assumed finite dimensional, in $\mathfrak{W}$ is a largest weight: $\vec{j}$. As F_k^+ would rise the eigenvalue by one we have:

(b) $$F_k^+ \mid j\rangle = 0.$$

The components must all be larger or equal to zero as largest $2m$ values of the "angular" momenta.

(c) The states of the form

$$F_{i_1}^- \dots F_{i_s}^- \mid j\rangle \equiv \Phi\,(i_1, \dots, i_s)$$

are eigenstates of G_k with eigenvalues $m_k = j_k - n_k$ where n_k is the number of $i_r = k$.

(d) The subspace $\mathfrak{M}' \subset \mathfrak{M}$ spanned by the $\{\Phi\,(i_1, \dots i_s)\}$ with a fixed state of highest weight is an invariant subspace. [According to (6, 2) we may commute an F_i^+ to the right until it annihilates $\mid j\,\rangle$ according to (b).]

(e) Therefore an irreducible $D\,(\mathfrak{L})$ has only one state of highest weight.

(f) Two representations with different highest weight are certainly inequivalent. On the other hand as the $F_k^{\pm}$, G_k generate the Lie algebra, the highest weight $\vec{j}$ defines uniquely the irreducible representation.

(g) From the representation theory of the angular momentum it follows that if $\vec{m} = (m_1, \dots, m_r, \dots m_l)$ is a weight, $\vec{m}' = (m_1, \dots, -m_r, \dots m_l)$ is a weight.

(h) Every l-tuple of nonnegative integers defines as highest weight a finite dimensional irreducible representation.

Most of these results will be illustrated by the discussion of the SU_3, i. e. the Lie algebra A_2 in the following lectures.

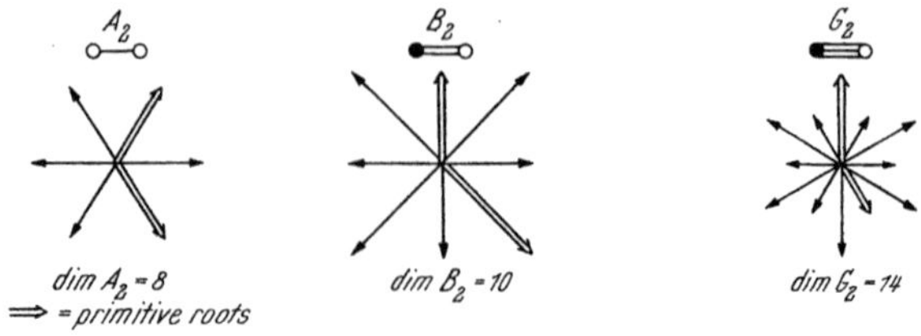

Fig. 1. Root and Dynkin diagrams of rank 2 groups

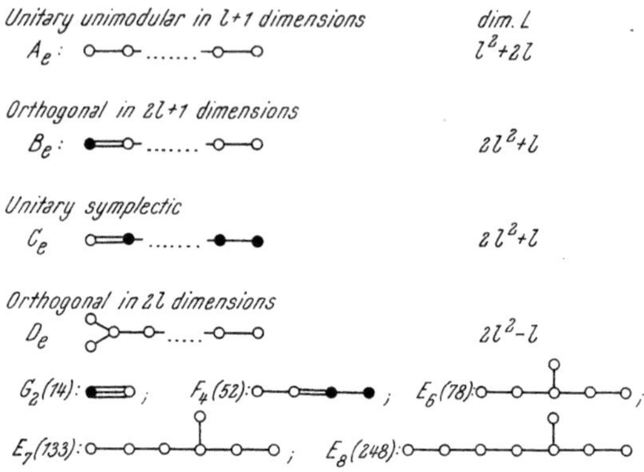

Fig. 2. Dynkin diagrams of the classical and exceptional simple, compact Lie algebras

The following literature was used for the preparation of these notes:

1. Behrends, R. E., J. Dreitlein, C. Fronsdal, and W. Lee, Rev. mod. Phys. **34**, 1 (1962).
2. Fronsdal, C., Brandeis Summer Institute 1962, Benjamin, New York, 1963.
3. Jacobson, N., Lie Algebras, Interscience Publishers, New York, London, 1962.
4. Pauli, W., CERN **56** — 31.
5. Pontrjagin, L. S., Topologische Gruppen, Bd. 2, B. G. Teubner, Leipzig, 1958.
6. Salam, A., Seminar on Theoretical Physics, Trieste, 1962, IAEA, Vienna, 1963.
7. Racah, G., Group theory and spectroscopy, CERN-reprint 61 — 8.

Unitary Symmetry of Strong Interactions*

By

Meinhard E. Mayer

Brandeis University, Waltham, Mass., U.S.A.

With 21 Figures

Introduction

We have become accustomed to the fact that theories in elementary-particle physics change with the season or the year, as women's fashions or new car models. Therefore, if a theory, or model, survives for longer than one or two years, one tends to become suspicious that either the theory really contains a grain of truth or — and this may well be the case with SU(3) symmetry — that the observed agreement between a model and experiment is in fact to some degree independent of the model and more a result of an accident of the dynamics of the problem.

During the last 3—4 years the classification of particles or states according to the quantum numbers associated with SU(3) has led to a few spectaculous results for the prediction and production of new particles, which seem in good agreement with the "broken SU(3) symmetry".

A recent compilation of Elementary particle data (Roos, Physics Letters, Nov. 1963) counts about 50 different states, as compared to the four (or five) "elementary particles" known thirty years ago.

We will deal in these lectures with only those states for which the baryon number $A = 0,1$, the electric charge is $Z = 0$, 1, 2, and will try to explain how the known states fit into the picture of a broken SU(3) symmetry.

We will first review the empirical situation classifying the strongly interacting particles according to the two special quantum conserved by S.I. namely isospin and hypercharge. Then we will show how to construct explicitly some of the representations of SU(3), illustrating some of the points for which Prof. Joos has outlined the general principles in his lectures. Finally, in the additional lecture scheduled for next week, we will not discuss weak interactions, but some recent model theories which are connected with SU(3). If time permits we will also briefly discuss the so-called "bootstrap" approach to symmetries and the connection between discrete (permutation) symmetry and continuous symmetries.

---

* These notes represent an abridged and amended form of notes of the 1963 Brandeis Summer Institute for Theoretical Physics, to be published shortly by Brandeis University.

1.1. Quantum Numbers for Particles

Tradition classifies the interactions between particles into strong, electromagnetic, and weak; and for all practical purposes and maybe also for some unknown deeper reason, one may as a first approximation treat strong interactions neglecting electromagnetic and weak ones (although this may be very dangerous). We deal almost exclusively in these lectures with strong interactions.

Strong interactions (S.I.) are characterized by scattering and production cross sections in the mb (10^{-27} cm^2) region, by lifetimes or characteristic times, derived via the uncertainty relation $\tau \sim \hbar/\Gamma$ from the width of the energy peak, of $\ll 10^{-20}$ sec (we do not know whether such a small number has any direct physical meaning; it would be better therefore to characterize S. I. in terms of the width), and somewhat more vaguely in terms of a Langrangian Field Theory with coupling constants $\gtrsim 1$.

The electromagnetic interactions are weaker than the S. I. by a factor of 100 (at least in the experimentally accessible region), i. e. the coupling constant is 1/137, the fine structure constant, and they have cross sections of the order of μb or less (with the exception of forward Coulomb scattering), and lifetimes of $\lesssim 10^{-16}$ sec.

The weak interactions are, in the low-momentum-transfer region we know, much weaker than the strong or the electromagnetic interactions; the coupling constant is of the order of 10^{-12} if we use the nucleon mass to make it dimensionless, cross sections as determined from neutrino experiments are $< 10^{-40}$ cm^2, and lifetimes $> 10^{-11}$ sec.

A nicer way to classify the interactions among particles is by the quantum numbers conserved in the process of interaction, or, since we have become used to associating conserved quantum numbers with invariances, by the invariance properties of the system.

The usual consensus is that the stronger an interaction is, the more quantum numbers it conserves. Let us start out with a review of the quantum numbers (or physical quantities) and their degree of conservation.

A. Conserved rigorously (except possibly for gravitational interactions, with which we do not deal).

1. Quantities connected with macroscopic space-time.

a) Energy-momentum: the Lorentz 4-vector $(E, \underset{\sim}{p}) = p_\mu$; in a reaction p^2 is invariant: $p_\mu p^\mu = p_\mu' p'^\mu = E^2 - \underset{\sim}{p}^2 = m^2 = E'^2 - \underset{\sim}{p}'^2 = m'^2 \geq 0$. Unprimed letters refer to particles before interaction, primed ones to particles after interaction. In classical or quantum field theory energy-momentum conservation is connected with space-time translations.

b) 4-angular momentum: $J_{\mu\nu}$ splits into the axial vector $\underset{\sim}{J}$, the angular momentum, and $\underset{\sim}{K}$, the centroidal momentum. The commutation rules (taken from correspondence with Lagrangian theory or postulated) correspond to those of the generators of rotation and proper Lorentz transformation, i. e., angular momentum and centroidal momentum, respectively. Physically acceptable values of $\underset{\sim}{J}$ are multiples of 1/2. In high energy physics one measures $\underset{\sim}{J} \cdot \underset{\sim}{p}/|\underset{\sim}{p}|$, the helicity, whose maximum eigenvalue is the spin.

2. Charges or "superselection rule" quantum numbers.

a) Electric charge: the number of charged particles minus the number of charged antiparticles is a constant. For an interaction with a macroscopic electric field, this is connected with the charge (in units of e), $Q = Ze$; we consider Z values of 0, ± 1, ± 2, higher charged systems are considered "less" elementary.

b) Baryon number: the number of heavy fermions minus the number of heavy anti-fermions denoted by A is a constant for all interactions. Again we restrict ourselves to systems with $A = 0$, ± 1, in analogy with charge. In nuclear physics this number along with Z is used to characterize a nucleus.

c) Lepton number L, muon number M: the conservation of these is not contradicted by experiment and they are important in weak interactions.

3. TCP invariance: formulated in various well-known ways — usually nowadays, as an analytic continuation property of amplitudes into their conjugates when 4-momenta are reversed [2].

B. Strongly conserved quantum numbers in addition to A.

1. Separate C or P invariance: we can assign a parity to strongly interacting particles (hadrons) at least within classes satisfying a superselection rule. This forces us to assign parity to one member of each fermion group carrying a specific charge, e. g.: since the neutron carries baryon number its parity is chosen $+1$, the proton has electric charge, we assign it arbitrarily parity $+1$, then there is the Λ which carries strangeness, so we assign it the parity $+1$. The rest of the baryons must now have determinate parities with respect to these three: so we know that the Σ's have parity $+1$ relative to p or Λ and hope that the J^P of Ξ will be $1/2^+$.

2. Isospin conservation: Because of the small neutron-proton mass difference, Heisenberg was led to treating p and n as different states of the more fundamental nucleon. This description turned out to be an important tool in particle physics, especially since we can use the formalism of angular momentum, i. e., a state is characterized by the total isospin I and its projection on the third axis I_3. [Group theoretically, it is preferable to view isospin not as 0(3), the group of all orthogonal matrices on a real 3-space, but rather as the locally isomorphic SU(2), the group of all unitary, unimodular transformation on a complex 2-space.] For a given I there are $2I + 1$ values of I_3, ranging from $-I$ to $+I$ in steps $\Delta I_3 = \pm 1$. For nonstrange particles we have evidently a connection between isospin and charge $Z = I_3 + A/2$ and hence electromagnetism which violates isospin, conserves, however, I_3.

3. Hypercharge: The discovery of heavy bosons and of the hyperons apparently led to conflict with the above formula. Thus the K's seemed

to be an isodoublet, but with charge 0, +1 (0, −1 for antiparticles) and Λ an isosinglet with $A=1$, but $Z=0$. In generalizing the expression for Z to include these particles, strangeness or better still hypercharge was introduced such that: $Z=\frac{Y}{2}+I_3=\frac{A}{2}+\frac{S}{2}+I_3$, where Y is hypercharge, S strangeness. K^0, K^+ has $Y=1$, $S=1$; $\bar{K}^0$, K^- has $Y=-1$, $S=-1$; Λ has $Y=0$, $S=-1$. Hypercharge may also be defined as twice the average charge of a multiplet: $Y=2\langle Z\rangle$, so that Y tells us where the center of charge for isomultiplet is located. For $n-p$ $\langle Z\rangle=1/2$; Λ and $\Sigma^{+,0,-}$, $\langle Z\rangle=0$; for $K-\bar{K}$ doublets $\langle Z\rangle=\pm 1/2$.

4. G-parity: For the treatment of totally neutral systems $A=Z=Y=0$, we combine C, the charge conjugation operator, and a rotation in I-spin space around the 2-axis by an angle π into $G=Ce^{i\pi I_2}$. Such a neutral system can decay into a set of n pions, for which $G|n\pi\rangle=(-1)^n|n\pi\rangle$ if it has the right G-parity. G-parity is thus a measure of "pionishness".

C. Electromagnetically conserved quantities: I_3 and (reportedly) C and P are conserved by electromagnetic interactions; isospin and hypercharge conservation are violated, however.

D. Weak interactions: the quantities L and M are conserved in weak interactions but the conservation of I, I_3, Y, P, C is violated.

E. Gravitation: the conservation of E, P, J, K requires revision if general relativity is to make any sense; this results from the fact that a Lorentz 4-vector is not conserved in its usual sense.

1.2 Classification of Particles

We have for the mesons $A=0$ and for the baryons $A=1$ where the mass differences of the order of, e. g., $\pi^+-\pi^0\sim 5$ Mev or $n-p\sim 2$ Mev are not drawn to scale:

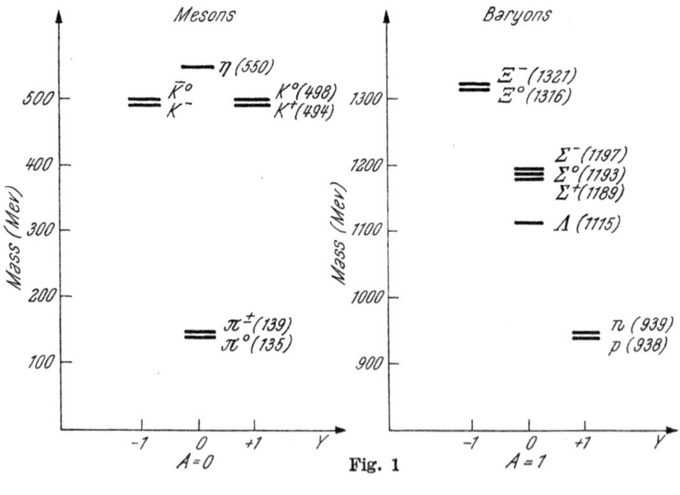

Fig. 1

However, rather than use this familiar way of drawing the mass-levels of states, we can use the fact that the strongly interacting particles conserve Y and I_3 and plot them in such a coordinate system. This allows us also to represent resonances in a convenient way. For the baryons $J^P = 1/2^+$:

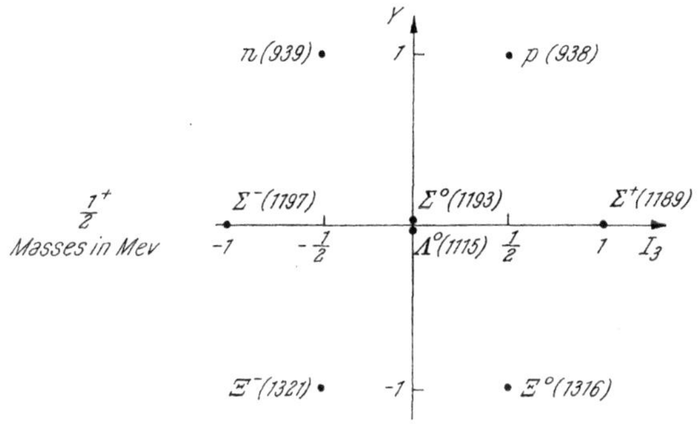

Fig. 2

Mass differences within a multiplet are accountable by electromagnetism. For the baryon resonances $J_R^P = 3/2^-$

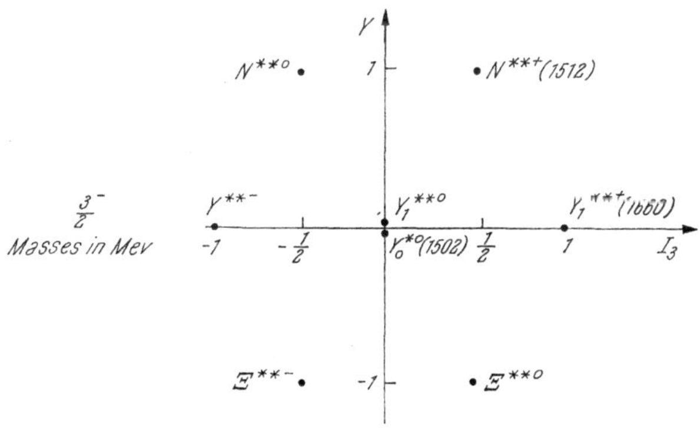

Fig. 3

The N^{xx} is identical to what the experimentalists called the 600 Mev πp resonance. It shows up as a bump in the $\pi - p$ cross section in the amplitude for $I = 1/2$, $J = 3/2$. The mass of the Y_1^{xx0} has not been actually determined nor has the parity of the Y_0^{x0}. The Ξ^{xx} has not been experimentally observed and the parity of the 1660 Y_1^{xx} resonance is more likely to be $+$. Therefore a slightly more complicated diagram appears when we look at baryon resonances, which have been established or believed to have $J^P = 3/2^+$, before February 1964.

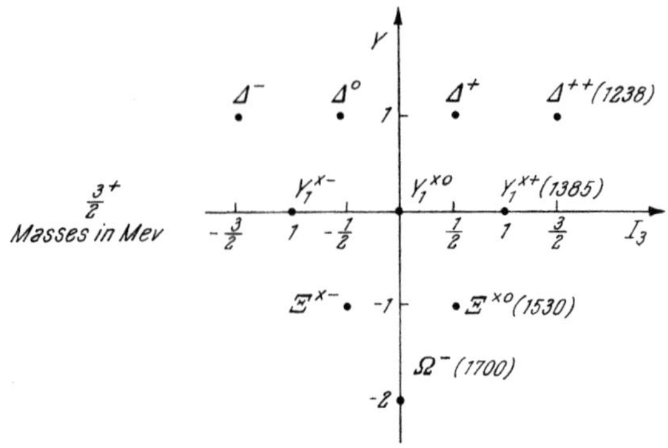

Fig. 4

If you believe in any symmetry of geometric figures then the vertex of the triangle (0, —2) should represent a particle, the Ω^- or Z^-, which is producted to have a mass of 1680 Mev. The prediction of the Ω^- from the symmetry of the diagram [or from a thorough analysis of SU(3)] was considered one of the crucial tests of the theory. Now about two weeks ago, an event which can be interpreted as an Ω^- has been found in an exposure of the Brookhaven 80″ hydrogen bubble chamber to a beam of 5 GeV K^--mesons. The topology of the event is the following (seen tracks are solid, unseen dotted!):

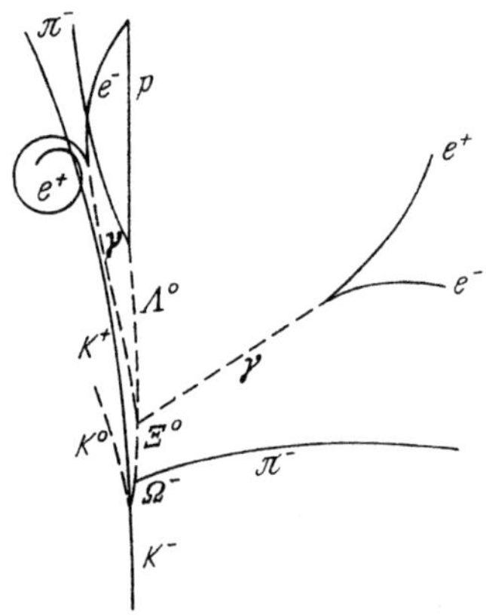

Fig. 5

An analysis of the event shows that

a) the missing mass of the K^0 particle corresponds to the expected value;

b) therefore Ω^- has $Y = -2$;

c) the mass is $\sim 1700 \pm 20$ MeV as compared to a predicted 1680 MeV.

Further, until recently, one used to group the $J^P = 3/2^-$ N^{**} (1512) resonance with the Y_0^{**} (1502) into an octet containing the Y_1^{**} (1660) resonance and an unobserved Ξ^{**} resonance. However recent data from Brookhaven have strengthened the doubts as to the parity assignment of the 1660 resonance — it seams likely to be a $3/2^+$ particle, and therefore it is premature to make any classification of these particles. There is also a singlet $J^P = 1/2^-$, the Y_0^* (1405 Mev). There are particles of higher spins; some people call these Regge recurrences. If we are willing to accept the idea of Regge trajectories, then the same multiplets of particles should have a recurrence if we go up by two units of angular momentum. Thus the $J^P = 1/2^+$ baryon octet should be mirrored by a $J^P = 5/2^+$ resonant octet. The N^{***} (1688 Mev or 900 Mev πp resonance) in both charge states and the Y_0^{***} (1850 Mev) have been seen, although the other members of the octet have not. An N^{****} (1920 Mev) $J^P = 7/2^+$ and a 9/2 resonance have also been seen. Other higher spin states have also been seen, but this completes the listing of BARKAS and ROSENFELD [1].

The bosons are simpler to treat this way. For the pseudoscalar mesons $J^P = 0^-$ (where appropriate, we also list G-parity J^{PG}).

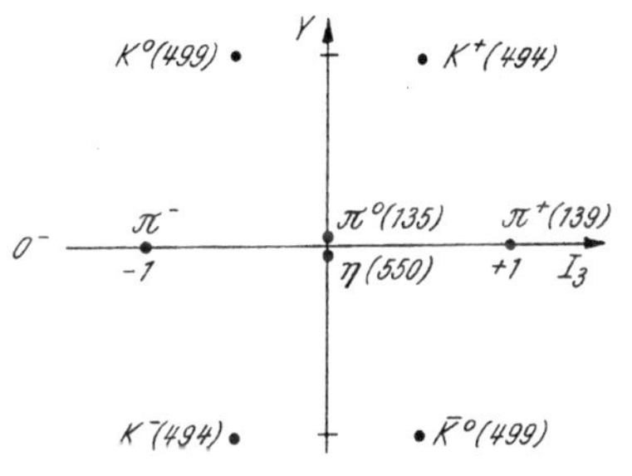

Fig. 6

We note that both particles and antiparticles belong to the same multiplet, this is because bosons do not carry baryon number. We have for the vector mesons $J^0 = 1^-$

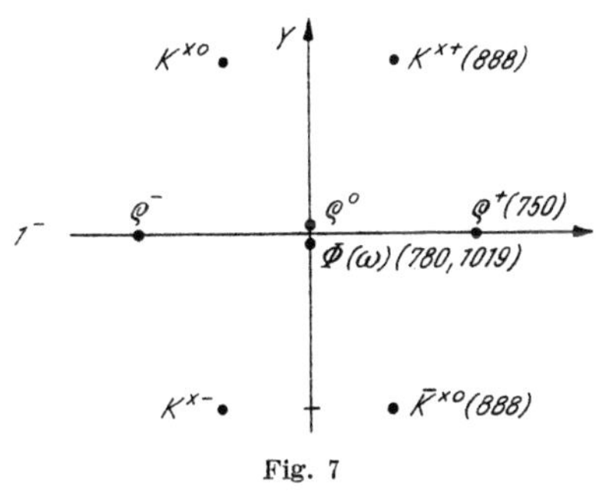

Fig. 7

There is also a doublet $K^{\times 0}$, $K^{\times +}$ at 725 MeV (sometimes called $\varkappa$) whose placement is obscure — probably 0^+. The singlet ω (φ) $J^{PG} = 1^{-+}$ and the φ (ω) $J^{PG} = 1^{--}$ are *mixed*. The ω has a mass of 780 MeV, the φ $\sim$ 1919 MeV; what combination is the singlet, what combination a part of the octet has not yet been answered. There is another boson singlet, the 2^{++} f^0 (1250 MeV).

These diagrams are obviously the weight diagrams of the algebra SU(3). Their symmetry properties will be studied in terms of Weyl reflections. Horizontal lines are related by I-symmetry lines with slope $\frac{Y}{I_3} = -2$ connect particles having the same charge, lines with slope $\frac{Y}{I_3} = +2$ are related by "U-symmetry".

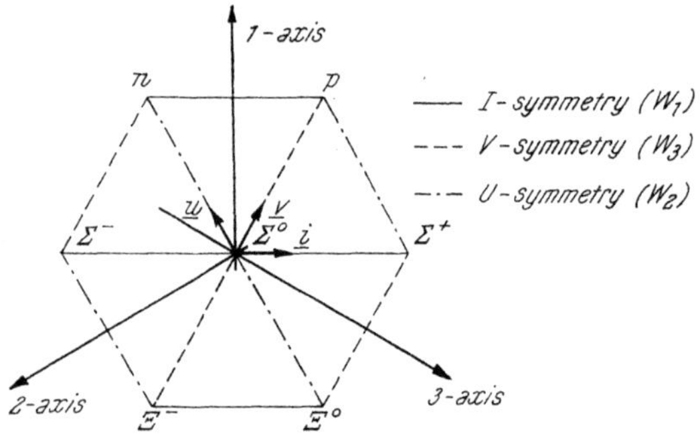

Fig. 8

There exists a proposal for a new notation for particle states by CHEW, GELL-MANN, and ROSENFELD [3]. For baryons they recommend using a capital Greek letter for states of different total isospin and hypercharge and a small Greek subscript for spins and parities:

	I	Y	*Subscripts*	J	P
N	1/2	1	α	1/2	+
	0	0	β	1/2	—
	1	0	γ	3/2	—
	1/2	—1	δ	3/2	+
	3/2	1	α_2	5/2	+
	0	—2		etc.	

α_2 is the first "recurrence" in the Regge sense. For bosons a small Greek letter for states of different total isospin and hypercharge should be used:

	I	Y		J	P
π	1	0	α	0	+
χ	0	0	β	0	—
$\varkappa$	1/2	1	γ	1	—
			δ	1	+

We have the correspondence with the "classical notation" for the baryons and baryon resonances:

$N \to N_\alpha$	$N^{x}_{1/2} \to N^{II}_\alpha$ (1688 Mev)
$\Lambda \to \Lambda_\alpha$	$\Xi^{x}_{1/2} \to N_\delta$ (1530 Mev)
$\Sigma \to \Sigma_\alpha$	$Y^{x}_{1} \to \Sigma_\gamma$ (1660 Mev)
$\Xi \to \Xi_\alpha$	$N^{xx}_{3/2} \to \Lambda_\gamma$ (1512 Mev)

for the pseudoscalar mesons:

$$\pi \to \pi_\beta$$
$$\eta \to \chi_\beta$$
$$K \to \varkappa_\beta$$

for the vector mesons (vectons):

$$\varrho \to \pi_\gamma$$
$$\omega \to \chi_\gamma$$
$$K^{x} \to \varkappa_\gamma$$

1.3 History of the Symmetry Approach to Strong Interactions

We take as the starting point the synthesizing of charge independence and charge symmetry into the isospin formalism. If we have any symmetry property in particle physics, then there is a correspondence with the symmetries of Lagrangian field theory. In field theory the various isospin multiplets, π's, K's, nucleons, are coupled in such a way that the interaction Lagrangian is invariant under isospin.

The generalizing of isospin as a rotation group led to the global symmetry schemes of Schwinger [4], Gell-Mann [5], and Salam and Polkinghorne [6] which all turned out to be in conflict with experiment.

The gauge theory of strong interactions had a comeback in 1960 with the work of Sakurai [7]. In its first form it was introduced by Yang and Mills [8] in 1954. It was based on the fact that whenever we have a Lagrangian field theory invariant under a group of transformations, there is a certain number of conserved currents. If one requires further that invariance be maintained if the transformations depend on the space-time point, one must couple these currents to vector fields (see e. g. ref. 9, for details, and further references). If we generalize to groups including isospin, then one must postulate the existence of at least 3 noncommuting vector fields. Gauge theory led to the conclusion that there should exist at least 5 vector mesons; 8 were known experimentally, encouraging people to take SU(3) seriously.

SU(3) was first introduced in the Sakata model [10]: all SIP's (strongly-interacting particles) can be built out of the p, n, and Λ. Transformations on these objects are what is called the unitary group in 3 dimensions. Wess [11], Ohnuki et al. [12] in 1960, all built on the Sakata model, taking p, n, and Λ as members of a triplet, not an octet. Mesons were built out of bound states on pn, $p\Lambda$, etc.

Recent revivals of global symmetry schemes appeared when Gell-Mann [13, 32], looking at invariance properties of currents in weak interactions, was led to the 8-fold way (this terminology goes back to Buddha). Independently Ne'eman [14] starting out from a generalization of the Sakata model was also led to the 8-fold way.

The 8-fold way is based on the symmetries of SU(3). One of the reasons for adopting SU(3) lies in the fact, that it is a group of rank 2; there exist 2 commuting operators, i. e., I_3, Y, out of the set of 8 independent, infinitesimal operators of the group. Other groups of rank 2 which were considered are G_2 with 14 parameters and B_2 with 10 parameters (see e. g. refs. [21, 27]).

1.4 The Meaning of Symmetry and Broken Symmetry

The meaning of symmetry in nonrelativistic quantum mechanics is well known: the Hamiltonian has a symmetry property if it commutes with the operator associated with the symmetry property of the system: $[H, T]=0$, so that the eigenvalues of H are not changed by T, and, at most, the eigenfunctions are permuted among themselves.

At the higher level of Lagrangian field theory the situation is not dissimilar: one deals with field operators $\psi_\alpha(x)$ which are subjected to linear transformations $\psi'_\alpha(x)=\sum_\beta M_{\alpha\beta}\,\psi_\beta(x)$ and the invariance of the Lagrangian under $\psi_\alpha(x)\to\psi'_\alpha(x)$ according to the generalization of Emmy Noether's theorem (see e. g. Ref. [15]) leads to conserved quantities. For continuous linear groups $M_{\alpha\beta}(\lambda_1, \lambda_2, \ldots \lambda_f)$, M is a function of f parameters. Invariance of the Lagrangian under the group means that there are f vector quantities $j_\mu{}^1, j_\mu{}^2 \ldots j_\mu{}^f$ such that $\partial_\mu j_\mu{}^i=0$ for all i, $i=1, 2, \ldots f$. With these conserved current densities we can associate what we call a charge. There are f conserved charges $Q^i = \int j_0{}^i\, d^3 x$.

Whenever we deal with transformations depending on parameters, we can introduce infinitesimal transformations, which for small λ_i

$$M_{\alpha\beta}(\lambda)=\delta_{\alpha\beta}+\sum_j i\, T^j_{\alpha\beta}\,\lambda_j+\ldots,$$

where T^j is an hermitian matrix. We have $T^j_{\alpha\beta}=\left(\frac{\partial M_{\alpha\beta}}{\partial\lambda_j}\right)_{\lambda_j=0}$. The matrices T^j determine the transformations completely. It can be proved that, if we know the T's, the infinitesimal generators, then the whole group can be constructed.

The commutation relations of the T's are

$$[T^i, T^j]=\sum_k c^{ij}_k\, T^k$$

and the charge operators satisfy the same commutation relations:

$$[Q^i, Q^j]=\sum_k c^{ij}_k\, Q^k.$$

c_k^{ij} are the structure constants of the Lie algebra of the infinitesimal generators. The subscripts i, j, k refer to the parameters; α, β refer to the field quantities.

One can go one step further (the Yang-Mills trick) by assuming that $\lambda_i = \lambda_i(x)$. This assumption leads with slightly forced arguments to the vector fields U_μ and the currents $j_\mu{}^i$ coupled to them (unfortunately obeying highly nonlinear equations of motion of the Yang-Mills type). This is the gauge theory approach to S.I. [7, 8, 9, 16].

Normally we think of symmetry as interchange. For protons and neutrons, kaons, and pions there is no symmetry in the primitive sense that we could replace a proton by a neutron or a kaon by a pion (not only is the set I, I_3, Y different from proton to neutron, kaon to pion, but also the mass). We consider super-multiplets of baryons and pseudoscalar mesons where the degenerate mass of all baryons is the same and of all pseudoscalar mesons the same. We then look for symmetry breaking interactions to account for the observed mass spectrum.

2. SU(3) and its Representations

2.1 The Sakata Model and SU (3) [17]

If we try to generalize I-spin to objects that also have hypercharge, we are led to models such as SAKATA's. SAKATA adds another particle with isospin 0 and hypercharge 0, the Λ, to the nucleon doublet.

We will use the Sakata model as an introduction to SU(3); we take the triplet $\begin{pmatrix} p \\ n \\ \Lambda \end{pmatrix} = \begin{pmatrix} \xi_1 \\ \xi_2 \\ \xi_3 \end{pmatrix}$, the doublet is $I = 1/2$, $Y = 1$, the singlet is $I = 0$, $Y = 0$. We subject this "vector" to unitary transformations as before [actually, in the Sakata model it is more appropriate to use the Strangeness S, rather than Y, since the latter (see below) is not integral].

$$\xi_i' = \sum_j U_{ij}\, \xi_j$$

$$U^+ = U^{-1},$$

consider det $U = 1$ for simplicity. Again, any unitary matrix can be written $U = e^{iH}$, H hermitian. Any matrix in 3-space can be written in terms of the unit matrix and 8 others. We take the set of 8 matrices, the λ matrices, traceless (corresponding to det $U = 1$).

The first 3 λ's, λ_1, λ_2, λ_3, will simply be the isospin operators in the doublet space (with $I_\pm = \frac{1}{2}(\lambda_1 \pm i\,\lambda_2), I_3 = \lambda_3$):

$$\lambda_1 = \begin{pmatrix} 0 & 1 & 0 \\ 1 & 0 & 0 \\ 0 & 0 & 0 \end{pmatrix}, \quad \lambda_2 = \begin{pmatrix} 0 & -i & 0 \\ i & 0 & 0 \\ 0 & 0 & 0 \end{pmatrix}, \quad \lambda_3 = \begin{pmatrix} 1 & 0 & 0 \\ 0 & -1 & 0 \\ 0 & 0 & 0 \end{pmatrix}.$$

A set that changes isospin and hypercharge (strangeness!) but does not change charge are U_+ and U_-,

$$U_+ = \begin{pmatrix} 0 & 0 & 0 \\ 0 & 0 & 1 \\ 0 & 0 & 0 \end{pmatrix}, \qquad U_- = \begin{pmatrix} 0 & 0 & 0 \\ 0 & 0 & 0 \\ 0 & 1 & 0 \end{pmatrix}$$

so that

$$\lambda_4 = U_+ + U_- = \begin{pmatrix} 0 & 0 & 0 \\ 0 & 0 & 1 \\ 0 & 1 & 0 \end{pmatrix}, \qquad \lambda_5 = -i\,U_+ + i\,U_- = \begin{pmatrix} 0 & 0 & 0 \\ 0 & 0 & -i \\ 0 & i & 0 \end{pmatrix}.$$

A set that changes isospin and hypercharge (strangeness) and also changes charge are V_+ and V_-,

$$V_+ = \begin{pmatrix} 0 & 0 & 1 \\ 0 & 0 & 0 \\ 0 & 0 & 0 \end{pmatrix}, \qquad V_- = \begin{pmatrix} 0 & 0 & 0 \\ 0 & 0 & 0 \\ 1 & 0 & 0 \end{pmatrix}$$

so that

$$\lambda_6 = V_+ + V_- = \begin{pmatrix} 0 & 0 & 1 \\ 0 & 0 & 0 \\ 1 & 0 & 0 \end{pmatrix}, \qquad \lambda_7 = -i\,V_+ + i\,V_- = \begin{pmatrix} 0 & 0 & -i \\ 0 & 0 & 0 \\ i & 0 & 0 \end{pmatrix}.$$

We want λ_8 to be connected with hypercharge (strangeness in the Sakata model proper) eigenvalues only and to commute with isospin. The only linearly independent matrix with vanishing trace is then:

$$\lambda_8 = \begin{pmatrix} a & 0 & 0 \\ 0 & a & 0 \\ 0 & 0 & -2a \end{pmatrix},$$

where we choose $a = 1/\sqrt{3}$ for normalization.

The commutation relations are similar to those for isospin: $[\lambda_i/2, \lambda_j/2] = 2i f_{ijk}\,\lambda_k/2$ where f_{ijk} is real antisymmetric with $f_{123} = +1$, $f_{147} = f_{246} = f_{257} = f_{345} = f_{571} = f_{367} = 1/2$, $f_{438} = f_{678} = \sqrt{3}/2$ and $f_{ijk} = 4\,i\,Tr\,\lambda_k\,[\lambda_i, \lambda_j]$.

The anticommutation relations are:

$\{\lambda_i/2, \lambda_j/2\} = \frac{1}{3}\,\delta_{ij} + d_{ijk}\,\lambda_k/2$ where d_{ijk} is real and symmetric with $d_{888} = -1/\sqrt{3}$, $d_{118} = d_{228} = d_{338} = 1/\sqrt{3}$, $d_{448} = d_{558} = d_{668} = d_{778} = -1/2\sqrt{3}$ and $d_{ijk} = 4\;Tr\,\lambda_k\,\{\lambda_i, \lambda_j\}$. We have further the normalization $Tr\,\lambda_i\,\lambda_j = 2\,\delta_{ij}$. We further define $\lambda_0 = \sqrt{2/3}\cdot 1$ so that we have a complete set of 9 matrices, i. e., any 3×3 matrix can be expressed in terms of the 9 matrices.

In the same way that one takes products of doublets in SU(2), we take products of triplets in SU(3). We label the representation in terms of spherical vector operators, i. e., we use the 3 sets of raising and lowering operators: $I_+, I_-, V_+, V_-, V_+, U_-$, and I_3, Y (or S). In $2\otimes 2 = 1 + 3$ the 3 representation is a vector representation the elements of which can be expressed in terms of τ_+, τ_-, τ_3 the spherical vector operators instead of by τ_1, τ_2, τ_3 cartesian vector operators. Similarly one can proceed with the "sakatons" p, n, Λ, and the λ_i-matrices, and construct "meson-states", i. e. states of integral spin and having the same isospin-hypercharge as the known mesons. There is no ambiguity for off-diagonal terms:

$$B = \begin{pmatrix} p \\ n \\ \Lambda \end{pmatrix}, \qquad \bar{B} = (\bar{p}\,\bar{n}\,\bar{\Lambda})$$

$$\begin{array}{ll} \bar{B} I_+ B = \bar{p} n \to \pi^- & \bar{B} I_- B = \bar{n} p \to \pi^+ \\ \bar{B} U_+ B = \bar{p} \Lambda \to K^- & \bar{B} U_- B = \bar{\Lambda} p \to K^+ \\ \bar{B} V_+ B = \bar{n} \Lambda \to \bar{K}^0 & \bar{B} V_- B = \Lambda n \to K^0 \end{array}$$

$$\bar{B} I_3 B = \frac{\bar{p} p - \bar{n} n}{\sqrt{2}} \to \pi^0 \quad \Big| \quad \bar{B} Y B = \frac{\bar{p} p + \bar{n} n - 2 \bar{\Lambda} \Lambda}{\sqrt{3}} \to \chi.$$

The $\bar{\Lambda}\Lambda$ term appears only with the particle χ, so that we expect $\bar{\Lambda}\Lambda = = \text{const.}\ \chi \,.\, \bar{p} p + \bar{n} n = -\bar{\Lambda}\Lambda$, if the trace vanishes. $\bar{p} p = \sqrt{2}\,\pi^0 + \bar{n} n$ or $\bar{p} p - \pi^0/\sqrt{2} = \bar{n} n + \pi^0/\sqrt{2}$. Let us add $+\chi/\sqrt{6}$ to both sides and equate to zero. Then $\bar{p} p = \pi^0/\sqrt{2} - \chi/\sqrt{6}$, $\bar{n} n = -\pi^0/\sqrt{2} - \chi/\sqrt{6}$ and $\bar{p} p + \bar{n} n = = -2/\sqrt{3}\,\chi$, therefore $\bar{\Lambda}\Lambda = \sqrt{2/3}\,\chi$. Or in matrix notation:

$$P = \begin{pmatrix} \frac{\pi^0}{\sqrt{2}} - \frac{\chi}{\sqrt{6}} & \pi^+ & K^- \\ \pi^- & -\frac{\pi^0}{\sqrt{2}} - \frac{\chi}{\sqrt{6}} & \bar{K}^0 \\ K^+ & K^0 & \sqrt{\frac{2}{3}}\,\chi \end{pmatrix}.$$

We can consider the 8 traceless λ matrices as the simplest representation of a set of more abstract quantities forming a matrix algebra, quantities which we call F_i, general operators in a finits dimensional space that obey the commutation relations $[F_i, F_j] = 2 f_{ijk}\, F_k$, $i, j, k = 1, 2, \ldots 8$. The Pauli spin matrices are just the simplest representation of the algebra of angular momentum operators, satisfying the commutation relation $[J_i, J_j] = i\,\varepsilon_{ijk}\, J_k$, $i, j, k = 1, 2, 3$. Any representation (i. e., mapping onto a set of linear transformations of a vector space) is characterized by the largest value of J_3, its spin or angular momentum. The J's are generators of continuous rotations. The F's are just the generalization of the J's to SU(3).

Some additional notions before we get into the tensor analysis of SU(3). SU(3) is what is called a compact group. A compact group is a closed manifold so that you can define averaging over the "volume" of the group. For compact groups you can make all representations unitary. SU(3) is also simple: it has no invariant subgroups except the trivial group itself and the identity. U(3) is not simple: it contains an invariant subgroup if det $U \neq 1$ (all matrices of type $e^{i\alpha}\,\delta_{kj}$). We wish to find representations of SU(3) (i. e., they obey the same commutation relations as the F's), but we are not interested in any representations except irreducible ones. If the representation is irreducible, then the vector space we consider does not admit subspaces that are invariant, that is, the elements of which transform among themselves.

The motivation for abandoning the Sakata model comes from the facts that although we can build the octets of pseudoscalar and vector mesons, the Σ's are considered bound states of $\Lambda\pi$; this gives the relative parity of $\Sigma\Lambda$ as minus, whereas experiment tells us it is plus. The Ξ doublet has completely different properties than what is expected under the Sakata model. And finally the annihilation of baryons and antibaryons into mesons is in contradiction with the predictions of this model. We can arrange the baryons into a matrix by analogy with the matrix of the pseudoscalar mesons:

$$B = \begin{Bmatrix} \frac{\Sigma^0}{\sqrt{2}} - \frac{\Lambda}{\sqrt{6}} & \Sigma^+ & \Xi^- \\ \Sigma^- & -\frac{\Sigma^0}{\sqrt{2}} - \frac{\Lambda}{\sqrt{6}} & \Xi^0 \\ p & n & \sqrt{\frac{2}{3}}\Lambda \end{Bmatrix}$$

i. e., we consider the 8 baryons as elements of the adjoint representation.

2.2. *Tensor Algebra of SU*(3)

The tensor algebra of SU(3) is a straightforward generalization from real vector spaces. A set of three complex numbers u^i, i. e. a 3-dimensional complex vector, transforms as:

$$u^j = \sum_i a_i{}^j\, u^i \qquad i, j = 1, 2, 3.$$

A transformation a is unitary if a^+ (hermitean conjugate) $= a^{-1}$ (inverse), i. e. $a_j{}^{i*} = (a^{-1})_i{}^j$. The complex conjugate vector transforms as:

$$(u'^j)^* = \sum_i (a_i{}^j)^*\,(u^i)^* = \sum_i (a^{-1})_j{}^i\,(u^i)^*.$$

We introduce the contragradient transformation $(a^{-1})_j{}^i$ and the vector $u'^{j*} = u'_j$, $u^{i*} = u_i$; u'_j, u_i are covariant vectors, whereas u'^j, u^i are contravariant vectors. In 2 dimensions the complex conjugate matrices $a_i{}^j$, $a_i{}^{j*}$ are equivalent, i. e., there exists a similarity transformation S such that $a^* = SaS^{-1}$; for 3 dimensions this is not so, and in general for even dimensional spaces a and a^* are equivalent, for odd dimensional spaces not. We have thus obtained two representations of SU(3): a representation by a, the $\{3\}$ and by a^*, the $\{3^*\}$. The simplest representation is the 1-dimensional one which is trivial.

We proceed as in the tensor calculus: take a product of p (upper indices) contravariant vectors and q (lower indices) covariant ones:

$$T^{ij\ldots k}_{rs\ldots t} = t^i u^j \ldots v^k x_r y_s \ldots z_t.$$

Call this a tensor of type (p, q). It transforms as (the tensor product of the vectors):

$$T^{i'j'\ldots k'}_{r's'\ldots t'} = a_i^{i'} a_j^{j'} \ldots a_k^{k'} \left(a^{-1}\right)^r_{r'} \left(a^{-1}\right)^s_{s'} \ldots \left(a^{-1}\right)^t_{t'} T^{ij\ldots k}_{rs\ldots t}$$

where all indices run from 1 to 3. Such an object is not what we call irreducible. Can we find subjects which transform among themselves? If we contract with respect to 2 indices (this is called taking a trace) then we have left objects $T^{ij\ldots k}_{is\ldots t}$ of $(p-1, q-1)$ indices which transform among themselves, i. e., that form an invariant subspace. Since the t's, u's, v's, etc. are just numbers we can permute them without changing T; therefore we should symmetrize T in all upper and lower indices. We finally arrive, after removing all subsets of lower indices and symmetrizing at both levels separately, at a tensor $T^{ij\ldots k}_{rs\ldots t}$ symmetric in all upper indices, symmetric in all lower indices, and traceless.

Let us see what dimension our tensors have. A representation is usually labeled by $D(p, q)$, the D coming from the German for representation: Darstellung. $D(p, q) = D(0, 0)$. $D(0, 0)$ is obviously the trivial representation of dimension 1, $\{n\} = \{1\}$. $D(1, 0) = \{3\}$. $D(0, 1) = \{3^*\}$. $D(1, 1)$ is the set of elements that transform as $t_j^i = u^i v_j$ and are traceless $t_j^i = u^i v_j - \frac{1}{3} \delta_j^i u^k v_k$ therefore $D(1, 1)$ has dimension 8. We call this the regular or adjoint representation, both are the same in this case. The 8 matrices are such quantities t_j^i.

In general for $D(p, q) = \{n\}$, $n = \frac{1}{2}(p+1)(q+1)(p+q+2)$.

Proof: We have an object with p upper and q lower indices: $T^{U_1\ldots U_p}_{V_1\ldots V_q}$.

Each index can take on only values 1, 2, 3; further since T is symmetric in all upper indices and in all lower ones, permutations of the upper or lower indices are irrelevant. We fix the number of upper indices equal to 1; call it s: $s = 0, 1, \ldots p$. There are $p-s$ indices equal to 2 or 3. There are $p-s+1$ ways of arranging the 2's and 3's hence

$$n_{\text{upper}} = \sum_{s=0}^{p} (p-s+1) = \tfrac{1}{2}(p+1)(p+2).$$

Similarly for the lower indices $n_{\text{lower}} = \frac{1}{2}(q+1)(q+2)$. The trace of a tensor (p, q) is a tensor $(p-1, q-1)$. How many independent traces are there or how many $(p-1, q-1)$ tensors? There are

$$[\tfrac{1}{2} p (p+1)] \times [\tfrac{1}{2} q (q+1)].$$

The dimension n of the representation

$$D(p, q) \text{ is } n = [\tfrac{1}{2}(p+1)(p+2)] \times [\tfrac{1}{2}(q+1)(q+2)] - [\tfrac{1}{2} p (p+1)] \times [\tfrac{1}{2} q (q+1)] = \tfrac{1}{2}(p+1)(q+1)(p+q+2).$$

We enumerate the irreducible representations of SU(3):

$D(0,0) = \{1\}$	
$D(1,0) = \{3\}$	$D(0,1) = \{3^x\}$
$D(2,0) = \{6\}$	$D(0,2) = \{6^x\}$
$D(1,1) = \{8\}$	
$D(3,0) = \{10\}$	$D(0,3) = \{10^x\}$
$D(2,1) = \{15\}$	$D(1,2) = \{15^x\}$
$D(4,0) = \{15\}$	$D(0,4) = \{15^x\}$
$D(5,0) = \{21\}$	$D(0,5) = \{21^x\}$
$D(3,1) = \{24\}$	$D(1,3) = \{24^x\}$
$D(2,2) = \{27\}$	

The underlined representations are physically significant. This is due to the requirement of having integral Y which obtains when $|p-q| = 3n$. [The other representations, like $D(1,0)$, for the Sakata model, have non-integral Y, but can be made to have integral S-strangeness.]

2.3 Infinitesimal Operators

For a general representation of SU(3) we have the F operators which obey the communication relation

$$[F_i, F_j] = 2\,i\,f_{ijk}\,F_k \qquad i, j, k = 1, 2, \ldots 8.$$

F_1, F_2, F_3 are the isospin operators I_1, I_2, I_3, $F_8 = \sqrt{3/2}\,Y$. Matrices of the transformations in the space of 3 dimensions are just the λ's, which, of course, satisfy the same commutation relations but are now not abstract operators but 3×3 matrices. One could use different ways of parametrizing the group; a particularly useful parametrization is such that for vanishing parameters one obtains the unit element. In terms of this parametrization

$$U_k^i = \delta_k^i + a_k^i$$

and the infinitesimal transformations are

$$A_k^i = \left(\frac{\partial U}{\partial a_i^k}\right)_{a_i^j = 0}.$$

There are 9 infinitesimal operators of the group of unitary transformations in 3-space

$$(A_k^{\,i})_\mu^{\;\nu} = \delta_\mu^{\;i}\,\delta_k^{\;\nu},$$

but we take out the trace [in order to have SU(3)], so that

$$(A_k^i)_\mu^{\;\nu} = \delta_\mu^{\;i}\,\delta_k^{\;\nu} - \tfrac{1}{3}\,\delta_k^i\,\delta_\mu^{\;\nu}.$$

Under hermitian conjugation the A's have the property under conjugation $(A_k^i)^+ = A_i^k$, and (this is not to be confused with tracelessness since the matrix element indices are not written out) $A_1^{\,1} + A_2^{\,2} + A_3^{\,3} = 0$. The commutation relations of these matrices are:

$$[A_k^{\,i}, A_l^{\,j}] = \delta_l^{\,i}\,A_k^{\,j} - \delta_k^{\,j}\,A_l^{\,i}$$

hence the structure constants of the group are $\delta_l^i \, \delta_k^j - \delta_k^j \, \delta_l^i$. The same notation is used for this representation in n dimensions.

Let us write out the connection between the A's and the F's.

$$A_1^{\,1} = F_3 + \frac{1}{\sqrt{3}} F_8 \qquad A_1^{\,2} = \frac{F_1 + i\,F_2}{2} \qquad A_1^{\,3} = \frac{F_4 + i\,F_5}{2} \qquad A_2^{\,3} = \frac{F_6 + i\,F_7}{2}$$

$$A_2^{\,2} = -F_3 + \frac{1}{\sqrt{3}} F_8 \qquad A_2^{\,1} = \frac{F_1 + i\,F_2}{2} \qquad A_3^{\,1} = \frac{F_4 - i\,F_5}{2} \qquad A_3^{\,2} = \frac{F_6 - i\,F_7}{2}$$

$$A_3^{\,3} = -\frac{2}{\sqrt{3}} F_8 .$$

Further let us write out the connection between the A's and the raising and lowering operators $I_\pm$, $U_\pm$, $V_\pm$:

$$A_1^{\,1} = I_3 + \frac{Y}{2} \quad A_1^{\,2} = I_+ \quad A_1^{\,3} = V_+ \quad A_2^{\,3} = U_+ \quad A_2^{\,2} = -I_3 + \frac{Y}{2}$$

$$A_2^{\,1} = I_- \qquad A_3^{\,1} = V_- \quad A_3^{\,2} = U_- \quad A_3^{\,3} = -Y .$$

For the 3×3 representation $\lambda_i = 2\,F_i$.

We want now to derive the communtation relations, particularly between $I_\pm$, $U_\pm$, $V_\pm$. For $[I_+, I_-]$ we should expect, by analogy with the angular momentum operators, to obtain $2\,I_3$. Indeed:

$$[I_+, I_-] = [A_1^{\,2}, A_2^{\,3}] = \delta_2^{\,2}\, A_1^{\,1} - \delta_1^{\,1}\, A_2^{\,2} = 2\, I_3 .$$

I_+, I_-, I_3 will generate an algebra (the SU(2) algebra) if we can verify that $[I_\pm, I_3] = \pm I_\pm$

$$[I_+, I_3] = \left[A_1^{\,2}, \frac{A_1^{\,1} - A_2^{\,2}}{2}\right] = \tfrac{1}{2}\,[A_1^{\,2}, A_1^{\,1}] - \tfrac{1}{2}\,[A_1^{\,2}, A_2^{\,2}] =$$
$$= -\tfrac{1}{2}\, A_1^{\,2} - \tfrac{1}{2}\, A_1^{\,2} = -I_+$$

$$[I_-, I_3] = \left[A_2^{\,1}, \frac{A_1^{\,1} - A_2^{\,2}}{2}\right] = \tfrac{1}{2}\,[A_2^{\,1}, A_1^{\,1}] - \tfrac{1}{2}\,[A_2^{\,1}, A_2^{\,2}] =$$
$$= \tfrac{1}{2}\, A_2^{\,1} + \tfrac{1}{2}\, A_2^{\,1} = +I_- .$$

Thus the I's form an algebra just like $\tau_\pm$, τ_3.

Let us look at the U's:

$$[U_+, U_-] = [A_2^{\,3}, A_3^{\,2}] = A_2^{\,2} - A_3^{\,3} = I_3 + \tfrac{3}{2}\, Y .$$

We define $U_3 = \frac{1}{2}\,(-I_3 + \frac{3}{2}\, Y)$. So we get another triplet of SU(2) operators. We can also do the same for $V_\pm$ defining $V_3 = \frac{1}{2}\,(I_3 + \frac{3}{2}\, Y)$. The sets U_+, U_-, U_3 and V_+, V_-, V_3 are thus also SU(2) algebras just like the set I_+, I_-, I_3 only rotated through angles of 120^0 and 240^0 respectively (look at I_3, Y diagram for any set of particles). So in some way SU(3) is a sum (although not a direct sum) of SU(2)'s. This aspect of SU(3) has been emphasized by Elliott [18], Edmonds [19], and Levinson, Lipkin, and Meshkov [20].

In the next two sections we will investigate the geometry of SU(3).

2.4 Weight Vectors

We introduce some concepts that illustrate the theory of groups of rank 2. Among the set of 8 operators only I_3 and Y commute. In our representation $\{3\}$ these 2 operators are already diagonal:

$$I_3 = \frac{1}{2}\begin{pmatrix} 1 & 0 & 0 \\ 0 & -1 & 0 \\ 0 & 0 & 0 \end{pmatrix} \qquad Y = \frac{2}{\sqrt{3}} F_8 = \frac{1}{3}\begin{pmatrix} 1 & 0 & 0 \\ 0 & 1 & 0 \\ 0 & 0 & -2 \end{pmatrix}$$

For higher representations we also want I_3 and Y to be diagonal. We are now in the $D(1, 0) = \{3\}$ representation. We can construct what is called a weight diagram, labeling the allowed eigenstates by the valus of I_3 and Y. The commuting operators for a group (their number is the rank of the group) are often denoted by $H_1, H_2, \ldots H_r$ where r is the rank. In our case $r = 2$ and we have $H_1 = I_3$, $H_2 = \frac{\sqrt{3}}{2} Y$.

We combine them into the vector $\underset{\sim}{H} = \left(I_3, \frac{\sqrt{3}}{2} Y\right)$. $\Bigl($Behrends et al. [21] use $\underset{\sim}{H} = \frac{1}{\sqrt{3}}\left(I_3, \frac{\sqrt{3}}{2} Y\right)\Bigr)$.

The eigenvalues of $\underset{\sim}{H}$ together form the weight vector $\underset{\sim}{W} = (W_1, W_2) = \left(I_3, \frac{\sqrt{3}}{2} Y\right)$ although when we draw weight vectors we plot I_3 along the X axis and $\sqrt{3}/2\, Y$ along the Y axis and label the vertical axis by the values of Y. The commutation relations become:

$$\begin{aligned} [\underset{\sim}{H}, I_+] &= \pm \underset{\sim}{i}\, I_+ & \underset{\sim}{i} &= (1, 0) \\ [\underset{\sim}{H}, U_\pm] &= \pm \underset{\sim}{u}\, U_\pm & \underset{\sim}{u} &= \left(-\frac{1}{2}, \frac{\sqrt{3}}{2}\right) \\ [\underset{\sim}{H}, V_+] &= \pm \underset{\sim}{v}\, V_\pm & \underset{\sim}{v} &= \left(\frac{1}{2}, \frac{\sqrt{3}}{2}\right) \end{aligned}$$

where the two-dimensional vectors $\underset{\sim}{i}$, $\underset{\sim}{u}$, $\underset{\sim}{v}$ are not themselves weight vectors since they characterize the operators and not the states. (They are called by mathematicians "root vectors".) The rest of the commutation relations may be written down in abbreviated form:

$$\begin{aligned} [I_+, I_-] &= 2 I_3 = 2 \underset{\sim}{i} \cdot \underset{\sim}{H} \\ [U_+, U_-] &= -I_3 + \tfrac{3}{2} Y = 2 \underset{\sim}{u} \cdot \underset{\sim}{H} \\ [V_+, V_-] &= I_3 + \tfrac{3}{2} Y = 2 \underset{\sim}{v} \cdot \underset{\sim}{H}. \end{aligned}$$

Further commutation relations are

$$\begin{aligned} [I_-, V_+] &= U_+ & [I_+, U_+] &= V_+ \\ [V_+, U_-] &= I_+ & [U_+, V_-] &= I_- \\ [U_-, I_-] &= V_- & [V_-, I_+] &= U_-. \end{aligned}$$

$[V_+, I_+] = 0$, as do all the other commutators of

$$I_\pm,\ U_\pm,\ V_\pm.$$

Let us plot $\underset{\sim}{i}$, $\underset{\sim}{u}$, $\underset{\sim}{v}$:

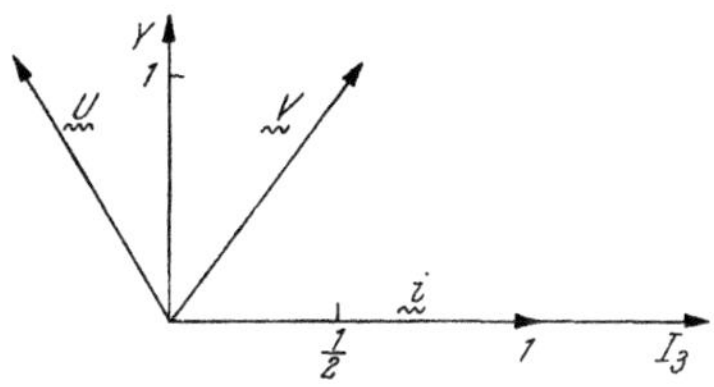

Fig. 9

Lines perpendicular to these vectors are of special significance and will figure prominently in the so-called Weyl reflections.

The representation $D(1, 1)$ contains, e. g., 8 states or particles. If we try to label these by I_3 and Y we will *not* get a complete identification as e. g. Λ and Σ^0 have the same I_3 and Y. A weight vector is not sufficient to uniquely identify a particle. We need another number. We label the particle also by the eigenvalue of $I^2 : I(I+1)$. If we add $(p \cdot q)$ we get a more than adequate labeling. An alternate way of identifying a representation is by labeling with the "Casimir operators" F^2 and G^3, where

$$F^2 = \sum_{i=1}^{8} F_i^2 = \tfrac{1}{2} \sum_{ij=1} A_j^{\,i} A_i^{\,j}$$

the square of "unitary momentum", and $G^3 = \sum_{ijk}^{3} A_j^{\,i} A_k^{\,j} A_j^{\,k}$ [22, 23].

We will study the I_3 and Y content of various irreducible representations. Each $T^{u_1 \ldots u_p}_{v_1 \ldots v_q} \in D(p, q)$ can be labeled by its weight. We can then plot the weights on an $I_3 - Y$ coordinate system, obtaining the weight diagram. The weight with the largest I_3 is the *highest weight*. $D(p \cdot q)$ is contained in $p\, D(1, 0) \otimes q\, D(0, 1)$ symmetrized and made traceless. Symmetrizing and taking out traces does not affect the weights. The highest weight of such a product is therefore $w_{\max}(p, q) = p\, w_{\max}(1, 0) + q\, w_{\max}(0, 1)$. For any irreducible representation the highest weight is always simple, i. e., never has 2 or more states with the same I_3 and Y. This state is labeled $|\,(p, q)\, I\, w_{\max}\rangle$.

Let us look at a representation that gives us non-integral values of Y. We introduce the triplet $\varphi = \begin{pmatrix} \varphi_1 \\ \varphi_2 \\ \varphi_3 \end{pmatrix}$. The corresponding weight vectors $\underset{\sim}{w}$, labeled by (I_3, Y) not $\left(I_3, \frac{\sqrt{3}}{2} Y\right)$ now, are $\underset{\sim}{w}_1 = \left(\frac{1}{2}, \frac{1}{3}\right)$, $\underset{\sim}{w}_2 = \left(-\frac{1}{2}, \frac{1}{3}\right)$ and $\underset{\sim}{w}_3 = \left(0, -\frac{2}{3}\right)$. The highest weight of $D(1, 0)$, which is the representation we are looking at, is $\underset{\sim}{w}_{\max} = \underset{\sim}{w}_1$. The weight diagram is:

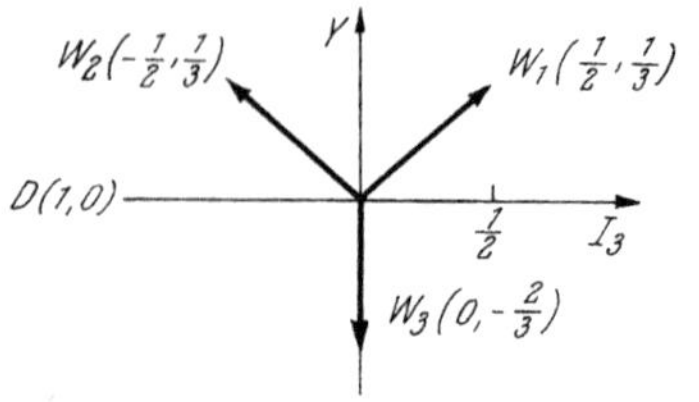

Fig. 10

Hypercharge is not a convenient number to label states here because it is not an integer. All representations for which $|p-q| \neq 3\,n$ have this defect (and, e. g., in the Sakata model, one uses strangeness in place of Y). We will use them only to construct the "physical" representation of the antiparticles of $D\,(1, 0)$. When we go over to antiparticles we take the charge conjugate which reverses the sign of the hypercharge. Therefore $\underset{\sim}{w}_1{}^{*} = (-\frac{1}{2}, -\frac{1}{3})$ and $\underset{\sim}{w}_2{}^{*} = (\frac{1}{2}, -\frac{1}{3})$ $\underset{\sim}{w}_3{}^{*} = (0, \frac{2}{3})$. The weight diagram is

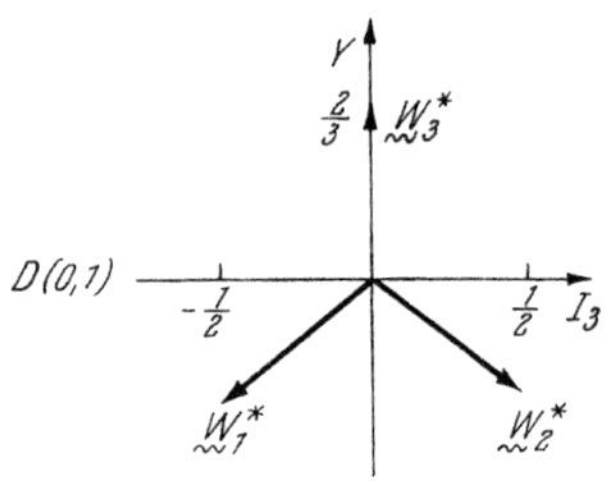

Fig. 11

We have $\underset{\sim}{w}^*_{\max} = \underset{\sim}{w}_{\max}\ (0, 1) = \left(\frac{1}{2}, -\frac{1}{3}\right) = \underset{\sim}{w}_2{}^{*}$. In general, therefore, $\underset{\sim}{w}_{\max}\ (p, q) = p\left(\frac{1}{2}, \frac{1}{3}\right) + q\left(\frac{1}{2}, -\frac{1}{3}\right) = \left(\frac{p+q}{2}, \frac{p-q}{3}\right)$. The highest isospin we can find for a representation is $\frac{p+q}{2}$. For physically meaningful results Y must be integral. In this way we find the highest weight from which we find the highest isospin state $I_{3_{\max}} = I$. From he vector with the highest value of isospin $\left|(p, q), I = \frac{p+q}{2}, I_3, Y = \frac{p-q}{3}\right\rangle$ we can generate whole sets of eigenvectors by applying the raising and lowering operators and choosing a set of appropriate phases.

From WEYL's "branching" theorem $\mathrm{SU}(3) = \sum_{\text{all combinations}} \mathrm{SU}(2)$ (see e. g. 22, 23, 24) one can derive that for any irreducible representation $D\,(p, q)$, for any pair of positive integers f, g, such that $p+q \geq$

$\geq f \geq q \geq g \geq 0$, there exists in $D(p, q)$ a set of states I, Y such that $I = \frac{1}{2}(f - g)$, $Y = f + g - \frac{2}{3}(p + 2q)$. From this theorem we derive if $q = 0$, $p = 3n$, then $I = \frac{1}{2}Y + \frac{1}{3}p$ and all weights are simple. This is important for the representation $D(3, 0) = \{10\}$.

The weights for the simplest representations can be found by the above method (we omit the $D(0, 0)$ which is trivial and $D(1, 0)$ and $D(0, 1)$ which we have seen):

$\underline{D(1, 1) = \{8\}}$

$$p + q \geq f \geq q \geq g \geq 0$$
$$2 \geq f \geq 1 \geq g \geq 0$$

If $f = 2$ $\quad g = 1$ or 0;
If $f = 1$ $\quad g = 1$ or 0;

$f = 2,\ g = 1,\ I = \frac{1}{2},\ I_3 = \frac{1}{2},\quad Y = 1,\ f = 1,\ g = 1,\ I = 0,\ I_3 = 0,\ Y = 0,$
$\qquad I_3 = -\frac{1}{2},\ Y = 1.$

$f = 2, g = 0, I = 1, I_3 = 1,\quad Y = 0, f = 1, g = 0, I = \frac{1}{2}, I_3 = \frac{1}{2},\quad Y = -1,$
$\qquad I_3 = 0,\quad Y = 0 \qquad\qquad I_3 = -\frac{1}{2}, Y = -1$
$\qquad I_3 = -1,\ Y = 0.$

The weight diagram is:

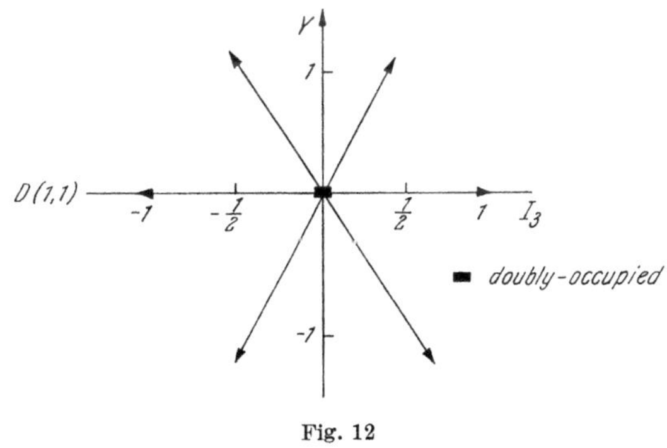

Fig. 12

For the $D(3, 0) = \{10\}$ we find the highest weight very easily by our general rule. We omit the details of findings the weights.

The next representation which is physically interesting, i. e., has integral Y value, and is found in the decomposition of $8 \otimes 8$ is $D(2, 2) = \{27\}$.

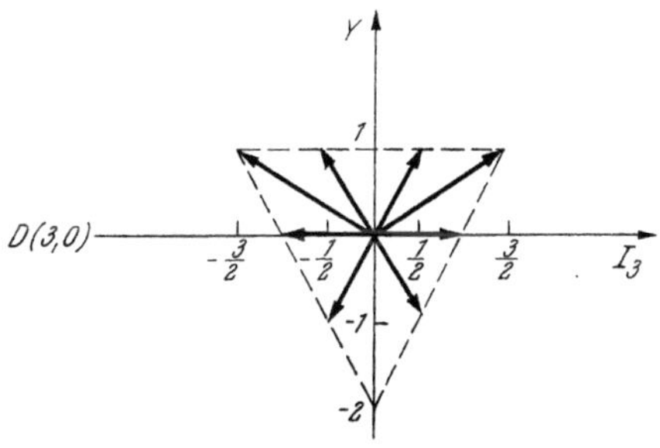

Fig. 13

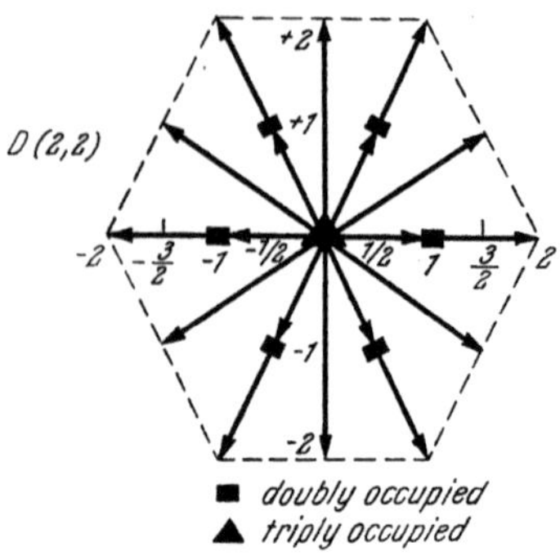

Fig. 14

We note that peripheral weights are always simple.

2.5. *Weyl Reflections*

$D(p, q) \to D(q, p)$ means going over to the contragredient representation. The matrices of the transformation U go over to their complex conjugate U^*.

$$U = e^{iH} = e^{i \sum_j a_j F_j}$$

$$U^* = e^{-iH^*} = e^{-i \sum_j a_j F_j^*} = e^{i \sum a_j F_j'}$$

$$U^{*T} = e^{-iH^{*T}} = e^{-i \sum_j a_j F_j^{*T}} = e^{-i \sum_j a_j F_j} = U^{-1}$$

$$F_j'^T = -F_j = -F_j^{*T}$$

$$\text{or } F_j' = -F_j^T = -F_j^*.$$

Thus $I_3' = -I_3$, $Y' = -Y$ since they are diagonal. Further $I_\pm' = -I_\mp$. This is unsatisfactory, for a phase change has been introduced, so that we fix by definition $I_\pm' = I_\mp$ and also $U_\pm = +U_\mp$. In order not to change the commutation relations one must define $V_\pm' = -V_\mp$. The state vectors undergo the change:

$$|(p, q)\, I, I_3\, Y> \to |(q, p)\, I, I_3, Y> =$$
$$= (-1)^{I_3 + \frac{Y}{2}} e^{i\alpha} (|(p, q)\, I, -I_3, -Y>)^*.$$

The symmetry axes of the diagrams are perpendicular to $\underset{\sim}{i}$, $\underset{\sim}{u}$, $\underset{\sim}{v}$ and are labeled 1, 2, 3 respectively. They were first studied by Cartan [25] and Weyl [24]. The 1-axis is related to charge symmetry, i. e., reflection through the 1-axis changes $I_3 \to -I_3$, so that it connects only the different isotopic states of an isomultiplet. Reflection through the 3-axis connects only states of the same charge. Reflection through the 2-axis has no self-evident physical meaning. WEYL reflections about these axes are denoted by W_1, W_2, W_3.

Other people use P_I, P_U, etc. We will adhere to the W notation. WEYL reflections give us no direct information about multiply-occupied states. W_1 is a true symmetry. W_2 and W_3 are broken symmetries to the extent that the masses are different: we will carry out our analysis as if they were not broken. For all α, $\alpha = 1, 2, 3$ $W_\alpha^{-1} = W_\alpha$. Further $W_3 = W_1 W_2 W_1$, W_2 and W_3 are conjugate elements, etc. In the Sakata model for the $D(1, 0)$ representation we can represent WEYL reflections by 3×3 matrices:

$$W_1 = -\begin{pmatrix} 0 & 1 & 0 \\ 1 & 0 & 0 \\ 0 & 0 & 1 \end{pmatrix}, \quad W_2 = -\begin{pmatrix} 1 & 0 & 0 \\ 0 & 0 & 1 \\ 0 & 1 & 0 \end{pmatrix}, \quad W_3 = -\begin{pmatrix} 0 & 0 & 1 \\ 0 & 1 & 0 \\ 1 & 0 & 0 \end{pmatrix}.$$

W interchanges p and n; W_2 interchanges n and Λ; W_3 interchanges p and Λ. We include the minus sign since we want, det $W_\alpha = 1$. There is one more way to represent WEYL reflections. Look at the isospin part of the G parity operator $e^{i\pi I_2}$; it changes the isospin projection by a sign. Therefore $W_1 = e^{i\pi I_2}$. The same thing can be done for the 2 and 3 axes. Since

$$[U_+, U_-] = -I_3 + \tfrac{3}{2} Y, \quad [V_+, V_-] = I_3 + \tfrac{3}{2} Y,$$

define $U_3 = -\dfrac{I_3}{2} + \dfrac{3}{4} Y$, $V_3 = +\dfrac{I_3}{2} + \dfrac{3}{4} Y$, making the U's and V's into full SU(2) algebras. By definition

$$U_1 = \frac{U_+ + U_-}{2}, \quad U_2 = \frac{U_+ - U_-}{2i},$$

similarly for V_1, V_2. The WEYL reflections W_2 and W_3 can now be written in analogy with charge symmetry as:

$$W_2 = e^{i\pi U_2}$$
$$W_3 = e^{i\pi V_2}.$$

Using the exponential form for W or looking at the $D(1, 0)$ representation of the operators (since we deal with an algebra, the relation we obtain must be true for any representation), we derive:

$$\begin{aligned} W_1 I_\pm W_1 &= -I_\mp \\ W_2 I_\pm W_2 &= V_\mp \\ W_3 I_\pm W_3 &= U_\mp. \end{aligned}$$

We can derive the properties of $\underset{\sim}{H} = \left(I_3, \dfrac{\sqrt{3}}{2} Y\right)$ under WEYL reflection.

$$\begin{aligned} W_1 \underset{\sim}{H} W_1 &= \underset{\sim}{H} - 2 \underset{\sim}{i} (\underset{\sim}{i} \cdot \underset{\sim}{H}) \\ W_2 \underset{\sim}{H} W_2 &= \underset{\sim}{H} - 2 \underset{\sim}{u} (\underset{\sim}{u} \cdot \underset{\sim}{H}) \\ W_3 \underset{\sim}{H} W_3 &= \underset{\sim}{H} - 2 \underset{\sim}{v} (\underset{\sim}{v} \cdot \underset{\sim}{H}). \end{aligned}$$

If we have a weight vector and use Weyl reflections, if the weight does not lie on an axis, we get 5 more vectors; if it does, we get only 2 more. Using ladder operators, we will then get all possible roots.

We list some results without proof. For further reference we have to agree on phases of eigenstates. We label an eigenstate by $|(p, q)\, I, I_3, Y\rangle$. When we operate with $I_\pm$, $U_\pm$, $V_\pm$ sometimes ambiguities in phase arise. For example, what phase do we choose for $I_\pm\,|(p, q)\, I, I_3, Y\rangle$? We use the phases of Condon and Shortley, and Racah (but not of Schiff's *Quantum Mechanics*).

$$I_+\,|(p, q)\, I, I_3, Y\rangle = \sqrt{(I-I_3)(I+I_3+1)}\,|(p, q)\, I, I_3+1, Y\rangle$$
$$I_-\,|(p, q)\, I, I_3, Y\rangle = \sqrt{(I+I_3)(I-I_3+1)}\,|(p, q)\, I, I_3-1, Y\rangle.$$

Since the U's form an algebra we can write $U^2 = U_1^2 + U_2^2 + U_3^2$. We can label the states by $|U, U_3, Z\rangle$ instead of $|I, I_3, Y\rangle$. Then

$$U_+\,|I, I_3, Y\rangle = \sum_I C_{I'}\,|I', I_3 - \tfrac{1}{2}, Y+1\rangle.$$

One can determine $C_{I'}$ from the commutation relation of $I_\pm$, $U_\pm$, $V_\pm$. Picking a set of phases, we take $C_\pm$ real and positive. Another method of fixing these phases is based on the fact that for any unitary group, and especially for SU(3) we can find a subgroup $SU(2) \otimes U(1) \subset SU(3)$. (See Biedenharn [26].) We write down Biedenharn's expression for the matrix elements of $V_\pm$ (we can find $U_\pm$ from $U_\pm = [I_\mp, V_\pm]$):

$$\left\langle I+\frac{1}{2}, I_3+\frac{1}{2}, Y+1 \right| V_+ \left| I, I_3, Y\right\rangle =$$
$$= \frac{\sqrt{3}}{18}\left\{(I+I_3+1)\left[p-q+3\left(I+\frac{Y}{2}+1\right)\right]\times\right.$$
$$\left.\times\left[p+2q+3\left(I+\frac{Y}{2}+2\right)\right]\left[2p+q-3\left(I+\frac{Y}{2}\right)\right]\right\}^{\frac{1}{2}}\{(I+1)(2I+1)\}^{-\frac{1}{2}},$$

$$\left\langle I-\frac{1}{2}, I_3+\frac{1}{2}, Y+1 \right| V_+ \left| I, I_3, Y\right\rangle =$$
$$= \frac{\sqrt{3}}{18}\left\{(I-I_3)\left[-p+q+3\left(I-\frac{Y}{2}\right)\right]\times\right.$$
$$\left.\times\left[p+2q-3\left(I-\frac{Y}{2}-1\right)\right]\left[2p+q+3\left(I-\frac{Y}{2}+1\right)\right]\right\}^{\frac{1}{2}}\{I(2I+1)\}^{-\frac{1}{2}}.$$

(Biedenharn uses a slightly different notation: $p_B = p + q$, $q_B = q$, since he uses Young symmetrizers and his p's and q's denots lenghts of the lines in the tableaux.)

2.6. *Kronecker Products*

In the physics of unitary symmetry we must construct direct products of representations as we have done to obtain the octet:

$$D(1,0) \otimes D(0,1) = D(1,1) \oplus D(0,0)$$
$$u^i v_j = u^i v_j - \tfrac{1}{3}\,\delta_j{}^i\, u^k v_k \oplus \tfrac{1}{3}\, u^k v_k.$$

All our known particles fit into the $D(1, 1)$ representation. We have a baryon octet, a pseudoscalar meson octet, and a vector meson octet. When we look at scattering, e. g., meson-baryon scattering, then since the mesons and baryons each belong to an octet, the allowed products belong to the decomposition of $8 \otimes 8$. The resonances will therefore group themselves into the allowed representations of $8 \otimes 8$, $8 \otimes 8 = 1 \oplus 8 \oplus 8 \oplus \oplus 10 \oplus 10 \oplus 27$. For example, in $\pi - N$ scattering we obtain an $N_{3/2, 3/2}$ resonance which falls into the decuplet. Baryon-baryon scattering will yield mesons and vectons which fall into the octets of the product decomposition.

When we look at Lagrangian field theory the fields must transform as irreducible representations of SU(3). For the baryons, e. g., we have 8 fields, one for each baryon. The Lagrangian is quadratic in the baryon fields, and any baryon may be transformed into any other baryon, two conditions which fix the group to be unitary and to have an irreducible representation of dimension 8. The simplest such group is the product representation 8 of $3 \otimes 3$ where $\{3\}$ or $D(1, 0)$ is the self-representation of SU(3).

For angular momenta $D(j') \otimes D(j'') = \sum_{|j'-j''|}^{|j'+j''|} \oplus D(j)$ where each $D(j)$ occurs only once. When we take the product of 2 octets the decomposition gives us some problems since we get two 8's in the reduction (the group is not "simply reducible"). There are therefore 2 invariant ways of coupling bosons to baryons; for one 8 we use symmetric coupling, for the other, antisymmetric coupling.

Let us sketch the reduction of $8 \otimes 8$. We use elementary tensor algebra. The representation of $D(1, 1)$ is $s_\beta{}^\alpha$, of $D'(1, 1)$: $t_j{}^i$. The direct product is $s_\beta{}^\alpha\, t_j{}^i$. There are 64 such independent products. We take away the traces in α, j, and i, β, and symmetrize giving us a symmetric traceless tensor and a trace. We have thus obtained a $D(2, 2)$ and $D(0, 0)$ representation. $D(2, 2) = \{27\}$, $D(0, 0) = \{1\}$, we have 36 more terms to find. Let us contract with $\delta_\alpha{}^j$ and $\delta_i{}^\beta$ separately.

$$\delta_\alpha{}^j\, s_\beta{}^\alpha\, t_j{}^i = v_\beta{}^i,$$

$$\delta_i{}^\beta\, s_\beta{}^\alpha\, t_j{}^i = v_j{}^\alpha.$$

Taking out traces of the V's we obtain two $D(1, 1)$ representations. $D(1, 1) = \{8\}$, so that we have 20 more elements to find. Let us contract with ε^{ijk} and ε_{ijk} (ε_{ijk} is the totally antisymmetric tensor of rank 3) instead of with $\delta_\alpha{}^j$ and $\delta_i{}^\beta$

$$\varepsilon^{kj\beta}\, s_\beta{}^\alpha\, t_j{}^i = W^{k\alpha i}, \quad \varepsilon_{ki\alpha}\, s_\beta{}^\alpha\, t_j{}^i = W_{k\beta j}.$$

$W^{k\beta i}$ is a contravariant tensor of rank 3, $W_{k\beta j}$ a covariant tensor of rank 3. These tensors are the representations $D(3, 0)$ and $D(0, 3)$ respectively. Since $D(3, 0) = 10$ and $D(0, 3) = 10$, we have accounted for all 64 components of $8 \otimes 8$ and found the representations to which they belong. See e. g. ref. [27] for general methods for treating the decomposition of a product.

We have not looked in the above decomposition of the $8 \otimes 8$ at what happens when we interchange s and t. This is an important operation when we deal with products of two representations of the same particles. For boson-boson scattering only symmetric states are allowed. The $D(0, 0)$ and $D(2, 2)$ are obviously symmetric representations. The two 10's are obviously antisymmetric ones since they result from contracting with ε's. The two 8's have undetermined symmetry. But if we take the sum and difference, we get symmetric and antisymmetric combinations. We take only the symmetric combination. It is important to do this with meson-baryon scattering too, since the two 8's also appear there and we will need either the symmetric or antisymmetric 8.

2.7. Generalized Clebsch-Gordan Coefficients*

The product representation of $D(p', q')$ and $D(p'', q'')$ may be written

$$D(p', q') \otimes D(p'', q'') = \sum_{p, q} \oplus M(p, q)\, D(p, q)$$

where $M(p, q)$ is the multiplicity, i. e., the number of times a given representation appears in the decomposition. M is normally 0 or 1 [but for SU(3) 2 also occurs]. We look at the conserved quantum numbers of $D(p', q')$ and $D(p'', q'')$ to find out what $D(p, q)$ occur. For the angular momentum group:

$$D(J') \otimes D(J'') = \sum_{J=|J'-J''|}^{|J'+J''|} D(J).$$

The relation between the base in the product space and the bases in $D(J')$ and $D(J'')$ is given by the Clebsch-Gordan coefficients. We form the product base

$$|I', I_3', I'', I_3''\rangle = |I', I_3'\rangle\, |I'', I_3''\rangle.$$

How is this related to $|I, I_3\rangle$, the base vector in the product space? It must be related by a unitary transformation whose matrix elements are just the $C-G$ coefficients for SU(2):

$$|I, I_3\rangle = \sum_{I_3', I_3''} |I', I_3', I'', I_3''\rangle\, \langle I', I_3', I'', I_3'' \,|\, I, I_3\rangle.$$

These coefficients all vanish unless $I = I' + I''$ and $I_3 = I_3' + I_3''$ and also satisfy well-known relations between rows and columns due to unitarity. A similar thing happens for SU(3) except labeling of states becomes more complicated.

We label by $|(p, q)\, I, I_3, Y\rangle$ and where no ambiguity arises by $|n, I, I_3, Y\rangle$ where n is the dimension $\{n\} = D(p, q)$. We may also

* Not to be confused with GORDON — one of the discovers of the relativistic wave equation. It may be noted that besides the wrong spelling Clebsch-Gordon-coefficients sometimes found in the literature, there exists one example where the Klein-Gordon equation is called the "Clebsch-Gordon" equation. One can also find Gordan replaced by Jordan (especially in Russian papers).

need to label whether we deal with symmetric or antisymmetric representations as for example 8_s or 8_a. For the product space we express $|n, I, I_3, Y\rangle$ in terms of the base vectors

$$|n', I', I_3', Y', n'', I'', I_3'', Y''\rangle = |n', I', I_3', Y'\rangle |n'', I'', I_3'' Y''\rangle.$$

We have

$$|n, I, I_3, Y\rangle = \sum_{\substack{I_3', I_3'', I', I'' \\ Y', Y''}} |n', I', I_3', Y', n'', I'', I_3'', Y''\rangle \langle n', I', I_3', Y', n'', I'', I_3'', Y'' | n, I, I_3, Y\rangle.$$

We can compute these directly from matrix elements of transformations of tensors. This has been done by several people and they are tabulated. (EDMONDS [19], TARJANNE [28], RASHID [29], DE SWART [23].)

SU(3) has one subgroup of special significance, the SU(2) group of isospin. There are restrictions on the $C-G$ coefficients not only by SU(3) symmetry but also by SU(2) symmetry. From conservation of isospin and also of hypercharge we can rewrite $C-G$ coefficients with an isoscalar factor removed

$$\langle n, I, I_3, Y | n', I', I_3', Y', n'', I'', I_3'', Y''\rangle$$
$$= \delta_{Y, Y'+Y''} \langle I, I_3 | I', I_3', I'', I_3''\rangle \times \langle n, I, Y | n', I', Y', n'', I'', Y''\rangle.$$

By unitarity the $C-G$ coefficients must satisfy relations of the same kind as do the rows and columns of the $C-G$ coefficients of SU(2). Besides using the I-spin subgroup we can also use the U, V subgroups with their appropriate $C-G$ coefficients. One can eventually reduce the isoscalar factor to some very simple combination of coefficients or for a particular problem one can evaluate the isoscalar factor directly by using the tables of DE SWART [23].

3. Physics of SU(3)

3.1. The Symmetry-Breaking Interactions and the Mass Formula

The baryon octet is:

$\cdot\, n$ (939)		$\cdot\, p$ (938 MeV)
$\cdot\, \Sigma^-$ (1197)	$\cdot\, \Sigma^0$ (1193)	$\cdot\, \Sigma^+$ (1189)
	$\cdot\, \Lambda^0$ (1115)	
	$\cdot\, \Xi^-$ (1321) $\cdot\, \Xi^0$ (1316)	

Under unitary symmetry all the baryons have degenerate masses. How do we break unitary symmetry and obtain the observed mass spectrum? First, keeping isospin, how do we account for breaking unitary symmetry between the 4 multiplets of the octet when the mass differences are of the order of the mass of π-mesons $\sim$200 MeV? Second, how, assuming the mass difference within each isomultiplet is due to electromagnetic interactions, do we account for its mass splitting?

First, in each horizontal line the masses are approximately equal, therefore, the mass operator must commute with I^2. Furthermore within each multiplet we want Y to have a constant value. The simplest choice is that the mass operator $M = \alpha Y = \beta F_8$. Using the Lagrangian formalism the mass splitting part of the Lagrangian would be (for spinor particles)

$$\delta L = \Delta m\, \psi_B\, F_8\, \psi_B.$$

Because the Lagrangian for Dirac particles is linear in m, $M = \Delta m F_8$ (it will be necessary after we derive the mass formula to replace m by m^2 if we wish to apply the formula to bosons; this corresponds to the fact that the Lagrangian for spin 0 or 1 particles depends on m^2 not m). We calculate the mass-splitting by sandwiching the mass operator between any baryon state.

$$\langle I, I_3, Y \mid M \mid I, I_3, Y \rangle = \langle I, I_3, Y \mid \Delta m\, F_8 \mid I, I_3, Y \rangle.$$

Nothing in the world tells us that this is the correct formula; the theoretical foundation for this method is very poor. Further there is no reason why the first order result should be as good as it is.

The mass operator will transform as the $T_3{}^3$ component of some tensor $T_\nu{}^\mu$ (it is important here to distinguish between the infinitesimal operators $A_\nu{}^\mu$ and the tensor operator $T_\nu{}^\mu$), since $M = \beta F_8$ and F_8 is proportional to $A_3{}^3$. Using Okubo's form for the matrix element of $T_\nu{}^\mu$ and redefining M so that it will include the additive constant that corresponds to the degenerate mass of the baryons:

$$\langle I, I_3, Y \mid M \mid I, I_3, Y \rangle =$$
$$= \langle I, I_3, Y \mid m_0 + m_1 Y + m_2 \left[I(I+1) - \frac{Y^2}{4} \right] \mid I, I_3, Y \rangle.$$

We derive this formula for $M \propto T_3{}^3 = a + b A_3{}^3 + c A_K{}^3 A_3{}^K$ by using the Casimir operator $F^2 = \sum_{i,j=1}^{3} A_j{}^i A_i{}^j = \text{const.}$ and the commutation relation of the A's. This yields for the baryons:

$$\begin{aligned} M_N &= m_0 + m_1 + \tfrac{1}{2} m_2 \\ M_\Lambda &= m_0 \\ M_\Sigma &= m_0 + 2 m_2 \\ M_\Xi &= m_0 - m_1 + \tfrac{1}{2} m_2, \end{aligned}$$

hence

$$\frac{M_N + M_\Xi}{2} = \frac{3 M_\Lambda + M_\Sigma}{4}.$$

For the pseudoscalar mesons:

$$m_K^2 = \frac{3 m_\eta{}^2 + m_\pi{}^2}{4}.$$

For the vector mesons:

$$m_{K^*}^2 = \frac{3 m_\Phi{}^2 + m_\varrho{}^2}{4},$$

where $m_\Phi = 940$ Mev, the mass of an "unsplit particle" Φ, the weighted average of the singlet φ and the $I = 0$ member of the octet ω (65% φ, 35% ω). We find excellent agreement with experiment for the baryons and the pseudoscalar mesons, but not for the vector mesons (unless we use the trick of mixing the φ and ω). If we apply the same method to the decuplet $D(3, 0)$ of baryon resonances where $I = \frac{1}{2}Y + \frac{1}{3}p$ i. e., $I = \frac{1}{2}Y + 1$, we have $M = m_0' + m_1'$. This gives a law of equal spacing for the decuplet and predicts the Ω mass to be 1670 Mev.

3.2. Relations Among Reaction Cross Sections

Here we consider another application of SU(3) symmetry: scattering reactions and decays. Treating SU(3) as an exact symmetry, we will sometimes relate amplitudes which are physically forbidden by conservation (mass conservation forbids e. g. $\Delta \to \Sigma + K$) with other allowed reactions from $D(3,0) \to D(1,1) \times D(1,1)$ like $Y^* \to \Sigma + K$ or $\Omega^- \to \Xi^- + K^0$ i. e., we do not worry about dynamics. Each member of the decuplet can decay into a baryon and a meson; if we take SU (3) symmetry seriously, the decuplet is built up from the meson and baryon octet. All these decays are related by Clebsch-Gordan coefficients to an isoscalar factor; some, however, as the above Δ decay, are unphysical. Using the general technique for computing any decay or scattering process one has to find the S-matrix element

$$\langle n, s, p^\mu, I, I_3, Y | S | n', s', p'^\mu; I', I_3', Y' \rangle.$$

Even for the simplest scattering problem $a + b \to c + d$, initial and final states must be built out of both particles using vector addition coefficients. For strong interactions I and Y are conserved so that the scattering amplitude factors into

$$\delta_{II'}\, \delta_{YY'}\, \delta(\underset{\sim}{p} - \underset{\sim}{p}')\, \delta(E - E') \langle n, I, I_3, Y | S | n', I', I_3', Y' \rangle$$

where $\langle n, I, I_3, Y|$ and $|n', I', I_3', Y'\rangle$ are built out of products of the oneparticle states. Usually the result is very lengthy and does not justify all the effort.

There are several shortcuts of which we will consider two: the Weyl reflection method and the generalized method of Shmushkevich. The Weyl reflection method allows simple derivations of relations between various reactions, especially reactions involving photons which conserve U-spin (thus the method is especially applicable to electromagnetic transitions) and uses both I-spin and U-spin conservation. The method of Shmushkevich is a very simple arithmetic of cross sections invariant under isospin which we generalize to SU(3) symmetry. For reference see LEVINSON, LIPKIN, and MESHKOV [20, 35], and MACFARLANE, SUDARSHAN, and DULLEMOND [36] for Weyl reflections and for the other method SHMUSHKEVICH [37, 38], LAPIDUS [39], MARSHAK and SUDARSHAN [40], and MARKOV [41], DULLEMOND, MACFARLANE, and SUDARSHAN [42] and unpublished work by the lecturer.

Weyl Reflection Method

Consider the effect of the Weyl reflection W_3 on I_3 and Y:

$$W_3\, I_3\, W_3 = -U_3$$
$$W_3\, Y\, W_3 = -\left(I_3 + \frac{Y}{2}\right) = -Z.$$

For the decuplet under W_3:

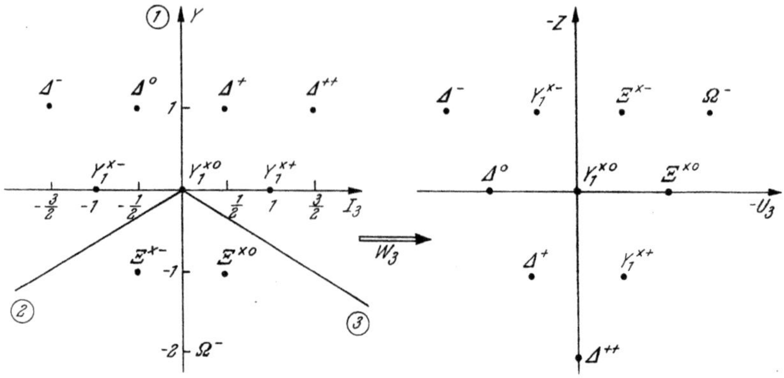

Fig. 15

Further we have

$$W_3 I_{\pm} W_3 = U_{\mp},\ U_3 = \frac{3}{4} Y - \frac{I_3}{2} \text{ and } U^2 = U_1^2 + U_2^2 + U_3^2 =$$
$$= \frac{1}{2}(U_+ U_- + U_- U_+) = U_3^2.$$

For the decuplet there is a one-to-one correspondence between states before and after W_3 reflection. What happens when we do not have a one-to-one correspondence between states?

Before answering this question, let us introduce a new notation. I-spinors are $|\, I\, I_3\rangle$, U spinors are $|\, U\, U_3)$ (note the different shape of the brackets). Further adding appropriate commuting variables to I and U-spinors, we get a sufficient notation to label states of any representation (octet, decouplet, etc.), i. e., $|\, I, I_3, Y\rangle$ and $|\, U, U_3, Z)$.

For the decuplet we can write down immediately the effect of W_3 on the states and if we fix the phase for the first member then all the other phases are determined.

$$\begin{aligned}
W_3 \,|\, \tfrac{3}{2}, \tfrac{3}{2}, 1\rangle &= |\, 0, 0, 2) \\
W_3 \,|\, \tfrac{3}{2}, \tfrac{1}{2}, 1\rangle &= |\, \tfrac{1}{2}, \tfrac{1}{2}, 1) \\
W_3 \,|\, \tfrac{3}{2}, -\tfrac{1}{2}, 1\rangle &= |\, 1, 1, 0) \\
W_3 \,|\, \tfrac{3}{2}, -\tfrac{3}{2}, 1\rangle &= |\, \tfrac{3}{2}, \tfrac{3}{2}, -1) \\
W_3 \,|\, 1, 1, 0\rangle &= |\, \tfrac{1}{2}, -\tfrac{1}{2}, 0) \\
W_3 \,|\, 1, 0, 0\rangle &= |\, 1, 0, 0) \\
W_3 \,|\, 1, -1, 0\rangle &= |\, \tfrac{3}{2}, \tfrac{1}{2}, -1) \\
W_3 \,|\, \tfrac{1}{2}, \tfrac{1}{2}, 1\rangle &= |\, 1, -1, 0) \\
W_3 \,|\, \tfrac{1}{2}, -\tfrac{1}{2}, 1\rangle &= |\, \tfrac{3}{2}, -\tfrac{1}{2}, 1) \\
W_3 \,|\, 0, 0, 2\rangle &= |\, \tfrac{3}{2}, -\tfrac{3}{2}, -1).
\end{aligned}$$

Therefore, for representations where all the weights (eigenstates) are simple we can use them directly to find relations for amplitudes.

For the octet under W_3:

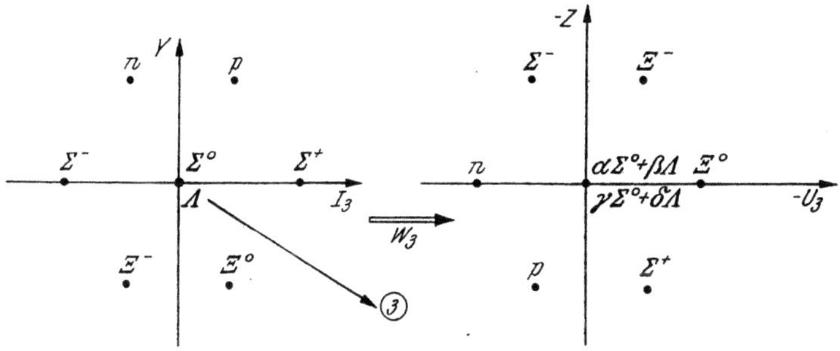

Fig. 16

There is no ambiguity, except for an overall phase, for the peripheral members of the octet.

$$\begin{aligned} W_3 \mid \tfrac{1}{2}, \tfrac{1}{2}, 1\rangle &= \mid \tfrac{1}{2}, \tfrac{1}{2}, 1) \\ W_3 \mid \tfrac{1}{2}, -\tfrac{1}{2}, 1\rangle &= \mid 1, 1, 0) \\ W_3 \mid 1, 1, 0\rangle &= \mid \tfrac{1}{2}, -\tfrac{1}{2}, 1) \\ W_3 \mid 1, -1, 0\rangle &= \mid \tfrac{1}{2}, \tfrac{1}{2}, -1) \\ W_3 \mid \tfrac{1}{2}, \tfrac{1}{2}, -1\rangle &= \mid 1, -1, 0) \\ W_3 \mid \tfrac{1}{2}, -\tfrac{1}{2}, -1\rangle &= \mid \tfrac{1}{2}, -\tfrac{1}{2}, -1). \end{aligned}$$

There is, however, an ambiguity in the doubly occupied $I=0$, $Y=0$ state.

$$\begin{aligned} W_3 \mid 1, 0, 0\rangle &= \alpha \mid 1, 0, 0) + \beta \mid 0, 0, 0) \\ W_3 \mid 0, 0, 0\rangle &= \delta \mid 1, 0, 0) + j \mid 0, 0, 0) \end{aligned}$$

or

$$W_3 \begin{pmatrix} \mid 1, 0, 0\rangle \\ \mid 0, 0, 0\rangle \end{pmatrix} = \begin{pmatrix} \alpha & \beta \\ \delta & \gamma \end{pmatrix} \begin{pmatrix} \mid 1, 0, 0) \\ \mid 0, 0, 0) \end{pmatrix}.$$

The matrix of coefficients α, β, δ, γ called C is obviously unitary and may be made unimodular. How do we find α and β? It is more convenient to find U-spinors in terms of I-spinors. Defining a new α and β ($C^+ W_3$ is itself unitary), we have

$$\begin{pmatrix} \mid 1\,0\,0) \\ \mid 0\,0\,0) \end{pmatrix} = \begin{pmatrix} \alpha & \beta \\ \beta^x & -\alpha \end{pmatrix} \begin{pmatrix} \mid 1\,0\,0\rangle \\ \mid 0\,0\,0\rangle \end{pmatrix}.$$

From the algebra of SU(2), we know that

$$U_\pm \mid U, U_3, Z) = \sqrt{(U \mp U_3)(U \pm U_3 + 1)} \mid U, U_3 \pm 1, Z).$$

The neutron $|n\rangle$ is $|\frac{1}{2}, -\frac{1}{2}, 1\rangle$ or $|1, 1, 0\rangle$

$$U_- |n\rangle = U_- |1, 1, 0\rangle = \sqrt{2}\,|1, 0, 0\rangle = \sqrt{2}\,(\alpha\,|1, 0, 0\rangle + \beta\,|0, 0, 0\rangle).$$

One could read off α and β from the Biedenharn [26] expression for the matrix elements of $V_\pm$, but it is simpler not to have to use them. Applying I_+ and using the commutation relation $[I_+, U_-] = 0$

$$I_+ U_- |n\rangle = \sqrt{2}\, I_+ |1, 0, 0\rangle = \sqrt{2}\,(\alpha\, I_+ |1, 0, 0\rangle + \beta\, I_+ |0, 0, 0\rangle) = \\ = 2\alpha\, |1, 1, 0\rangle$$

$$U_- I_+ |n\rangle = U_- |p\rangle = U_- |\tfrac{1}{2}, \tfrac{1}{2}, 1\rangle = |\tfrac{1}{2}, -\tfrac{1}{2}, 1\rangle = |1, 1, 0\rangle.$$

Hence

$$\alpha = \tfrac{1}{2}.$$

By normalization $|\beta| = \sqrt{3}/2$ and we choose as phase $+1$. (This choice of phase agrees with Biedenharn's phase conventions for the $D(1, 1)$ representation.)

For the triplet state $|1, 0, 0\rangle = \frac{1}{2}|\Sigma^0\rangle + \frac{\sqrt{3}}{2}|\Lambda^0\rangle$ and for the singlet state $|0, 0, 0\rangle = \frac{\sqrt{3}}{2}|\Sigma^0\rangle - \frac{1}{2}|\Lambda^0\rangle$. Thus $|1, 0, 0\rangle$ goes under W_3 into $\frac{1}{2}|1, 0, 0\rangle + \frac{\sqrt{3}}{2}|0, 0, 0\rangle$ and $|0, 0, 0\rangle$ goes into $\frac{\sqrt{3}}{2}|1, 0, 0\rangle - \frac{1}{2}|0, 0, 0\rangle$. We may draw the octet after W_3 reflection as

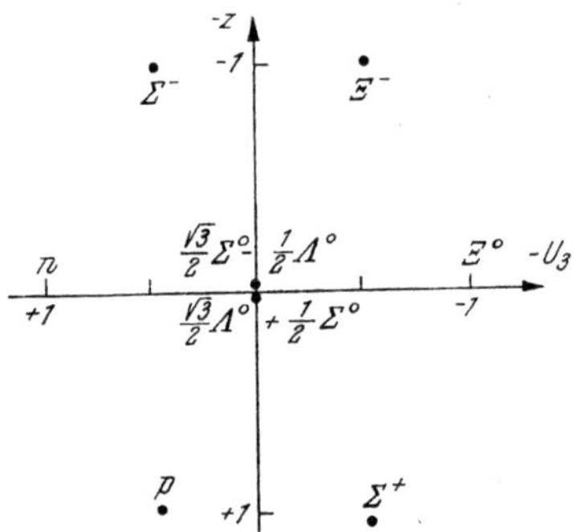

Fig. 17

with an analogous diagram for the pseudoscalar mesons. Using the same method, we may see how the multiply occupied states of the 27 representation transform under Weyl reflection (cf. 42).

As a simple example of this technique, let us derive SU(3) relations for the decay of a baryon resonance from $D(3, 0)$ into a baryon and a meson, both being in octets; symbolically;

$$D(3, 0) \to D(1, 1) + D(1, 1)$$
$$R \to B + P.$$

Full-fledged SU(3) would require computing Clebsch-Gordan coefficients $\langle 10, I, I_3, Y \mid 8', I', I_3', Y', 8'', I'', I_3'', Y'' \rangle$ which multiply the reduced S-matrix element

$$S^{10}_{8', 8''}, \; {}^{\frac{3}{2}+}_{\frac{1}{2}+, 0-} .$$

Working this out can be a very long calculation. However, using U-spin, we have for the quartet of states with charge-1 decaying into a charged baryon and a neutral meson and into a neutral baryon and a charged meson (only triplet states contribute because U-spin is conserved).

$$(\Delta^-, Y^{*-}, \Xi^{*-}, \Omega^-) \begin{array}{l} \nearrow (\Xi^-, \Sigma^-) + \left(K^0, \bar{K}^0, \frac{1}{2}\pi^0 + \frac{\sqrt{3}}{2}\eta\right) \\ \searrow \left(\frac{1}{2}\Sigma^0 + \frac{\sqrt{3}}{2}\Lambda^0, n, \Xi^0\right) + (\pi^-, K^-) \end{array}$$

where we are to find the appropriate pairings. We denote U-spin $C-G$ coefficients by $(|)$, I spin by $\langle | \rangle$. We examine all decays that satisfy the conservation laws. They are all related horizontally by U-spin and vertically by I-spin. For the first mode, charged baryon-neutral meson, calling the reduced matrix element S_1 the amplitude depending only on total quantum numbers I and Y:

$$\langle \Delta^- \mid \Sigma^- K^0 \rangle = \left(\frac{1}{2}, \frac{1}{2}, 1, 1 \,\middle|\, \frac{3}{2}, \frac{3}{2}\right) S_1$$

$$\langle Y_1^{x-} \mid \Xi^- K^0 \rangle = \left(\frac{1}{2}, -\frac{1}{2}, 1, 1 \,\middle|\, \frac{3}{2}, \frac{1}{2}\right) S_1$$

$$\langle Y_1^{x-} \mid \Sigma^- n \rangle = \frac{\sqrt{3}}{2}\left(\frac{1}{2}, \frac{1}{2}, 1, 0 \,\middle|\, \frac{3}{2}, \frac{1}{2}\right) S_1$$

$$\langle Y_1^{x-} \mid \Sigma^- \pi^0 \rangle = \frac{1}{2}\left(\frac{1}{2}, \frac{1}{2}, 1, 0 \,\middle|\, \frac{3}{2}, \frac{1}{2}\right) S_1$$

$$\langle \Xi^{x-} \mid \Sigma^- \bar{K}^0 \rangle = \left(\frac{1}{2}, \frac{1}{2}, 1, -1 \,\middle|\, \frac{3}{2}, -\frac{1}{2}\right) S_1$$

$$\langle \Xi^{x-} \mid \Xi^- \pi^0 \rangle = \frac{1}{2}\left(\frac{1}{2}, -\frac{1}{2}, 1, 0 \,\middle|\, \frac{3}{2}, -\frac{1}{2}\right) S_1$$

$$\langle \Xi^{x-} \mid \Xi^- n \rangle = \frac{\sqrt{3}}{2}\left(\frac{1}{2}, -\frac{1}{2}, 1, 0 \,\middle|\, \frac{3}{2}, -\frac{1}{2}\right) S_1$$

$$\langle \Omega^- \mid \Xi^- \bar{K}^0 \rangle = \left(\frac{1}{2}, -\frac{1}{2}, 1, -1 \,\middle|\, \frac{3}{2}, -\frac{3}{2}\right) S_1 .$$

For the second mode, neutral baryon and charged meson:

$$\langle \Delta^- \mid n \pi^- \rangle = \left(1, 1, \frac{1}{2}, \frac{1}{2} \,\middle|\, \frac{3}{2}, \frac{3}{2}\right) S_2$$

$$\langle Y_1^{x-} \mid \Sigma^0 \pi^- \rangle = \frac{1}{2}\left(1, 0, \frac{1}{2}, \frac{1}{2} \,\middle|\, \frac{3}{2}, \frac{1}{2}\right) S_2$$

$$\langle Y_1^{x-} \mid \Lambda^0 \pi^- \rangle = \frac{\sqrt{3}}{2}\left(1, 0, \frac{1}{2}, \frac{3}{2} \,\middle|\, \frac{3}{2}, \frac{1}{2}\right) S_2$$

$$\langle Y_1^{x-} \mid n K^- \rangle = \left(1, 1, \frac{1}{2}, -\frac{1}{2} \,\middle|\, \frac{3}{2}, \frac{1}{2}\right) S_2$$

$$\langle \Xi^{x-} \mid n\,\pi^- \rangle = \left(1,\ 1,\ \frac{1}{2},\ \frac{1}{2} \middle| \frac{3}{2},\ -\frac{1}{2}\right) S_2$$

$$\langle \Xi^{x-} \mid \Sigma^0\, K^- \rangle = \frac{1}{2}\left(1,\ 0,\ \frac{1}{2},\ -\frac{1}{2} \middle| \frac{3}{2},\ -\frac{1}{2}\right) S_2$$

$$\langle \Xi^{x-} \mid \Lambda^0\, K^- \rangle = \frac{\sqrt{3}}{2}\left(1,\ 0,\ \frac{1}{2},\ -\frac{1}{2} \middle| \frac{3}{2},\ -\frac{1}{2}\right) S_2$$

$$\langle \Omega^- \mid \Xi^0\, K^- \rangle = \left(1,\ -1,\frac{1}{2},\ -\frac{1}{2} \middle| \frac{3}{2},\ -\frac{3}{2}\right) S_2.$$

However, S_1 and S_2 are not independent but related by I-spin, which must also be conserved. Ω^- has $I=0$. What combination of $|\,\Xi^- K^0\rangle$ and $|\,\Xi^0 K^-\rangle$ has $I=0$? It is easy to see that the right combination is $|\,\Xi^- K^0\rangle - |\,\Xi^0 K^-\rangle$. The amplitude for $\langle \Omega^- \mid \Xi^- K^0\rangle$ and for $\langle \Omega^- \mid \Xi^0 K^-\rangle$ must appear with equal weight but opposite in sign. Therefore $S_1 = -S_2$. We can check by looking at another channel. The complete set of 16 numbers gives us the desired information on relative cross section or frequency.

The real advantage of using U-spin shows itself in electromagnetic processes, since the electric charge $Z = I_3 + \frac{Y}{2}$ commutes with all U operators, i. e., electromagnetic interactions are U-scalar and the expectation value for any electromagnetic operator is the same for all members of a U-multiplet. Photons are U-spin zero particles. Thus for any electromagnetic quantity ε (Q)

$$\langle p \mid \varepsilon \mid p \rangle = \langle \Sigma^+ \mid \varepsilon \mid \Sigma^+ \rangle$$

$$\langle \Sigma^- \mid \varepsilon \mid \Sigma^- \rangle = \langle \Xi^- \mid \varepsilon \mid \Xi^- \rangle$$

$$\langle n \mid \varepsilon \mid n \rangle = \langle \Xi^0 \mid \varepsilon \mid \Xi^0 \rangle =$$

$$= \frac{1}{4}(1\,0 \mid \varepsilon \mid 1\,0) + \frac{3}{4}(0\,0 \mid \varepsilon \mid 0\,0) + \sqrt{3}(1\,0 \mid \varepsilon \mid 0\,0)$$

$$\langle \Xi^0 \mid \varepsilon \mid \Lambda^0 \rangle = \frac{\sqrt{3}}{4}[(1\,0 \mid \varepsilon \mid 1\,0) - (0\,0 \mid \varepsilon \mid 0\,0)]$$

$$\langle \Lambda^0 \mid \varepsilon \mid \Lambda^0 \rangle = \frac{3}{4}(1\,0 \mid \varepsilon \mid 1\,0) + \frac{1}{4}(0\,0 \mid \varepsilon \mid 0\,0) - \sqrt{3}(1\,0 \mid \varepsilon \mid 0\,0)$$

$$\langle \Sigma^0 \mid \varepsilon \mid \Sigma^0 \rangle = \frac{1}{4}(1\,0 \mid \varepsilon \mid 1\,0) + \frac{3}{4}(0\,0 \mid \varepsilon \mid 0\,0) + (1\,0 \mid \varepsilon \mid 0\,0).$$

We can obtain the electromagnetic mass difference and magnetic moments from the above expressions.

Whenever photons are involved, it is more convenient to use U-spin rather than I-spin. We have $|\,\gamma\, p\rangle$ with $U=1/2$, $U_3=1/2$ and $|\,\gamma\, n\rangle$ with $U=1$, $U_3=1$. Relations for photoproduction are:

$$\frac{\langle \gamma\, p \mid \Delta^0\, \pi^+ \rangle}{\langle \gamma\, p \mid Y_1^{*0},\, K^+ \rangle} = \frac{\langle \gamma\, p \mid \Delta^0\, \varrho^+ \rangle}{\langle \gamma\, p \mid Y_1^{*0},\, K^{*+} \rangle} = \frac{(\frac{1}{2}, \frac{1}{2} \mid 1, 1, \frac{1}{2}, -\frac{1}{2})}{(\frac{1}{2}, \frac{1}{2} \mid 1, 0, \frac{1}{2}, \frac{1}{2})} = -\sqrt{2}.$$

The branching ratio is then 2/1. Similar relations for neutrons give a branching ratio of 3/1. We have for the electromagnetic decays allowed by U-spin conservation $\langle Y_1^{*0} | \Sigma^+ \gamma \rangle = \langle \Delta^+ | p \gamma \rangle$ and $\langle \Delta^0 | n \gamma \rangle = 2 \langle Y^{*0} | \Sigma^0 \gamma \rangle$ whereas $\langle Y_1^{*0} | \Sigma^- \gamma \rangle$ is forbidden by U-spin conservation.

Just adding U-spin conservation to what we usually do with I-spin gives us a simple means of calculating many quantities of physical interest. U-spin is not rigorously true unless SU(3) is exactly correct. $SU_I(2) \otimes SU_U(2)$ gives the same results as SU(3) (but is not identical to it). If SU(3) is broken with I and Y exact, U is broken. It may be simpler to break U-spin rather than the whole SU(3) symmetry. Experimentally we do know that U-spin is not exactly conserved.

Bibliography

On Group Representations

BOERNER, H., Representations of Groups, North-Holland Publ. Co. (1963).

HAMERMESH, M., Group Theory and Its Application to Physical Problems, Addison-Wesley, Reading, Massachusetts (1962).

WEYL, H., The Classical Groups, Princeton University Press, Princeton, New Jersey (1946).

WIGNER, E. P., Group Theory and Its Application to the Quantum Mechanics of Atomic Spectra, Academic Press, New York (1959).

On Lie Groups

Ankara Summer School Notes 1962

BEHRENDS, R. E., J. DREITLEIN, C. FRONSDAL, and B. W. LEE, Reviews of Modern Physics **34**, 1, 1 (Jan. 1962).

FRONSDAL, C., 1962 Brandeis Lectures, Vol. 1, W. A. Benjamin, New York (1963).

RACAH, G., Group Theory and Spectroscopy, CERN Report 61-8 (1961).

Theoretical Physics — Lectures Presented at the Seminar on Theoretical Physics, Trieste, Summer 1962, International Atomic Energy Agency, Vienna (1963).

Varenna Summer School Notes 1962.

References

1. BARKAS, H., and A. ROSENFELD, University of California Report UCRL — 8030 Rev. VII. (April 1963).
2. JOST, R., Helvetia Physica Acta **30**, 409 (1957).
3. ROSENFELD, A., Proceedings of the 1962 International Conference on High-Energy Physics at CERN, CERN, Geneva p. 325.
4. SCHWINGER, J., Annals of Physics **2**, 407 (1957).
5. GELL-MANN, M., Physical Review **106**, 1296 (1957).
6. SALAM, A., and J. C. POLKINGHORNE, Nuovo cimento **2**, 685 (1955).
7. SAKURAI, J. J., Annals of Physics **11**, 1 (1960).
8. YANG, C. N., and R. L. MILLS, Physical Review **96**, 191 (1954).
9. MAYER, M. E., Nuovo cimento **11**, 760 (1959).
10. SAKATA, S., Progress of Theoretical Physics **16**, 686 (1956).
11. WESS, J., Nuovo cimento **15**, 52 (1960).
12. IKEDA, M., S. OGAWA, and Y. OHNUKI, Progress of Theoretical Physics **22**, 715 (1959).
13. GELL-MANN, M., Physical Review **125**, 1067 (1962).
14. NE'EMAN, Y., Nuclear Physics **26**, 222 (1961), Unpublished thesis, Imperial College 1961.

15. BOGOLIUBOV, N. N., and D. V. SHIRKOV, Introduction to the Theory of Quantized Fields, Interscience, New York, 1959, p. 19.
16. GLASHOW, S., and M. GELL-MANN, Annals of Physics **15,** 437 (1961).
17. GELL-MANN, M., Lecture Notes on Special Topics in Relativistic Quantum Theory, MIT, Spring 1963.
18. ELLIOTT, J. P., Proceedings of the Royal Society of London, Series A, **245,** 128, 562 (1958).
19. EDMONDS, A. R., Proceedings of the Royal Society of London, Series **A, 268,** 567 (1962).
20. LEVINSON, C. A., H. J. LIPKIN, and S. MESHKOV, Nuovo cimento **23,** 231 (1962).
21. BEHRENDS, R. E., J. DREITLEIN, C. FRONSDAL, and B. W. LEE, Reviews of Modern Physics **34,** 1 (1962).
22. OKUBO, S., Progress of Theoretical Physics **27,** 949 (1962).
23. SWART, J. J. DE, CERN Preprint (May 1963).
24. WEYL, H., The Classical Groups, Princeton University Press, Princeton (1941).
25. CARTAN, E., Thèse (Paris, 1894; 2nd ed., 1933).
26. BIEDENHARN, L. C., Physics Letters **3,** 69 (1962).
27. SPEISER, D., and J. TARSKI, Journal of Mathematical Physics **4,** 588 (1963).
28. TARJANNE, P., Carnegie Tech. Preprint, NYO — 9290 A (April 1963).
29. RASHID, M. A., Nuovo cimento **26,** 118 (1962).
30. DIU, B., Nuovo cimento **28,** 466 (March 1963).
31. GINIBRE, J., Orsay Preprint Th. 19 (1962).
32. GELL-MANN, M., California Institute of Technology Synchroton Laboratory Report No. CTSL — 20 (1961).
33. COLEMAN, S., and S. GLASHOW, Physical Review Letters **6,** 423 (1961).
34. CABIBBO, N., and R. GATTO, Nouvo cimento **21,** 832 (1961).
35. LEVINSON, C. A., H. J. LIPKIN, and S. MESHKOV, Physics Letters **1,** 44, 125, 307 (1962); Physical Review Letters **10,** 361 (1962).
36. MACFARLANE, A. J., E. C. G. SUDARSHAN, and C. DULLEMOND, University of Rochester Preprint NYO — 10266.
37. SHMUSHKEVICH, J. M., Doklady Akad. Nauk (SSR) **103,** 235 (1955).
38. DUSHKIN, N., and J. M. SMUSHKEVICH, Soviet Physics Doklady **1,** 94 (1956).
39. LAPIDUS, L. J., Soviet Physics JETP **4,** 740 (1957).
40. MARSHAK, R. E., and E. C. G. SUDARSHAN, Introduction to Elementary Particle Physics, Interscience, New York (1961).
41. MARKOV, M. A., Hyperonen und K-Mesonen, VEB, Berlin (1960).
42. DULLEMOND, C., A. J. MACFARLANE, and E. C. G. SUDARSHAN, Physical Review Letters **10,** 423 (May 1963).
43. CERULUS, F., Supp. Nuovo cimento **15,** 402 (1960).
44. ZACHARIASEN, F., and C. ZEMACH, Physical Review **128,** 849 (1962).
45. CAPPS, R. H., Physical Review Letters **10,** 312 (April 1963).
46. CUTKOSKY, R. E., Physical Review **131,** 1888 (August 1963).
47. MARTIN, A. W., and K. C. WALI, Argonne National Laboratory Preprint ANL 6729 (1962).
48. SAKURAI, J. J., Physical Review Letters **10,** 446 (May 1963).
49. CARPS, R. M., CERN Preprint Th. 360 (1963).
50. CABIBBO, N., Physical Review Letters **10,** 531 (June 1963).
51. CORNWALL, J. M., and V. SINGH, Physical Review Letters **10,** 551 (June 1963).

Appendix A.

3.2. Electromagnetic Mass Differences

Secondly in order to calculate the electromagnetic mass difference we want something that leaves I_3 and Y invariant although it must not commute with I spin. We know that $Z = I_3 + \frac{Y}{2}$ is a strictly conserved quantum number even for interactions that violate I-spin. Let us introduce the electromagnetic current j^μ where

$$j^\mu = j_3{}^\mu + \frac{1}{\sqrt{3}} j_8{}^\mu \text{ and } Q = \int j^0 \, d^3 x = e Z.$$

If we look at the mass splittings $\Xi^- - \Xi^0$, $\Sigma^+ - \Sigma^0$, $\Sigma^0 - \Sigma^-$, $p - n$, then these mass differences occur between a charged and an uncharged particle. If we assume that the mass difference is an electromagnetic effect then the lowest non-vanishing term of the perturbation expansion for the interaction of a charged and uncharged particle is of second-order. Mass differences should then depend on $j^\mu j_\mu$.

We want the current to represent the baryons, so we expect j_μ to have 8 components. We will have something like:

$$j_k{}^\mu = \bar{B}_i J^\mu_{ijk} B_j.$$

However, this will require looking at the product representation $8 \otimes 8$ whose decomposition includes two 8 representations. We will need, therefore, two independent matrices J to couple the baryons together. We take an antisymmetric coupling $F_3 + \frac{1}{\sqrt{3}} F_8$ and a symmetric coupling $D_3 + \frac{1}{\sqrt{3}} D_8$ by using Clebsch-Gordan coefficients. The F and D matrices were first introduced in Gell-Mann's original work [32]. In order to find the matrix elements of the F's and D's, it will be necessary to consider the invariant couplings of baryons to pseudoscalar mesons.

We write out the 8 baryons with their quantum numbers and express them in terms of the infinitesimal operators $A_j{}^i$ (we are in the regular representation since $t_j{}^i$ and $A_j{}^i$ have the same commutation relations). We can find all the baryons from $\Lambda = |0\,0\,0\rangle$ and $\Sigma^0 = |1\,0\,0\rangle$ which correspond to the center of the weight diagram of the octet and which depend only on I_3 and Y, i. e., $A_1{}^1$, $A_2{}^2$, $A_3{}^3$ in appropriate combinations (we will apply the conditions that the trace of the baryon tensor which we form vanish).

Λ does not carry isospin so that it will depend only on Y, i. e., on $A_3{}^3$. $p = |\tfrac{1}{2}, \tfrac{1}{2}, 1\rangle$. To go from Λ to p we raise Y by 1 unit, I by 1/2, I_3 by 1/2, therefore we need $V_+ = A_1{}^3$. $n = |\tfrac{1}{2}, -\tfrac{1}{2}, 1\rangle$.

The matrix which raises Y and lowers I_3, we know to be $U_+ = A_2{}^3$. All the rest of the baryons are almost determined. $\Xi^- = |\tfrac{1}{2}, -\tfrac{1}{2}, -1\rangle$ and $\Xi^0 = |\tfrac{1}{2}, \tfrac{1}{2}, -1\rangle$ will clearly be related to $V_- = A_3{}^1$ and $U_- = A_3{}^2$, respectively. The Σ's, $\Sigma^+ = |1, 1, 0\rangle$, $\Sigma^0 = |1, 0, 0\rangle$, $\Sigma^- = |1, -1, 0\rangle$

do not change hypercharge so we expect them to involve only isospin. Σ^+ must be $I_+ = A_1{}^3$, Σ^-, $I_- = A_3{}^1$, and Σ^0 which leaves I_3 invariant must be related to $I_3 = \frac{A_1{}^1 - A_2{}^2}{2}$. We may write the baryon tensor B as a 3×3 matrix using the set of λ's (previously defined) as a basis. For example

$$V^- = \frac{\lambda_4 + i\,\lambda_5}{2} = \begin{pmatrix} 0 & 0 & 1 \\ 0 & 0 & 0 \\ 0 & 0 & 0 \end{pmatrix}, \quad V_+\,p = \begin{pmatrix} 0 & 0 & p \\ 0 & 0 & 0 \\ 0 & 0 & 0 \end{pmatrix}$$

where p is a field operator. The tensor B is:

$$B = \sqrt{2}\begin{pmatrix} \frac{\Sigma^0}{\sqrt{2}} - \frac{\Lambda^0}{\sqrt{6}} & \Sigma^+ & p \\ \Sigma^- & -\frac{\Sigma^0}{\sqrt{2}} - \frac{\Lambda^0}{\sqrt{6}} & n \\ \Xi^- & \Xi^0 & \sqrt{\frac{2}{3}}\Lambda^0 \end{pmatrix},$$

the $\sqrt{2}$ is introduced so that the nucleon-nucleon-meson coupling constant $g_{NN\pi}$ will appear without any extraneous factors. We may rewrite B as $B = \sum_i \lambda_i B_i$ if we define B_i as the proper field operators. We may do the same thing for the antiparticles where, since B behaves as $A_j{}^i$, $\bar{B}$ will behave as $A_i{}^j$.

$$\bar{B} = \sqrt{2}\begin{pmatrix} \frac{\bar{\Sigma}^0}{\sqrt{2}} - \frac{\bar{\Lambda}^0}{\sqrt{6}} & \bar{\Sigma}^- & \bar{\Xi}^- \\ \bar{\Sigma}^+ & \frac{-\bar{\Sigma}^0}{\sqrt{2}} - \frac{\bar{\Lambda}^0}{\sqrt{6}} & \bar{\Xi}^0 \\ \bar{p} & \bar{n} & \sqrt{\frac{2}{3}}\bar{\Lambda}^0 \end{pmatrix}$$

and for the pseudoscalar mesons, by analogy

$$P = \begin{pmatrix} \frac{\pi^0}{\sqrt{2}} - \frac{\eta}{\sqrt{6}} & \pi^+ & K^+ \\ \pi^- & -\frac{\pi^0}{\sqrt{2}} - \frac{\eta}{\sqrt{6}} & K^0 \\ K^- & \bar{K}^0 & \sqrt{\frac{2}{3}}\eta \end{pmatrix}.$$

Defining $\bar{B}_i$ and P_i as the appropriate field operators we rewrite $\bar{B} = \sum_i \lambda_i \bar{B}_i$ and $P = \sum_i \lambda_i P_i$.

How do we form an invariant Lagrangian from B, $\bar{B}$, and P? There are two independent ways of coupling:

$$\mathcal{L}_1 = Tr\,\bar{B}\,\gamma_5\,B\,P \quad \text{and} \quad \mathcal{L}_2 = Tr\,\bar{B}\,\gamma_5\,P\,B$$

(there is no third independent way since the trace is not changed by a cyclic permutation of the matrices).

From our definition of B, $\bar{B}$, and P in terms of the λ matrices, we derive (using the summation convention)

$$\mathcal{L}_1 = Tr\, \lambda_i\, \bar{B}_i\, \lambda_j\, B_j\, \lambda_k\, P_k = \bar{B}_i\, B_j\, P_k\, Tr\, \lambda_i\, \lambda_j\, \lambda_k$$
$$\mathcal{L}_2 = Tr\, \lambda_i\, \bar{B}_i\, \lambda_k\, P_k\, \lambda_j\, B_j = \bar{B}_i\, B_j\, P_k\, Tr\, \lambda_i\, \lambda_k\, \lambda_j.$$

Defining antisymmetric and symmetric Lagrangians L_1 and L_2:

$$L_1 = 4\, i\, (\mathcal{L}_1 - \mathcal{L}_2) = 4\, i\, \bar{B}_i\, B_j\, P_k\, Tr\, \lambda_i\, [\lambda_j, \lambda_k]$$
$$L_2 = 4\, (\mathcal{L}_1 + \mathcal{L}_2) = 4\, \bar{B}_i\, B_j\, P_k\, Tr\, \lambda_i\, \{\lambda_j, \lambda_k\}.$$

However, $f_{ijk} = 4\, i\, Tr\, \lambda_i\, [\lambda_j, \lambda_k]$, $d_{ijk} = 4\, Tr\, \lambda_i\, \{\lambda_j, \lambda_k\}$
so that

$$L_1 = \bar{B}_i\, B_j\, P_k\, f_{ijk} = \bar{B}_i\, (f_k)_{ij}\, B_j\, P_k$$
$$L_2 = \bar{B}_i\, B_j\, P_k\, d_{ijk} = \bar{B}_i\, (d_k)_{ij}\, B_j\, P_k.$$

The set of matrices F_k are defined as $(F_k)_{ij} = f_{kij}$ and their matrix elements are just the structure constants of the algebra of SU(3) and they satisfy the same commutation relations as the λ's. They form the regular representation. The set D_k are defined as $(D_k)_{ij} = d_{kij}$ and they form an adjoint representation defined by

$$[F_i, D_j] = i\, f_{ijk}\, D_k,$$

which can be proved from the Jacobi identity.

In the beginning of the SU(3) symmetry approach the hope was that only F or D coupling would occur in nature. There was introduced the operation of R symmetry or the interchange $\Xi \to N$ which would exclude one of these couplings. It was discovered that this was not true so that both F and D couplings are needed.

The electromagnetic currents may now written as

$$j_F^{\mu} = \bar{B}_i \left(F_3 + \frac{1}{\sqrt{3}} F_8\right)_{ij} \gamma^{\mu}\, B_j$$

and

$$j_D^{\mu} = \bar{B}_i \left(D_3 + \frac{1}{\sqrt{3}} D_8\right)_{ij} \gamma^{\mu}\, B_j;$$

since the currents should depend linearly on Q, $Q = I + Y/2 = F_3 + {}$ $+ 1/\sqrt{3}\, F_8$ or $D_3 + 1/\sqrt{3}\, D_8$. The mass differences due to the electromagnetic interaction may now be computed. These mass differences will depend on the algebraic quantities:

$$\left(F_3 + \frac{1}{\sqrt{3}} F_8\right)^2, \quad \left(F_3 + \frac{1}{\sqrt{3}} F_8\right)\left(D_3 + \frac{1}{\sqrt{3}} D_8\right),$$

and

$$\left(D_3 + \frac{1}{\sqrt{3}} D_8\right)^2;$$

since we are looking at the graph

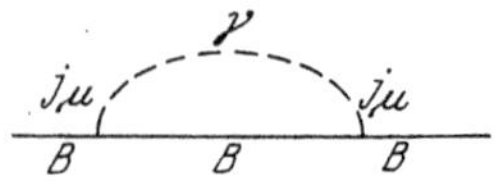

Fig. 18

where there are 2 electronmagnetic couplings on this one line, we need $j^{\mu} j_{\mu}$. However, $\left(D_3 + \frac{1}{\sqrt{3}} D_8\right)^2 = \frac{1}{9} + \frac{1}{3} \quad \left(F_3 + \frac{1}{\sqrt{3}} F_8\right)^2$, so that $\left(D_3 + \frac{1}{\sqrt{3}} D_8\right)^2$ is not independent of the first two products. We introduce a third independent matrix Q, since we have 4 arbitrary constants; the exact choice of Q is to some extent arbitrary. We choose non-vanishing matrix elements only for the Σ^0, Λ^0, $\Sigma^0 \leftrightarrow \Lambda^0$, $\Sigma^0 = 1$, $\Lambda^0 = 1/3$, $\Sigma^0 \leftrightarrow \Lambda^0 = 1/\sqrt{3}$. Then using first order perturbation theory (it is much safer to use perturbation theory for the electromagnetic interaction than for the strong interactions to break degeneracy) we evaluate the mass operator M

$$M = M_0 + A\left(F_3 + \frac{1}{\sqrt{3}} F_8\right)^2 + B\left(F_3 + \frac{1}{\sqrt{3}} F_8\right)\left(D_3 + \frac{1}{\sqrt{3}} D_8\right) + CQ$$

between $\langle B \mid \mid B \rangle$ to find the mass differences. We have 4 independent mass differences for the octet and using them with the matrix elements of the F's and D's, which are tabulated in the next section, and eliminating the 3 undetermined constants A, B, and C gives us the prediction (see the Appendix B for a different derivation of this result).

$$M_n - M_p + M_{\Xi^-} - M_{\Xi^0} = M_{\Sigma^-} - M_{\Sigma^+}.$$

We have the values of the masses (see Barkas and Rosenfeld [1])

$$\begin{aligned}
M_p &= 938.213 \pm 0.01 \text{ Mev} \\
M_n &= 939.507 \pm 0.01 \\
M_\Lambda &= 1115.36 \pm 0.14 \\
M_{\Sigma^+} &= 1189.35 \pm 0.15 \\
M_{\Sigma^0} &= 1193.2 \pm 0.7 \\
M_{\Sigma^-} &= 1197.6 \pm 0.5 \\
M_{\Xi^-} &= 1321.0 \pm 0.5 \\
M_{\Xi^0} &= 1315.8 \pm 0.5
\end{aligned}$$

so that we find for the above relation:

Left side = 1.3 Mev + 5.2 Mev = 6.5 Mev,
Right side = 8.2 Mev.

Older values of the masses gave almost exact agreement. When we compute the current for mesons, the D terms vanish, since the coupling is derivative, i. e., mesons have only F type coupling. Since we have only two independent mass differences, we do not obtain any prediction of electromagnetic mass differences for mesons.

3.3. *Magnetic Moments of the Baryons*

First we need a table of matrix elements for the physically interesting operators $F_3 + \frac{1}{\sqrt{3}} F_8$, $D_3 + \frac{1}{\sqrt{3}} D_8$, all of which are diagonal with respect to baryon states except for D_3 which mixes Λ^0 and Σ^0.

	F_3	F_8	D_3	D_8	$F_3 + \frac{1}{\sqrt{3}} F_8$	$D_3 + \frac{1}{\sqrt{3}} D_8$
p	$\frac{1}{2}$	$\frac{\sqrt{3}}{2}$	$\frac{1}{2}$	$-\frac{1}{2\sqrt{3}}$	1	$\frac{1}{3}$
n	$-\frac{1}{2}$	$\frac{\sqrt{3}}{2}$	$-\frac{1}{2}$	$-\frac{1}{2\sqrt{3}}$	0	$-\frac{2}{3}$
Λ	0	0	0	$-\frac{1}{\sqrt{3}}$	0	$-\frac{1}{3}$
Σ^+	1	0	0	$\frac{1}{\sqrt{3}}$	1	$\frac{1}{3}$
Σ^0	0	0	0	$\frac{1}{\sqrt{3}}$	0	$\frac{1}{3}$
Σ^-	-1	0	0	$\frac{1}{\sqrt{3}}$	-1	$\frac{1}{3}$
Ξ^0	$\frac{1}{2}$	$-\frac{3}{2}$	$-\frac{1}{2}$	$-\frac{1}{2\sqrt{3}}$	0	$-\frac{2}{3}$
Ξ^-	$-\frac{1}{2}$	$-\frac{\sqrt{3}}{2}$	$\frac{1}{2}$	$-\frac{1}{2\sqrt{3}}$	-1	$\frac{1}{3}$
$\Sigma^0 \to \Lambda^0$	0	0	$\frac{1}{\sqrt{3}}$	0	0	$\frac{1}{\sqrt{3}}$

The form factors and magnetic moments of the baryons are first order in j_μ. We can calculate the anomalous magnetic moments by:

$$\mu_B = \langle B \mid \theta \left(F_3 + \frac{1}{\sqrt{3}} F_8\right) + \xi \left(D_3 + \frac{1}{\sqrt{3}} D_8\right) \mid B \rangle$$

$$\mu_p = \theta + \tfrac{1}{3}\xi$$

$$\mu_n = -\tfrac{3}{2}\xi.$$

All the other magnetic moments of the baryons can be expressed in terms of μ_p or μ_n.

$$\mu_\Lambda = \tfrac{1}{2} \cdot \mu_n \quad \text{(the Sakata model predicts } \mu_{\Lambda^0} = \mu_n)$$

$$\mu_{\Sigma^+} = \mu_p, \; \mu_{\Xi^0} = -\mu_n, \; \mu_{\Sigma^0 \to \Lambda^0} = -\frac{\sqrt{3}}{2}\mu_n$$

$$\mu_{\Sigma^0} = -\tfrac{1}{2}\mu_n, \; \mu_{\Xi^-} = \mu_{\Sigma^-} = -\mu_p - \mu_n$$

$$\mu_{\Sigma^-} = -\mu_p - \mu_n.$$

$\mu_{\Sigma^0 \to \Lambda^0}$ is responsible for the electromagnetic decay $\Sigma^0 \to \Lambda^0 + \gamma$. The important magnetic moments are those that can be experimentally studied best: μ_Λ and $\mu_{\Sigma^0 \to \Lambda^0}$. For reference see Coleman and Glashow [33], Cabibbo and Gatto [34]. As an aside to this method of attack, we note that the Okubo mass formula is just $M = a + b\, F_8 + c\, D_8$ where F_8 is the Y term, D_8 is the $I\,(I+1) - \frac{Y^2}{4}$ term of Okubo.

Appendix B

Another Derivation of the Electromagnetic Mass Difference of the Baryons by *Lance Heiko, Brandeis U.*

The mass difference between isomultiplets is found by calculating the matrix elements of a mass operator which transforms as Y. In analogy we would expect the mass differences within an isomultiplet to be dependent on a mass operator which transforms as I_3. But $I_3 = \frac{A_1{}^1 - A_2{}^2}{2}$, where $A_1{}^1$ and $A_2{}^2$ are two of the infinitesimal operators of the group. We would then expect M to be of the form:

$$M = \frac{T_1{}^1 - T_2{}^2}{2}\,.$$

In Okubo's form for the tensor operator, when calculating matrix elements, we have $T_\nu{}^\mu = a\,\delta_\nu{}^\mu + b\,A_\nu{}^\mu + c\,A_\lambda{}^\mu\,A_\nu{}^\lambda$. Therefore

$$T_1{}^1 = a + b\,A_1{}^1 + c\,A_\mu{}^1\,A_1{}^\mu$$
$$T_2{}^2 = a^1 + b^1\,A_2{}^2 + c^1\,A_\mu{}^2\,A_2{}^\mu.$$

For simplicity let us assume $a = a^1$, $b = b^1$, $c = c^1$, then

$$M = b\,I_3 + c\,(A_\mu{}^1\,A_1{}^\mu - A_\mu{}^2\,A_2{}^\mu),$$

where we redefine c to include the additional $1/2$.

Let us evaluate the second term:

$$A_\mu{}^1\,A_1{}^\mu - A_\mu{}^2\,A_2{}^\mu = A_1{}^1\,A_1{}^1 + A_2{}^1\,A_1{}^2 + A_3{}^1\,A_1{}^3 - A_1{}^2\,A_2{}^1 - A_2{}^2\,A_2{}^2 - A_3{}^2\,A_2{}^3.$$

We break this sum up into 3 terms which we evaluate separately. First,

$$A_1{}^1\,A_1{}^1 - A_2{}^2\,A_2{}^2 = \left(I_3 + \frac{Y}{2}\right)\left(I_3 + \frac{Y}{2}\right) - \left(-I_3 + \frac{Y}{2}\right)\left(-I_3 + \frac{Y}{2}\right) = 2\,I_3\,Y.$$

Second,

$$A_2{}^1\,A_1{}^2 - A_1{}^2\,A_2{}^1 = -\,[A_1{}^2\,A_2{}^1] = -2\,I_3,$$

which we absorb into the b term. Lastly,

$$A_3{}^1\,A_1{}^3 - A_3{}^2\,A_2{}^3 = V_-\,V_+ - U_-\,U_+.$$

But $U_-\,U_+ \mid U, U_3, Z) = (U - U_3)\;(U + U_3 + 1) \mid U, U_3, Z)$ and if we look at baryon states which are not degenerate, i. e., peripheral members of the octet, then a baryon state can be looked at as an I-spinor, a U-spinor or a V-spinor, whichever is most convenient. Then,

$$A_3^1 A_1^3 - A_3^2 A_2^3 = V^2 - V_3^2 + V - V_3 - (U^2 - U_3^2 + U - U_3).$$

However,

$$-V_3^2 + U_3^2 = -(\tfrac{3}{4}Y + \tfrac{1}{2}I_3)(\tfrac{3}{4}Y + \tfrac{1}{2}I_3) + (\tfrac{3}{4}Y - \tfrac{1}{2}I_3)(\tfrac{3}{4}Y - \tfrac{1}{2}I_3) = \\ = -\tfrac{3}{2}I_3 Y,$$

which contributes to the $A_1^1 A_1^1 - A_2^2 A_2^2$ term so that we have left $+\frac{1}{2} I_3 Y$. Further we have $-V_3 + U_3 = -I_3$ which we add to the b term. Then neglecting these subtractions

$$A_3^1 A_1^3 - A_3^2 A_2^3 = V(V+1) - U(U+1).$$

The mass operator can now be written:

$$M = bI_3 + c\left(\tfrac{1}{2} I_3 Y + V(V+1) - U(U+1)\right).$$

Under Weyl reflection W_3, I-spinors become U-spinors. The weight diagram becomes:

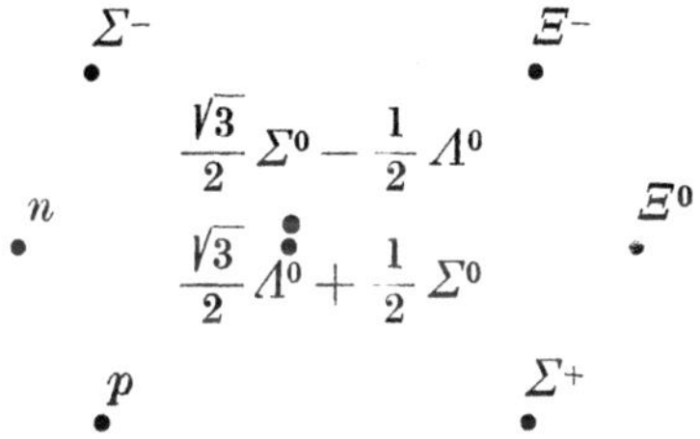

Analogously with W_2

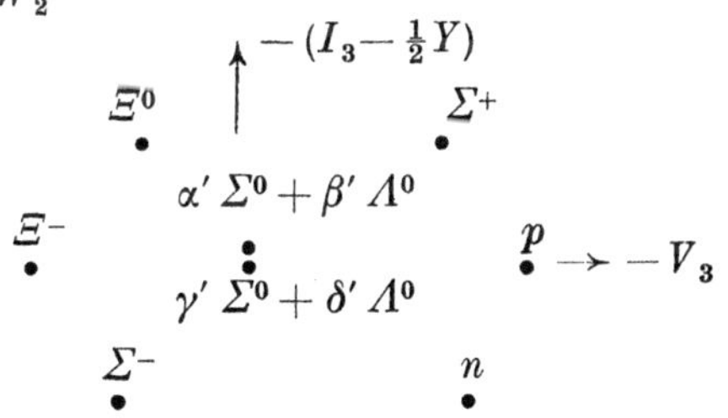

We can read off from the diagrams the U and V values for any baryon except Σ^0 and Λ^0 which are mixtures of $U=0$ and $U=1$ in one case and $V=0$ and $V=1$ in the other. The matrix elements are:

$$\begin{aligned} M_n &= -\tfrac{1}{2}b - \tfrac{3}{2}c \\ M_p &= \tfrac{1}{2}b + \tfrac{3}{2}c \\ M_{\Xi^-} &= -\tfrac{1}{2}b + \tfrac{3}{2}c \\ M_{\Xi^0} &= \tfrac{1}{2}b - \tfrac{3}{2}c \\ M_{\Sigma^-} &= -b \\ M_{\Sigma^+} &= b. \end{aligned}$$

Therefore:

$$M_n - M_p + M_{\Xi^-} - M_{\Xi^0} = M_{\Sigma^-} - M_{\Sigma^+}.$$

Appendix C.

Some more Recent and Speculative Developments

1. The "tadpole" model by Coleman and Glashow
(Harvard University preprint, Dec. 1963)

Four kinds of fundamental interactions of elementary particles are assumed:

1) Very strong interactions invariant under SU(3).
2) Strong interactions.
3) Electromagnetic interactions.
4) Weak interactions.

The interactions 2), 3), and 4) are symmetry breaking interactions. The departure from the exact SU(3) symmetry is the strongest for the strong interactions and becomes weaker for the other ones. The main assumption of the model is that there exists an octet of scalar mesons, which allows the existence of some Feynman diagrams, which have non-vanishing vacuum expectation values: $\langle\psi\rangle \neq 0$. Such diagrams (tadpoles)

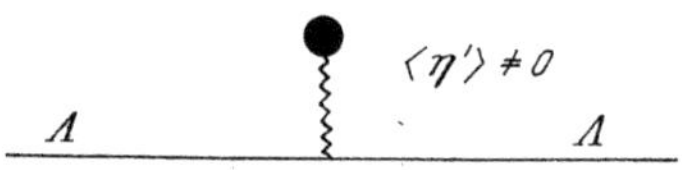

Fig. 19

had been used originally by SCHWINGER in an attempt to explain the muon mass [SCHWINGER, Ann. Phys. (N. Y.) **2**, 407 (1957)] and also by SALAM and J. C. WARD [Phys. Rev. Letters **5**, 390 (1960)]. The origin of this idea is of course a consideration of effects which had been found from unitary symmetry, namely

1) the mass formula; the mass splitting interaction transforms like the component with $Y=0$ and $I_3=0$ of an octet.

2) The electromagnetic mass corrections satisfy approximately a $\Delta I = 1$ rule, at least for baryons. (The $\Delta I = 2$ is about 10% of the $\Delta I = 1$ term.) The mass splitting operator behaves like I_3.

3) The non-leptonic weak interactions satisfy an approximate $\Delta I = 1/2$ rule. The Lagrangian for the non-leptonic-decays behaves like a member with $\Delta I = 1/2$, $Y = \pm 1$ of an octet (like V or U). According to the assumption there exists an octet of scalar mesons. These particles are called η', π', K', $\overline{K}'$ in analogy to the pseudoscalar mesons. The breaking of symmetry makes the vacuum expectation values of neutral primed fields non-vanishing.

$\langle\eta'\rangle \neq 0$ leads to the mass formula.

$\langle\pi^{0\prime}\rangle \neq 0$ produces the electromagnetic mass splitting.

$\langle K_1{}'\rangle \neq 0$ is responsible for the non-leptonic weak interactions. Furthermore

$\langle K_1\rangle \neq 0$ may also exist, because of parity violation.

If we have exact SU(3) symmetry, all these expectation values must vanish.

For the electromagnetic mass splitting there are also non-tadpole contributions necessary (this is a modernization of the FEYNMAN-SPEISMAN calculations). The agreement with experiment is astonishing.

Arguments for the existence of tadpoles:

The K' may possibly be identified with the $\varkappa$ (725 MeV), 0^+ or 1^- because of K^* (885)$\to \varkappa + \pi$.

The scalar pion π' has odd G-parity, and couples to at least 5 pions or $\eta\pi$. If it is lighter than $\eta + \pi$, it can decay electromagnetically into $2\pi + \gamma$. A possible candidate is the ζ (2π peak at 570 MeV).

The η' with $J^{PG} = 0^{++}$ should show up as a $I = 0$, s-wave pion-pion resonance.

The empirical fact, which is explained after all by the tadpole theory,

$$\varrho - \pi = K^* - K = \widehat{\omega} - \eta$$

where

$$\varrho = (m_\varrho)^2, \ \ldots \quad \widehat{\omega} = \tfrac{1}{2}\,[(m_K)^2 + (m_\varphi)^2]$$

leads one to apply this "rule" to the masses of the conjectured scalar mesons. With K' at 730 MeV one obtains η' at 770 MeV and π' at 560 MeV.

Assuming a Yukawa type coupling, one gets the following SU(3) invariant Lagrangian:

$$\mathscr{L}' = d\, Tr\,\{\psi, \psi\}\,\varphi' + f\, Tr\,[\psi, \psi]\,\varphi' + g\, Tr\,\psi\,\psi\,\psi' + \ldots$$

D-coupling F-coupling D-coupling

The dots represent further couplings: ψ baryons

D-type	$V\,V\,\varphi'$	φ	p. s. mesons
F-type	$V\,\varphi'\,\varphi'$	φ'	s. mesons
D-type	$\varphi'\,\varphi'\,\varphi'$ ...	V	v. mesons

If one writes down the Lagrangian explicitly and considers the self-energy operators one gets by assumption of tadpole dominance 13 mass relations, e. g.:

$$p = m_0 + (3f - d)\,\langle\eta'\rangle/\sqrt{3} + (d + f)\,\langle\pi^{0'}\rangle$$
$$n = m_0 + (3f - d)\,\langle\eta'\rangle/\sqrt{3} - (d + f)\,\langle\pi^{0'}\rangle$$
$$\Xi^- = m_0 - (3f + d)\,\langle\eta'\rangle/\sqrt{3} + (d - f)\,\langle\pi^{0'}\rangle$$
$$\Xi^0 = m_0 - (3f + d)\,\langle\eta'\rangle/\sqrt{3} - (d - f)\,\langle\pi^{0'}\rangle$$
$$\Sigma^+ = m_0 + 2d\,\langle\eta'\rangle/\sqrt{3} + 2f\,\langle\pi^{0'}\rangle$$
$$\Sigma^0 = m_0 + 2d\,\langle\eta'\rangle/\sqrt{3}$$
$$\Sigma^- = m_0 + 2d\,\langle\eta'\rangle/\sqrt{3} - 2f\,\langle\pi^{0'}\rangle$$
$$\Lambda = m_0 - 2d\,\langle\eta'\rangle/\sqrt{3}$$
$$K^+ = \mu_0^2 - g\,\langle\eta'\rangle/\sqrt{3} + g\,\langle\pi^{0'}\rangle$$
$$K^0 = \mu_0^2 - g\,\langle\eta'\rangle/\sqrt{3} - g\,\langle\pi^{0'}\rangle$$
$$\pi^+ = \mu_0^2 + 2g\,\langle\eta'\rangle/\sqrt{3}$$

$$\pi^0 = \mu_0^2 + 2g\,\langle\eta'\rangle/\sqrt{3}$$
$$\eta = \mu_0^2 - 2g\,\langle\eta'\rangle/\sqrt{3}.$$

The masses, resp. the (masses)^2 depend on 7 parameters, one of them gives only the scale. Eliminating the parameters gives 3 classes of sum rules:

Intermultiplet sum rules

$\frac{1}{2}(N + \Xi) = \frac{1}{4}(3\Lambda + \Sigma)$ (i. e. the Gell-Mann Okubo mass formula)

$$K = \tfrac{1}{4}(3\eta + \pi).$$

Intramultiplet sum rules (electromagnetic mass splitting)

$$\Xi^- - \Xi^0 = \Sigma^- - \Sigma^+ + p - n$$
$$\Sigma^0 = \tfrac{1}{2}(\Sigma^+ + \Sigma^-)$$
$$\pi^0 = \pi^+;$$

this last one is not so good, but it can be corrected by the non-tadpole contributions.

Hybrid sum rules

$$\frac{\Sigma^- - \Sigma^+}{\Xi - N} = \frac{2}{3}\,\frac{\Xi^- - \Xi^0 + p - n}{\Sigma - \Lambda} \qquad \text{(A)}$$
$$\frac{K^0 - K^+}{K - \pi} = \frac{\frac{1}{2}(n + \Xi^0) - \frac{1}{2}(p + \Xi^-)}{\frac{1}{2}(\Xi + N) - \Sigma}. \qquad \text{(B)}$$

From the experimental values one gets: an agreement at least in the order of magnitude:

	left hand side	right hand side
(A)	0.021	0.038
(B)	0.017	0.038

Non-tadpole contributions arises from ordinary electromagnetic interactions, e. g.

Fig. 20

By combining the experimental nucleon form factors and SU(3) one gets the non-tadpole contributions which are listed in the following table, together with the tadpole contributions and the experimental values of the mass differences*:

Non-tadpole Contribution	Tadpole	Calculated Splitting	Observed
0.9 MeV	−2.6 MeV	−1.7 MeV	−1.3 MeV
0.7	−3.9	−3.2	−3.6±0.5
0.0	3.9	3.9	4.5±0.4
0.0	5.2	5.2	5.6±1.4
2280 (MeV)^2	−4880 (MeV)^2	−2.6 MeV	−3.9±0.6 MeV
1140	0	4.2	4.6

* The non-tadpole contributions have been calculated by S. Coleman and H. Schnitzer (unpublished) for the baryons and by R. Socolow for the fermions.

The non-leptonic decays should be dominated by the following diagrams:

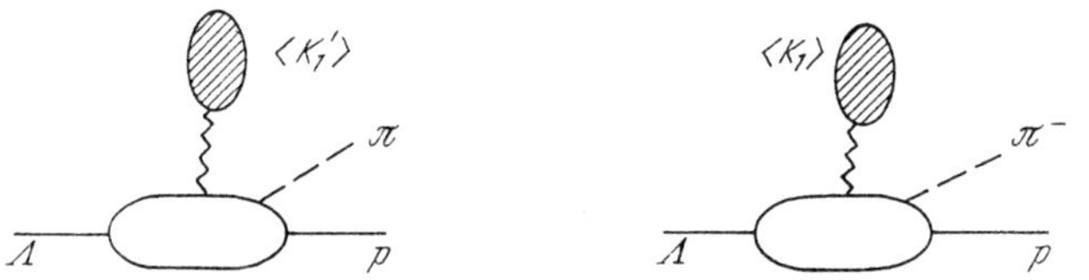

Fig. 21

But due to some difficulties which arise in the treatment of the parity-conserving decays (the theory seems to work well for parity violating decays) we will not discuss it further here, pending clearifications from the authors.

2. The new Schwinger model
(Harvard University preprint, January 1964)

In his paper [Ann. phys. N. Y. 2, 407 (1957)], SCHWINGER proposed a classification of leptons into a triplet of multiplicity 2×3: $\mu^{\pm}$, $e^{\pm}$, ν, $\bar{\nu}$, where the electron and the muon are coupled to different helicity states of the neutrino. The interactions are mediated by the electromagnetic field A^{μ}, which is coupled to the exactly conserved charge Q, and by the vector field Z which exchanges the charge and is responsible for weak interactions.

Schwingers new idea is to construct a theory of baryons along the same lines. He supposes that there exists a symmetry between leptons and baryons, namely that there exist fundamental (unobservable) baryon fields

$$N=1:\begin{pmatrix}\psi_0{}^{+}\\ \psi^{0}\\ \psi_0{}^{0}\end{pmatrix}\qquad\qquad N=-1:\begin{pmatrix}\overline{\psi^{+}}\\ \overline{\psi^{0}}\\ \overline{\psi_0{}^{0}}\end{pmatrix}$$

similar to the Sakaton. The conservation of baryon number N implies by gauge arguments the existence of a vector field B which is strongly coupled to ψ and should have no zero mass state. Besides this vector field Schwinger assumes the existence of a second vector field V, which carries baryon number ± 2

$$\begin{pmatrix}V^{-}\\ V^{0}\\ V^{00}\end{pmatrix}$$

B is very strongly coupled to ψ and V, but there is no direct coupling between ψ and V.

Now suppose you can subject ψ and V independently to U_3 transformations. The direct product $U_3\times U_3$ is called W_3 by *Schwinger*. It corresponds to various bound states $\psi\,\psi\to$ bosons $0\pm$, $1\pm$; $\psi\, V\to$ fermions $\frac{1}{2}\pm$, $\frac{3}{2}\pm$. If one takes $\overline{\psi}_a\,\psi_b$, $\overline{V_a}\,V_b$, this corresponds to $3^*\times 3 =$

$=8+1$. So we get octets and singlets of mesons. If you take $\bar{\psi}_a V_b$ $(N=1)$ or $\bar{V}_a \psi_b$ $(N=-1)$, this corresponds to $3\times 3=9$, giving 9 physical baryons, which are still degenerate.

However if you look at the effective Lagrangian, you will see, that there are only 2 possible charge conserving terms $\psi_a \psi_b \bar{V}_c$ and $\bar{\psi}_a \bar{\psi}_b \psi_c$, invariant under W_3. The effective Lagrangian is not invariant under general U_3 transformations but you can minimalize the noninvariance by taking a preferred axis in the 3-dimensional unitary space, the 3-axis or the Y-axis, and this gives the usual splitting of U_3 into the (I-spin and Y) conserving part. Let us take $a=3$, which must be associated with an electrically neutral component of 1, there is still a choice to be made between a symmetrical and antisymmetrical combination P_n of $\psi_a \psi_b$, before forming a scalar product with $\bar{V}_c$ and setting $a=3$. If the antisymmetrical combination is used, V_3 and $\bar{V}_3$ will not appear in the coupling term. Such a theory is invariant under independent phase transformations of the field V_3, which implies another conservation law of the same structure level as hypercharge and isospin conservation. Lacking evidence for an additional selection rule, that possibility is rejected and the symmetrical combination adopted. The ψ V interaction is pseudovector.

Second order perturbation theory leads to destruction of symmetry between the 9 baryon states

$$\bar{\psi}_a V_b\ [N,\ \Xi,\ \Lambda,\ \Sigma,\ Y^0;\ Y=0,\ I=0].$$

You will not get a simple mass formula, unless you separate the not seen Y^0 from the other ones. If one adds the condition that the mass of the nucleon should be independent of the mass of Y^0, one obtains the 2 formulas:

$$\tfrac{1}{2}(N+\Xi)=\tfrac{1}{4}(3\Lambda+\Sigma)$$

and

$$Y^0=3\Sigma-2\Lambda.$$

The mass predicted for the ninth baryon is about 1350 MeV. The closest candidate according to Schwinger is Y_0 (1405 MeV). Naturally it should have $1/2^+$, but experiments prefer $1/2^-$ or $3/2^+$.

Then he has treated the vector mesons and pseudoscalar mesons by simplifying the model. He obtained a formula for the (mass)^2

$$m_\varrho^2=\varrho,\ m_\pi^2=\pi,\ m_\omega^2=\omega \text{ etc.}$$

for vectormesons:

$$(\varphi-\varrho)(\omega-\varrho)=\tfrac{4}{3}(K^*-\varrho)(\varphi+\omega-2K^*).$$

This gives $\sqrt{\varrho}=752\pm 5$ MeV (experimentally: 755 ± 3 MeV). If the ninth 0^- meson is called δ, the (mass)^2 relation

$$(\delta-\pi)(\eta-\pi)=\tfrac{4}{3}(K-\pi)(\delta+\eta-2K)$$

yields $\sqrt{\delta}\sim 1.5$ GeV.

Further, $K-\pi=K^*-\varrho$ as in the tadpole model.

Finally as a result of the W_3 invariance one is able to predict relations between the widths of the decays of various particles.

For example, the pion coupling between $Y_0{}^*$ and Σ is related to the pion-nucleon coupling which gives

$$\Gamma_\pi (Y_0{}^*) = \tfrac{4}{3} (f_{\pi N}/4\pi) (p^3/\mu^2)_\pi \sim 30 \text{ MeV}.$$

Other predictions: from the ϱ width, which is known, one gets:

$$\Gamma (K^*) = 35 \pm 3, \ \Gamma (\varphi) = 2,4 \pm 0,2.$$

Therefore the very small width of the φ, compared with the large width of the other members of the octet is a consequence of the dynamics of the theory.

3. The Quarks of Gell-Mann
(Physics Letters, February 1, 1964)

Another model, which has been discussed recently, the so-called "quarks" of Gell-Mann and the practically identical model, the "Aces" proposed by Zweig, starts with the fact, that one has never seen other multiplets than 1, 8, and 10. A model, which would lead only to these representations, would be to consider not the original Sakata model, where we started out from 3×3, but to take 3 Sakatons or "quarks" and take the direct product $3 \times 3 \times 3 = 1 + 8 + 8 + 10$. Now what kind of animals should these three be? Obviously they should be fermions $1/2^+$, however, to get an integral baryon number out of three identical objects, we have to assume the baryon number 1/3. The electric charge turns out to be 1/3, 1/3, −2/3. Out of three of theses triplets you can build up all the octets and decuplets of baryons without any difficulty. The meson octets are to be constructed out of quark-antiquark pairs, as in the Sakatian model. A simple calculation shows that one gets objects with baryon number 0 and, of course, integral or zero charge. Gell-Mann went so far, as to suggest that experimentalists should look for such particles. In Brookhaven people found none.

Dr. Kummer remarked that at CERN there was also a search for these particles. But it turned out that, if such particles exist, the cross section would be down a factor 100 to the cross sections corresponding to strong interactions.

It is of course not difficult to invent variations on similar themes, and one can foresee a flux of papers of this kind in the near future.

SU(3) Invariant Static Model

By

J. E. Wess, Vienna

With 2 Figures

The static model seems to be useful for studying the influence of SU_3 invariance on the dynamics of a baryon — meson system. It is well known that the eightfold way is the most successful assignment of particles to irreducible representations of SU_3. Baryons and mesons are assigned to the 8-dimensional regular representation. Therefore, only the representations which can be obtained by direct products of the 8-representation are of physical interest.

Starting from the 3 by 3 representation of the SU_3 we find that the invariant subgroup:

$$\zeta = \begin{pmatrix} e^{2\pi i \frac{n}{3}} & & \\ & e^{2\pi i \frac{n}{3}} & \\ & & e^{2\pi i \frac{n}{3}} \end{pmatrix} \quad n = 0, 1, 2 \tag{1}$$

is represented by unity in the 8-representation. The 8-representation' therefore, is a faithful representation of the factorgroup SU_3/Z_3 and the eightfold way deals with the representations of this group. If we label the representations by their highest weight μv (for the highest weight vector we have $\vec{H} \mid \mu v\rangle = (\mu \vec{\alpha} + v \vec{\beta}) \mid \mu v\rangle$, $\vec{\alpha}$ and $\vec{\beta}$ being the simple roots), we find from the condition that Z_3 has to be represented by the unity that μ and v have to be integers. These are exactly the representations for which the hypercharge $Y = 2 H_2$ has integer eigenvalues. (The situation is thus similar to the rotation group in three dimensions, which is a factor group of SU_2 and which leads to representations with integer spin). According to the above, it is clear that we will have to decompose the $8 \otimes 8$ representation into its irreducible components when we are dealing with a baryon — meson system. We know that $8 \otimes 8 = 1 + 8 + 8' + 10 + \overline{10} + 27$.

It is easy to write down the combinations which transform like 8 and 8′. We know that there are two independent 8 — vector operators among the 8 by 8 matrices, Z^i and Y^i. The Z's are simply the generators of the group. We want to choose a representation in which the Z's are antisymmetric matrices and the Y's symmetric ones. To find the matrices Z^k_{ij} and Y^k_{ij} we start from the fundamental representation:

$$X^1 = \frac{1}{6}\begin{pmatrix}1 & & \\ & 1 & \\ & & -2\end{pmatrix} \qquad X^2 = \frac{1}{2\sqrt{3}}\begin{pmatrix}1 & & \\ & -1 & \\ & & 0\end{pmatrix} \tag{2}$$

$$X^3 = \frac{1}{2\sqrt{3}}\begin{pmatrix}0 & 1 & 0\\ 1 & 0 & 0\\ 0 & 0 & 0\end{pmatrix} \quad X^4 = \frac{i}{2\sqrt{3}}\begin{pmatrix}0 & 1 & 0\\ -1 & 0 & 0\\ 0 & 0 & 0\end{pmatrix} \quad X^5 = \frac{1}{2\sqrt{3}}\begin{pmatrix}0 & 0 & 1\\ 0 & 0 & 0\\ 1 & 0 & 0\end{pmatrix}$$

$$X^6 = \frac{i}{2\sqrt{3}}\begin{pmatrix}0 & 0 & 1\\ 0 & 0 & 0\\ -1 & 0 & 0\end{pmatrix} \quad X^7 = \frac{1}{2\sqrt{3}}\begin{pmatrix}0 & 0 & 0\\ 0 & 0 & 1\\ 0 & 1 & 0\end{pmatrix} \quad X^8 = \frac{i}{2\sqrt{3}}\begin{pmatrix}0 & 0 & 0\\ 0 & 0 & 1\\ 0 & -1 & 0\end{pmatrix}$$

It is easy to show that

$$\begin{aligned} 6\, i\, T r\, [X^i\, X^j]_-\, X^k &= Z^k_{ij} \\ 6\, T r\, \{X^i\, X^j\}_+\, X^k &= Y^k_{ij}\,. \end{aligned} \tag{3}$$

The fact that they are 8-vectors can be expressed through:

$$\begin{aligned} [Z^i\, Z^k] &= g^{ikl}\, Z^l \\ [Z^i\, Y^k] &= g^{ikl}\, Y^l \end{aligned} \qquad g^{ikl} = (Z^i)_{lk}. \tag{4}$$

Thus, if we have two sets of 8-vectors $\mathfrak{x}_i$, $\mathfrak{y}_i$ the combinations:

$$Z^i_{rs}\, \mathfrak{x}_r\, \mathfrak{y}_s \qquad Y^i_{rs}\, \mathfrak{x}_r\, \mathfrak{y}_s \tag{5}$$

belong to the 8 and 8′ representation, respectively. The invariant formed from two vectors is simply $\mathfrak{x}_i\, \mathfrak{y}_i$.

To find the other combinations we would have to compute the Clebsch-Gordan coefficients, but for our purpose it is enough to know the projection operators which project into the invariant subspaces.

To derive this projection operators we use the Casimir-operators; these are the two invariants that can be formed from the group generators. According to the above

$$\begin{aligned} I_1 &= Z^i\, Z^j\, Z^l\, Z^l_{ij} = \tfrac{1}{2}\, Z^i\, Z^i \\ I_2 &= Z^i\, Z^j\, Z^l\, Y^l_{ij} \end{aligned} \tag{6}$$

are invariants. For the representations in question they have the following eigenvalues:

Dim	1	8	10	$\overline{10}$	27
I^1	0	$^1/_2$	1	1	$^4/_3$
I^2	0	0	-1	1	0
μ, ν	0, 0	1, 1	1, 2	2, 1	2, 2

(7)

The projection operators, which project into the 10-representations therefore are:

$$P^{10} = \tfrac{1}{2}\, I_2\, (I_2 - 1)$$

$$P^{\overline{10}} = \tfrac{1}{2}\, I_2\, (I_2 + 1) \tag{8a}$$

and

$$P^{27} = \tfrac{27}{5}\, I_1\, (I_1 - \tfrac{1}{2})\, (I_1 - 1).$$

As we have already the proper combinations for the other representations, the corresponding projection operators are easily written down:

$$\begin{aligned} P^{1}_{ir,js} &= \tfrac{1}{8}\, \delta_{ir,js} \\ P^{8}_{ir,js} &= -\, Z^{k}_{ir}\, Z^{k}_{js} \\ P^{8'}_{ir,js} &= \tfrac{9}{5}\, Y^{k}_{ir}\, Y^{k}_{js}. \end{aligned} \tag{8b}$$

The normalization is chosen such that $P^2 = P$.

The static model is defined by the following Hamiltonian:

$$\begin{aligned} H^{\circ} &= \tfrac{1}{2} \sum_{\alpha} \int d^3 x\, \{\Phi^{\alpha}\, \Phi^{\alpha} + \nabla \Phi^{\alpha}\, \nabla \Phi^{\alpha} + m^2\, \Phi^{\alpha}\, \Phi^{\alpha}\} \\ H' &= \int d^3 x\, \varrho\, (r)\, \vec{\sigma}\, \vec{\nabla}\, \Phi^{\alpha}\, \{f_1\, Z^{\alpha} + f_2\, Y^{\alpha}\} \end{aligned} \tag{9}$$

Φ^{α} describes the meson field and therefore transforms like an 8-vector. H' is the most general invariant that can be formed from Φ^{α} and a set of 8×8 matrices.

We introduced a notation where the Z^{i}'s are antisymmetric. Thus, we are dealing with a subgroup of the SO_8 and the Φ^{α}'s can be taken to be real $\Phi^{\alpha +} = \Phi^{\alpha}$. This notation is of advantage for calculational purposes; to compare with measurements we have to transform to a coordinate system adapted to the particle notation. This is achieved by:

$$\begin{aligned} X_k &= U^{*}_{kj}\, U\, Z^{j}\, U^{+} \\ Y_k &= U^{*}_{kj}\, U\, Y^{j}\, U^{+} \end{aligned} \tag{10}$$

where

$$U = \begin{pmatrix} 1 & 0 & & & \\ 0 & 1 & & & \\ & & \begin{pmatrix} 1/\sqrt{2} & i/\sqrt{2} \\ 1/\sqrt{2} & -i/\sqrt{2} \end{pmatrix} & & \\ & & & (\quad) & \\ & & & & (\quad) \end{pmatrix} \tag{10'}$$

The assignment is as follows:

$$\begin{aligned} &\eta\ \pi^{0}\ \pi^{+}\ \pi^{-}\ \overline{K^{-}}\ K^{-}\ K^{0}\ \overline{K^{0}} \\ &\Lambda\ \Sigma^{0}\ \Sigma^{+}\ \Sigma^{-}\ P\ \Xi^{-}\ N\ \Xi^{0}. \end{aligned}$$

The relation $Z^{iT} = -Z^{i}$, $Y^{iT} = Y^{i}$ becomes $A\, X_i\, A^{-1} = -X_i^{T}$, $A\, Y_i\, A^{-1} = Y_i^{T}$ where $A = U\, U^{T}$. This transformation is called R-conjugation and it would leave H' invariant if either f_1 or f_2 were zero. As we shall see, the R-invariance is not realized in nature; therefore we have to use both couplings and determine the ratio of the couplings by comparison with experiment. The Z's correspond to the F-type and the Y's to the D-type coupling introduced by Gell-Mann.

The Hamiltonian leads to the following equations of motion:

$$(m^2 + \Box)\, \Phi^\alpha = \vec{\sigma}\, \vec{\nabla}\, \varrho(x)\, \{f_1 Z^\alpha + f_2 Y^\alpha\}\,. \tag{11}$$

We can develop Φ into a momentum or angular momentum representation respectively:

$$\begin{aligned}\Phi^\alpha(r) &= \int \frac{d^3 k}{(2\pi)^{3/2}} \frac{a^\alpha(\vec{k})\, e^{i \vec{k}\vec{r}} + a^{\alpha +}(\vec{k})\, e^{-i\vec{k}\vec{r}}}{\sqrt{2\omega}} \\ &= \int \frac{dk}{\sqrt{2\omega}} \sum U_K^l(r)\, \left[a_{lm}^\alpha(k)\, Y_l^m(\vartheta,\varphi) + a_{lm}^{\alpha +}(k)\, Y_l^{m*}(\vartheta,\varphi)\right].\end{aligned} \tag{12}$$

The angular momentum representation is especially convenient as only the $l = 1$ component is coupled to the source.

The field equations can be integrated with the help of the Green's function to give:

$$a_K(t) = a_K^{in}(t) - i \int_{-\infty}^{t} dt'\, e^{i\omega(t-t')}\, V_K(t')$$

where

$$V_K(t) = \frac{k^2 \varrho(k)}{\sqrt{12\,\omega\,\pi^2}}\, \sigma^i(t)\, \{f_1 Z^\alpha(t) + f_2 Y^\alpha(t)\}. \tag{13}$$

K denotes the spin variable i and the SU_3 variable a of the meson as well as the variable k.

Our aim is to calculate an S-matrix element

$$\langle \zeta' |\, a_{K'}^{\mathrm{out}}\, a_K^{+\mathrm{in}}\, | \zeta \rangle = S_{K'\zeta', K\zeta} \tag{14}$$

again, ζ denoting the spin variable and SU_3 variable of the baryon.

We express a^{in} in terms of a^{out}:

$$a_K^{+\mathrm{in}}(0) = a_K^{+\mathrm{out}}(0) - i \int_{-\infty}^{+\infty} dt\, e^{-i\omega t}\, V_K(t) \tag{15}$$

and get

$$S_{K'\zeta', K\zeta} = \delta_{\zeta'\zeta}\, \delta_{K'K} - 2\pi i\, \delta(\omega - \omega')\, T_{K'\zeta', K\zeta} \tag{16}$$

where, after some calculations

$$T_{K'\zeta', K\zeta} = -i \int_0^\infty \langle \zeta' |\, [V_{K'}(t),\ V_K(0)]\, | \zeta \rangle\, e^{i\omega t}\, dt. \tag{17}$$

To darive this equation, we made use of the fact that we can restrict ourselves to $\omega = \omega'$.

To calculate this expression further we introduce a complete set of intermediate states $|out, n\rangle$. We use

$$\begin{aligned}\langle \zeta' |\, V_K(t)\, |out, n\rangle &= \langle \zeta' |\, e^{iHt}\, V_K(0)\, e^{-iHt}\, |out, n\rangle \\ &= \langle \zeta' |\, V_K(0)\, |out, n\rangle\, e^{-iE_n t} = e^{-iE_n t}\, T^*_{n, K'\zeta'}\end{aligned} \tag{18}$$

and we obtain:

$$T_{K'\zeta', K\zeta} = -\mathop{S}_{n} \left\{ \frac{T^*_{n, K'\zeta'}\, T_{n, K\zeta}}{E_n - \omega - i\,\varepsilon} + \frac{T^*_{n, K'\zeta'}\, T_{n, K'\zeta}}{E_n + \omega} \right\} \tag{19}$$

It is useful to introduce the t-matrix:

$$t_{\zeta' K', \zeta K}(z) = -\mathop{S}_{n} \left\{ \frac{T^*_{n, \zeta' K'}\, T_{n, \zeta K}}{E_n - z} + \frac{T^*_{n, \zeta' K'}\, T_{n, \zeta K'}}{E_n + z} \right\}. \tag{20}$$

The t-matrix is to be considered a function of the complex variable z, and has simple analytic properties. From the energy spectrum of the intermediate states it follows that t has a pole for $z = 0$ (ground state) and a cut from ± 1 to $\pm\infty$. (The meson mass is assumed to be 1).

Thus:

$$t_{\zeta' K', \zeta K}(z) = \frac{R_{\zeta' K', \zeta K}}{z} + \int_1^\infty dE \left[\frac{F_{\zeta' K', \zeta K}(E)}{E - z} + \frac{G_{\zeta' K', \zeta K}(E)}{E + z} \right]. \tag{21}$$

From the definition of t follows:

$$T_{\zeta' K', \zeta K} = \lim_{z \to \omega + i\varepsilon} t_{\zeta' K', \zeta K}. \tag{22}$$

Moreover, we have a simple theorem of analytic continuation, following directly from the definition:

$$t_{K'\zeta', K\zeta}(z) = t^*_{K\zeta, K'\zeta'}(z^*) \tag{23a}$$

$$t_{\zeta' K', K\zeta}(z) = t_{\zeta' K, \zeta K'}(-z) \tag{23b}$$

(23b) is referred to as crossing symmetry.

Let us evaluate the pole-term first. This term also gives insight into a more realistic theory, as it corresponds to the renormalized Born approximation of a relativistic theory.

From (17) follows:

$$R_{\zeta' K', \zeta K} = \mathop{S}_{\xi} \{ \langle \zeta' |\, V_{K'} | \zeta \rangle \langle \xi | V_K | \zeta \rangle - \langle \zeta' | V_K | \xi \rangle \langle \xi | V_{K'} | \xi \rangle \}. \tag{24}$$

To evaluate this expression further, we use the Wigner-Eckart-theorem. We know that V_K is an operator of the form σZ, σY i. e., an 8-vector in the SU_3 space. There are exactly two 8-dimensional representations of such an operator:

$$\begin{aligned}\langle \zeta' | \sigma^i Z^\alpha | \xi \rangle &= \mathfrak{r}_1 (\sigma^i Z^\alpha)_{\zeta'\xi} + \mathfrak{r}_2 (\sigma^i Y^\alpha)_{\zeta'\xi} \\ \langle \zeta' | \sigma^i Y^\alpha | \xi \rangle &= \mathfrak{s}_1 (\sigma^i Z^\alpha)_{\zeta'\xi} + \mathfrak{s}_2 (\sigma^i Y^\alpha)_{\zeta'\xi}.\end{aligned} \tag{25}$$

We define the renormalized coupling constants:

$$g_1 = \mathfrak{r}_1 f_1 + \mathfrak{s}_1 f_2 \qquad g_2 = f_1 \mathfrak{r}_2 + f_2 \mathfrak{s}_2. \tag{26}$$

Eq. (24) can be written as follows:

$$R_{\zeta' K', \zeta K} = \frac{\varrho^2(k)\, k^4}{12\,\omega\,\pi^2} \{g_1{}^2 [\sigma^{i'} Z^{\alpha'} \sigma^i Z^\alpha - \sigma^i Z^\alpha \sigma^{i'} Z^{\alpha'}]_{\zeta'\zeta}$$
$$+ g_2{}^2 [\sigma^{i'} Y^{\alpha'} \sigma^i Y^\alpha - \sigma^i Y^\alpha \sigma^{i'} Y^{\alpha'}]_{\zeta'\zeta} \quad (27)$$
$$+ g_1\, g_2 [\sigma^{i'} Z^{\alpha'} \sigma^i Y^\alpha + \sigma^{i'} Y^{\alpha'} \sigma^i Z^\alpha - \sigma^i Z^\alpha \sigma^{i'} Y^{\alpha'} - \sigma^i Y^\sigma \sigma^{i'} Z^{\alpha'}]_{\zeta'\zeta}\}.$$

We remember that ζ stands for the spin and the SU_3 variables of the nucleon, i denotes the meson spin, α the meson SU_3 variable. Thus, R is altogether a complicated 384 by 384 matrix. To get any information we shall have to diagonalize R. To do this we shall have to transform from the one meson state $|\zeta\, k\rangle$ to a representation where J^2, J_z and I_1, I_2, X_1, X_2 and T^2 are diagonal. As the interaction conserves angular momentum and "SU_3 spin", the scattering matrix should be diagonal in such a representation, i. e., we expect S to take the following form:

$$\langle \alpha\, I_1', I_2', X_1', X_2', T', J', J_z' |\, S\, | I_1, I_2, X_1, X_2, T, J, J_z\, \alpha' \rangle$$
$$= \delta_{I_1 I_1'}\, \delta_{I_2 I_2'}\, \delta_{X_1 X_1'}\, \delta_{X_2 X_2'}\, \delta_{TT'}\, \delta_{JJ'}\, \delta_{J_z J_z'}\, e^{i 2 \delta_{I_1 I_2 J \alpha \alpha'}}\, \delta(E - E') \frac{dE}{dk}. \quad (28)$$

The phase shift δ should depend only on the total angular momentum and on the representation of SU_3 and on α, α', two parameters that label the states in case of a degeneracy. We know that for one-meson scattering in the static model, where only the $l = 1$ meson term is coupled to the source, meson and nucleon can form two irreducible angular momentum representations: 1/2 and 3/2, and 6 irreducible representations of SU_3: $8 \times 8 = 1 + 8 + 8' + 10 + 10^* + 27$. Therefore, we expect 12 independent phase shifts, corresponding to the 12 independent channels.

Thus, we should project the T-matrix into the corresponding subspaces. We have already determined the projection operators for the SU_3 representations. There we have chosen symmetric and antisymmetric combinations for the 8 and 8′ representation, respectively. However, there is no reason — as long as we do not have R-invariance — why T should be diagonal in these combinations. Actually, the right combination will depend on the mixing of g_1/g_2 and will be different for $J = 1/2$ and $J = 3/2$ states. To take this into account we introduce the following operators:

$$P^+_{ir,js} = \tfrac{3}{\sqrt{5}} Z^k_{ir}\, Y^k_{js}$$
$$P^-_{ir,js} = -\tfrac{3}{\sqrt{5}} Y^k_{ir}\, Z^k_{js}. \quad (29)$$

Together with P^8 and $P^{8'}$, they form a set of operators isomorphic to:

$$P^8 \sim \begin{pmatrix} 1 & 0 \\ 0 & 0 \end{pmatrix} \quad P^{8'} \sim \begin{pmatrix} 0 & 0 \\ 0 & 1 \end{pmatrix} \quad P^+ \sim \begin{pmatrix} 0 & 1 \\ 0 & 0 \end{pmatrix} \quad P^- \sim \begin{pmatrix} 0 & 0 \\ 1 & 0 \end{pmatrix}$$

The projection operators into the angular momentum representations are simply:

$$P^{1/2} = \tfrac{1}{3}\left(\tfrac{15}{4} - L^2\right) \qquad P^{3/2} = -\tfrac{1}{3}\left(\tfrac{3}{4} - L^2\right).$$

The projection operators onto the different channels now are simply the direct product of the projection operators for angular momentum and SU_3 representations.

Let us turn to the decomposition of R into these subspaces

$$R_{\zeta' K', \zeta K} = -\frac{\varrho^2 k^4}{12\,\omega\,\pi^2} \sum_{\substack{j = 1/2, 3/2 \\ \nu = 1, 8, 8', 10, 10^*, 27}} \left\{ {}^j P^\nu_{\zeta' K', \zeta K}\, \lambda_j^\nu + {}^j P^+\, \lambda_j^+ + {}^j P^-\, \lambda_j^- \right\}. \tag{30}$$

It takes some effort to calculate the λ's from (27), but only matrix multiplication is involved. We find:

$$\lambda^{1/2} = \begin{pmatrix} 1 \\ -7/2 \\ 1/2 \\ 0 \\ 0 \\ -1/3 \end{pmatrix} g_1^2 + \begin{pmatrix} -5/9 \\ 5/18 \\ -3/2 \\ -2/9 \\ -2/9 \\ -1/9 \end{pmatrix} g_2^2 + \begin{pmatrix} 0 \\ 0 \\ 0 \\ 2/3 \\ -2/3 \\ 0 \end{pmatrix} g_1 g_2 \qquad \lambda^-_{1/2} - \lambda^+_{1/2} - \sqrt{5}\, g_1 g_2$$

$$\lambda^{3/2} = \begin{pmatrix} -2 \\ 1 \\ -1 \\ 0 \\ 0 \\ 2/3 \end{pmatrix} g_1^2 + \begin{pmatrix} 10/9 \\ -5/9 \\ -1/3 \\ 4/9 \\ 4/9 \\ 2/9 \end{pmatrix} g_2^2 + \begin{pmatrix} 0 \\ 0 \\ 0 \\ -4/3 \\ 4/3 \\ 0 \end{pmatrix} g_1 g_2 \qquad \lambda^-_{3/2} = \lambda^+_{3/2} = 0. \tag{31}$$

For the $J = 3/2$ channel R is already diagonal for all values of g_1 and g_2. For the $J = 1/2$ channel R is a symmetric matrix which we can diagonalize by the orthogonal transformation:

$$\left.\begin{aligned} P^8 &= \cos^2\varphi\, Q^1 + \sin^2\varphi\, Q^2 - \sin\varphi\cos\varphi\, Q^+ - \sin\varphi\cos\varphi\, Q^- \\ P^{8'} &= \sin^2\varphi\, Q^1 + \cos^2\varphi\, Q^2 + \sin\varphi\cos\varphi\, Q^+ + \sin\varphi\cos\varphi\, Q^- \\ P^+ &= \sin\varphi\cos\varphi\, Q^1 - \sin\varphi\cos\varphi\, Q^2 + \cos^2\varphi\, Q^+ - \sin^2\varphi\, Q^- \\ P^- &= \sin\varphi\cos\varphi\, Q^1 - \sin\varphi\cos\varphi\, Q^2 - \sin^2\varphi\, Q^+ + \cos^2\varphi\, Q^- \end{aligned}\right\} \tag{32}$$

The transformation (32) diagonalizes R if

$$\operatorname{tg}\varphi = -\frac{1}{\alpha\sqrt{5}} \left(-2 + \tfrac{8}{9}\alpha^2 - \sqrt{4 + \tfrac{13}{9}\alpha^2 + \tfrac{64}{81}\alpha^4} \right) \quad \alpha = g_2/g_2. \tag{33}$$

The new projection operators are denoted by ${}^{1/2}Q$ (also for the 1, 10, 10*, and 27 channels where they are identical with the P's).

$$R_{\zeta' K', \zeta K} = -\frac{\varrho^2 k^4}{12\,\omega\,\pi^2} \sum_{\nu = 1, 8, 8', 10, 10^*, 27} \left\{ \tilde{\lambda}^\nu_{1/2}\, {}^{1/2}Q^\nu_{\zeta' K', \zeta K} + \lambda^\nu_{3/2}\, {}^{3/2} P^\nu_{\zeta' K', \zeta K} \right\}.$$

$$\tilde{\lambda}^\nu_{1/2} = g_1^2 \begin{pmatrix} 1 - 5/9\,\alpha^2 \\ -3/2 - 11/8\,\alpha^2 + \sqrt{4 + 13/9\,\alpha^2 + 64/81\,\alpha^4} \\ -3/2 - 11/8\,\alpha^2 - \sqrt{4 + 13/9\,\alpha^2 + 64/81\,\alpha^4} \\ 2/3\,\alpha - 2/9\,\alpha^2 \\ -2/3\,\alpha - 2/9\,\alpha^2 \\ -1/3 - 1/9\,\alpha^2 \end{pmatrix}; \quad \lambda^\nu_{3/2} = g_1^2 \begin{pmatrix} -2 + 10/9\,\alpha^2 \\ 1 - 5/9\,\alpha^2 \\ -1 - 1/3\,\alpha^2 \\ 4/9\,\alpha^2 - 4/3\,\alpha \\ 4/9\,\alpha^2 + 4/3\,\alpha \\ 2/3 + 2/9\,\alpha^2 \end{pmatrix} \quad (3$$

From scattering theory it is known that positive and negative pole terms correspond to attraction and repulsion, respectively. As we are interested in possible resonances, we have plotted all terms which can become positive as a function of α in Fig. 1. We note that for a wide range of α the 3/2 singlet term is dominant, followed by the 3/2 decublet term. In order to discuss the consequences of this result, we have to determine α by experiment.

Before doing so we show that the transformation (32) diagonalizes not only the pole term but also the entire Low equation: We have seen that $t(z)$ has a pole at $z=0$ and a cut from ± 1 to $\pm\infty$. Thus, by Cauchy's theorem:

$$t_{\zeta' K', \zeta K}(z) = \frac{R_{\zeta' K', \zeta K}}{z} + \frac{1}{2\pi i}\int\limits_{> \; <} dE\, \frac{t_{\zeta' K', \zeta K}(E)}{E - z}. \tag{35}$$

This is equation (21) with

$$F_{\zeta' K', \zeta K}(E) = G_{\zeta' K, \zeta K}(E) = \frac{1}{2\pi i} \lim_{z \to E + i\varepsilon} (t_{\zeta' K', \zeta K}(z) - t_{\zeta' K', \zeta K}(z^*)). \tag{36}$$

Finally, we project t into irreducible subspaces:

$$\begin{aligned} t(\omega) = -\sum_{\nu} \frac{\varrho^2(k^2)\, k^4}{12\,\omega\,\pi^2} \{ {}^{1\!/\!2} Q^{\nu} h^{\nu}(\omega) + {}^{3\!/\!2} P^{\nu} g^{\nu}(\omega) \} \\ + {}^{1\!/\!2} Q^{+} h^{+} + {}^{1\!/\!2} Q^{-} h^{-} + {}^{3\!/\!2} P^{+} g^{+} + {}^{3\!/\!2} P^{-} g^{-}. \end{aligned} \tag{37}$$

As the projection operators are symmetric we can use the hermiticity condition (23a) to write the Low equation (35) as follows:

$$\begin{aligned} &\sum_{\nu} \{ Q^{\nu} h^{\nu} + P^{\nu} g^{\nu} + Q^{\pm} h^{\pm} + P^{\pm} g^{\pm} \}_{\zeta' K', \zeta K} \\ &= \sum_{\nu} \Big\{ \tilde{\lambda}^{\nu} Q^{\nu} + \lambda^{\nu}_{3/2} P^{\nu} + \\ &+ \frac{1}{\pi} \int_1^{\infty} \frac{d\omega'}{\omega' - \omega - i\varepsilon} Jm (Q^{\nu} h^{\nu} + P^{\nu} g^{\nu} + Q^{\pm} h^{\pm} + P^{\pm} g^{\pm}(\omega')\Big\}_{\zeta' K', \zeta K} \\ &+ \sum_{\nu} \Big\{ \frac{1}{\pi} \int_1^{\infty} \frac{d\omega'}{\omega' + \omega} Im (Q^{\nu} h^{\nu} + P^{\nu} g^{\nu} + Q^{\pm} h^{\pm} + P^{\pm} g^{\pm}(\omega')\Big\}_{\zeta' K, \zeta K'}. \end{aligned} \tag{38}$$

The projection operators $Q_{\zeta' K, \zeta K'}$ and $P_{\zeta' K, \zeta K'}$ can be expressed by linear combinations of the $Q_{\zeta' K', \zeta K}$ and $P_{\zeta' K', \zeta K}$ with the meson indices exchanged. This leads to the Low equation in its final form:

$$\begin{aligned} \binom{h}{g} &= \frac{1}{z}\binom{\tilde{\lambda}}{\lambda} + \frac{1}{\pi}\int_1^{\infty} d\omega \left\{ \frac{1}{\omega - z} Jm \binom{h}{g} + \frac{1}{\omega + z} Jm A \binom{h}{g} + \frac{1}{\omega + z} Jm \binom{h^{+-}}{g^{+-}} \right\} \\ h^{+} + h^{-} &= \frac{1}{\pi}\int_1^{\infty} \frac{d\omega}{\omega - z} Jm (h^{+} + h^{-}) + \frac{1}{\pi}\int_1^{\infty} \frac{d\omega}{\omega + z} Jm\, f \\ g^{+} + g^{-} &= \frac{1}{\pi}\int_1^{\infty} \frac{d\omega}{\omega - z} Jm (g^{+} + g^{-}) + \frac{1}{\pi}\int_1^{\infty} \frac{d\omega}{\omega + z} Jm\, \tilde{f}. \end{aligned} \tag{39}$$

The functions f and $\tilde{f}$ are linear combinations of the h^ν and g^ν. They satisfy the equations:

$$f = \frac{1}{\pi}\int_1^\infty \frac{d\omega}{\omega - z} Jm\, f + \frac{1}{\pi}\int_1^\infty \frac{d\omega}{\omega + z}\frac{4}{5}\sin^2\varphi\cos^2\varphi\, Jm\, f$$

$$\tilde{f} = \frac{1}{\pi}\int_1^\infty \frac{d\omega}{\omega - z} Jm\,\tilde{f}.$$

Thus $h^\pm = 0$, $g^\pm = 0$, $f = 0$, $\tilde{f} = 0$ is consistent with the Low equation and, therefore, we have diagonalized the t—matrix.

The crossing matrix A is a rather complicated 12 by 12 matrix. It has the properties

$$A^2 = 1$$

$$A\begin{pmatrix}\tilde{\lambda}\\ \lambda\end{pmatrix} = -\begin{pmatrix}\tilde{\lambda}\\ \lambda\end{pmatrix}. \tag{40}$$

This leads to the crossing relation

$$\begin{pmatrix}h\,(z)\\ g\,(z)\end{pmatrix} = A\begin{pmatrix}h\,(-z)\\ g\,(-z)\end{pmatrix}. \tag{41}$$

To get a better insight into the meson — baryon system described by the static model, we are going to study its electromagnetic interactions. We can calculate magnetic moments and photoproduction cross sections and compare the results with experiments.

The photons interact with the charge and current of the total system. Charge and current contain contributions from the nucleons and mesons. According to the definition of the group generators we have:

$$Q = T_3 + \tfrac{1}{2}Y = \sqrt{3}\,H_1 + H_2 = X^1 + \sqrt{3}X^2.$$

Thus
$$\begin{aligned}
&Q_N = \varrho\,(r)\,(X^1 + \sqrt{3}\,X^2)\\
&Q_\pi\,(x) = i\Phi\,(Z^1 + \sqrt{3}\,Z^2)\,\dot{\Phi}\\
&\vec{j}_\pi(x) = -i\Phi\,(Z^1 + \sqrt{3}\,Z^2)\,\vec{\nabla}\Phi \quad (\Phi \text{ are hermitian fields})\\
&\vec{j}_N = \vec{\nabla}\times\mu_N\,(r).
\end{aligned} \tag{42}$$

The nucleon current arises from the normal moment of the nucleon. We assume that it is divergence-free and, moreover, that it can be expressed in terms of bare-nucleon variables only, i. e. it should commute with meson creation and annihilation operators.

For the total charge and current we demand the continuity equation:

$$\dot{Q} + \vec{\nabla}\vec{j} = 0.$$

To calculate $\dot{Q}_\pi + \vec{\nabla}\vec{j}_\pi$ we remember that the Z's are antisymmetric matrices:

$$\begin{aligned}
\dot{Q}_\pi + \vec{\nabla}\vec{j}_\pi &= i\Phi\,(Z^1 + \sqrt{3}\,Z^2)\,\{\ddot{\Phi} - \Delta\Phi\} =\\
&= i\vec{\sigma}\vec{\nabla}\,\varrho\,(x)\,\Phi^\alpha\,(Z^1 + \sqrt{3}\,Z^2)_{\alpha\nu}\,(f_1 Z^\nu + f_2\,Y^\nu).
\end{aligned}$$

Furthermore, $\vec{\nabla}\vec{j}_N=0$

$$\dot{Q}_N=i\,[H,Q_N]=i\varrho(x)\int d^3y\,\vec{\sigma}\vec{\nabla}_y\,\Phi^\alpha(y)\,(Z^1_{\alpha\nu}+\sqrt{3}\,Z^2_{\alpha\nu})\,(f_1 Z^\nu+f_2 Y^\nu).$$

We find that

$$\begin{aligned}&\dot{Q}_N+\dot{Q}_\pi+\operatorname{div}(j_N+j_\pi)=-\operatorname{div} j_I \qquad \text{if } \varrho(x)=\delta(x)\\ &j_I=-i\vec{\sigma}\,\delta^3(\vec{x})\Phi^\alpha\,(Z^1_{\alpha\nu}+\sqrt{3}\,Z^2_{\alpha\nu})\,(f_1,Z^\nu+f_2Y^\nu).\end{aligned} \tag{43}$$

Thus, in order to satisfy the continuity equation, we have to introduce the additional current j_I. The total current has to be defined as follows:

$$\vec{j}=\vec{j}_N+\vec{j}_\pi+j_I.$$

The interaction with the photons is now described by the following interaction Hamiltonian:

$$H_{\text{int}}=-e\int d^3x\,\vec{A}(\vec{x})\cdot\vec{j}(\vec{x}). \tag{44}$$

$\vec{A}$ is the electromagnetic vector potential. With this interaction Hamiltonian we can describe the following photoproduction process:

$$\gamma+N\to N+\pi.$$

To lowest order in e this process is described by the matrix element:

$$\langle\,\text{out},N+\pi|\,H_{\text{int}}\,|p+\gamma\rangle.$$

If we assume that the incoming photon has momentum $\vec{k}$, polarization $\vec{\varepsilon}$ and is normalized to one particle per unit volume, we have:

$$\langle 0\,|\,\vec{A}(r)\,|\,\gamma\rangle=e^{i\vec{k}\vec{r}}\,\frac{\vec{\varepsilon}}{\sqrt{2k}\,(2\pi)^{3/2}}.$$

Our matrix element becomes:

$$\begin{aligned}\langle\,\text{out},N+\pi\,|\,H_{\text{int}}\,|\,p+\gamma\rangle&=-\int\frac{d^3x}{(2\pi)^{3/2}}\,\frac{e^{i\vec{k}\vec{x}}}{\sqrt{2k}}\,\vec{\varepsilon}\,\langle\text{out},N+\pi|\,\vec{j}(\vec{x})\,|N\rangle\\ &=\vec{\varepsilon}\,\langle\text{out},N+\pi\,|\,\vec{j}(-\vec{k})\,|\,N\rangle\frac{1}{\sqrt{2k}}.\end{aligned} \tag{45}$$

Where we have used the momentum transformed current

$$\vec{j}(\vec{k})=\frac{1}{(2\pi)^{3/2}}\int d^3r\,e^{-i\vec{k}\vec{x}}\,\vec{j}(\vec{x}).$$

To evaluate the matrix element $\langle\text{out},N+\pi|\,\vec{j}(\vec{x})\,|\,N\rangle$ we proceed in a manner similar to the derivation of the Low equation:

$$\begin{aligned}\langle\zeta',p\alpha\,|\,\vec{j}(-\vec{k})\,|\,\zeta\rangle&=\langle\zeta'\,|\,[a^{\text{out}}_\alpha(\vec{p})\,\vec{j}(-\vec{k})]\,|\,\zeta\rangle\\ &=\langle\zeta'|\left[a_\alpha(p)+\int_0^\infty V_\alpha(p)\,e^{i\omega t}\,dt,\vec{j}(-\vec{k})\right]|\zeta\rangle=\\ &=\langle\zeta'|\,[a_\alpha(p),j(-k)]\,|\zeta\rangle+\\ &+\mathop{S}_n\left[\frac{\langle\zeta'|\,V_\alpha(p)\,|n\rangle\langle n|\,j(-k)\,|\zeta\rangle}{E_n-\omega-i\varepsilon}+\frac{\langle\zeta'|\,j(-k)\,|n\rangle\langle n|\,V_\alpha(p)\,|\zeta\rangle}{E_n+\omega}\right].\end{aligned} \tag{46}$$

We are going to treat the strong interaction in Born approximation only, i. e., we take only the ground state contribution of the sum over the intermediate states:

$$\langle\zeta'|\left[a_\alpha(p),\vec{j}\left(-\vec{k}\right)\right]|\zeta\rangle+\frac{1}{\omega}\left(\langle\zeta'|\,j\,(-k)\,|\zeta''\rangle\langle\zeta''|\,V_\alpha(p)\,|\zeta\rangle - \langle\zeta'|\,V_\alpha(p)\,|\zeta''\rangle\langle\zeta''|\,j\,(-k)\,|\zeta\rangle\right). \tag{47}$$

The contributions of the first term are easily calculated using the canonical commutations relations and the Wigner-Eckart theorem:

$$\langle\zeta'|\left[a_\alpha(\vec{p}),j_I\left(-\vec{k}\right)\right]|\zeta\rangle = = \frac{\vec{\sigma}}{(2\pi)^3}\frac{\varrho(p-k)}{\sqrt{2\omega_p}}(-i)\left(Z^1_{\alpha\nu}+\sqrt{3}\,Z^2_{\alpha\nu}\right)(g_1Z^\nu+g_2Y^\nu)_{\zeta'\zeta}$$

$$\langle\zeta'|\left[a_\alpha(\vec{p}),\vec{j}_\pi\left(-\vec{k}\right)\right]|\zeta\rangle = = \frac{2\vec{p}-\vec{k}}{\sqrt{2\omega_p}}\frac{i\,\varrho(p-k)\,\vec{\sigma}\left(\vec{p}-\vec{k}\right)}{(2\pi)^3[(p-k)^2+\mu^2]}\left(Z^1_{\alpha\nu}+\sqrt{3}Z^2_{\alpha\nu}\right)(g_1Z^\nu+g_2Y^\nu)_{\zeta'\zeta}$$

$$\langle\zeta'|\left[a_\alpha(\vec{p}),\vec{j}_N\left(-\vec{k}\right)\right]|\zeta\rangle=0. \tag{48}$$

As we do not know much about the nucleon current $\vec{j}_N\left(\vec{k}\right)$, we cannot evaluate the matrix element $\langle\zeta'|\,j\,(-k)\,|\zeta\rangle$ completely within the framework of the static model. But we can make an "ansatz" and relate it to other observable quantities, i. e., the magnetic moment.

The matrix element has to be a polar vector and can depend only on the vectors $\vec{k}$ and $\vec{\sigma}$. In the SU_3 space it has to transform like the $Z^1+\sqrt{3}Z^2$ component of an 8-vector, and it follows from the continuity equation that $\vec{k}\,\langle\zeta|\,\vec{j}\left(\vec{k}\right)|\zeta'\rangle=0$.
Thus

$$\langle\zeta'|\,\vec{j}\left(-\vec{k}\right)|\zeta\rangle=i\,\frac{\vec{\sigma}\times\vec{k}}{(2\pi)^{3/2}}\left[F_1(k^2)\left(Z^1+\sqrt{3}\,Z^2\right)+F_2(k^2)\left(Y^1+\sqrt{3}\,Y^2\right)\right]. \tag{49}$$

It can be easily verified that the contributions of j_π and j_I are consistent with this ansatz. We want to relate the form factors F_1 and F_2 to the magnetic moment.

$$\mathfrak{M}=\lim_{k\to0}\frac{1}{2}\int d^3x\left[\vec{p}\times\vec{j}\left(\vec{x}\right)\right]e^{i\vec{k}\vec{x}}=-\frac{i}{2}\lim_{k\to0}\nabla_k\times\int j\,(r)\,e^{ikr}\,d^3r \tag{50}$$

$$\langle\zeta|\,\mathfrak{M}\,|\zeta\rangle=\tfrac{1}{2}\,\vec{\sigma}\langle\zeta|\left[F_1(0)\left(Z^1+\sqrt{3}\,Z^2\right)+F_2(0)\left(Y^1+\sqrt{3}\,Y^2\right)\right]|\zeta\rangle.$$

This gives:

$$\begin{aligned}\mathfrak{M}_p&=F_1(0)+\tfrac{1}{3}F_2(0) &\qquad F_1(0)&=\mathfrak{M}_p+\tfrac{1}{2}\mathfrak{M}_N\\ \mathfrak{M}_N&=-\tfrac{2}{3}F_2(0) & F_2(0)&=-\tfrac{3}{2}\mathfrak{M}_N\end{aligned} \tag{51}$$

F_1 and F_2 are of the same order of magnitude. The form factors $F_1(0)$ and $F_2(0)$ have thus been related to the proton and neutron magnetic moments. From other experiments (electron proton scattering) we learn that we can expect F_1 and F_2 to be approximately constant for $k<2$ meson masses. Thus, we replace $F(k_2)$ by $F(0)$ in equ. (49).

Summing up all the terms we finally get:

$$\langle \text{out}\ \zeta', \alpha\, p \mid H_{\text{int}} \mid \zeta, \gamma_{k\varepsilon} \rangle =$$

$$= -\frac{\vec{\varepsilon}}{(2\pi)^3} \Big\{ \frac{\varrho(p-k)}{2\,\omega_p}\, i \left[Z^1_{k\nu} + \sqrt{3}\, Z^2_{\alpha\nu} \right] [g_1 Z^\nu +$$

$$+ g_2 Y^\nu] \left[\vec{\sigma} - \frac{2p}{(p-k)^2 + \mu^2} \left(\vec{\sigma}, \vec{p} - \vec{k} \right) \right] + \frac{\varrho(p)}{2\,\omega_p{}^2} \Big[i\, \vec{\sigma}\, \vec{p}\ \{ g_1 Z^\nu + g_2 Y^\nu \},$$

$$i \vec{\sigma} \times \vec{k} \{ (\mathfrak{M}_p + \tfrac{1}{2}\, \mathfrak{M}_N) \left(Z^1 + \sqrt{3} Z^2 \right) - \tfrac{3}{2}\, \mathfrak{M}_N \left(Y^1 + \sqrt{3}\, Y^2 \right) \} \Big] \Big\} .$$

Near threshold, $(p \to 0)$, the first term, being proportional to $\vec{\sigma}$, is leading. It gives rise to a creation of mesons in an S-state mainly. The second term, proportional to $\vec{p}$, contains all angular momenta due to the denominator $1/[(p-k)^2 + \mu^2]$. Both terms describe the creation of charged mesons only, whereas the third term describes the creation of neutral mesons, too. The three terms correspond to the three different contributions to the current.

The first term arises from $\vec{j}_I$ and corresponds to the photons being absorbed in the process of the meson creation.

The second term arises from j_π, and corresponds to the photons being absorbed in the cloud.

The third term arises from j_N; and corresponds to the photons interacting with the magnetic moment of the nucleon.

We now want to compare our results with photoproduction experiments. As we want to get information beyond the isotopic spin group, we have to look for processes where particles with strangeness different from zero are produced. There are measurements for the processes

$$\gamma + p \to \Lambda + K^+$$
$$\gamma + p \to \Sigma^0 + K^+.$$

Experiments indicate that near treshold S-wave production is dominant. Therefore, we keep only the leading term proportional to $\vec{\sigma}$ and we find:

$$\lim_{\vec{p} \to 0} \frac{\langle \Lambda\, K^+ | H_{\text{int}} | p\, \gamma \rangle}{\langle \Sigma^0\, K^+ | H_{\text{int}} | p\, \gamma \rangle} = \sqrt{3}\, \frac{g_1 + \frac{1}{3}\, g_2}{g_1 - g_2} .$$

The dependence of the cross sections on $\alpha = g_2/g_1$ is shown in Fig. 2.

To compare it with the actual measurements, we have to remember that by assuming SU_3 invariance we have neglected the mass differences between the mesons. Thus we cannot just compare the measurements at the same energies. If we evaluate the cross section from the interaction current j_I, we find $\sigma_I \sim p$. Therefore, we should rather compare the measurements at the same energy above threshold. The ratio of the S-wave Λ/Σ^0 production then is about 2. From Fig. 2 we find that $\alpha \sim 3.5$. This is in rather good agreement with the value obtained by J. J. Swart and C. K. Iddings from low energy $\Sigma - p$, $\Sigma - d$ interaction.

From Fig. 1 we learn that the ${}^{3/2}P^{\overline{10}}$ and ${}^{3/2}P^{1}$ terms in the Low equation are by far the dominant terms for $\alpha \sim 3.5$. Thus, resonances have to be assumed for these representations. The $\overline{10}$ representation allows the well-known assignment N_1^*, Y_1^*, Ξ^*, Ω. Moreover, a singlet seems to be predicted. But the Low equation favors the $3/2\ p^{10}$ resonance. This is due to the crossing matrix which becomes

$$\frac{1}{3}\begin{pmatrix} 1/8 - 5/4 \\ -1/8 + 1/4 \end{pmatrix}$$

for these two channels. The singlet would have quantum numbers $S = -1$ $Q = 0$, $T = 0$, $N = 1$ spin 3/2, p-wave. Provided that Y_0^* (1815 MeV) turns out to have spin 3/2 it could be assigned to this singlet.

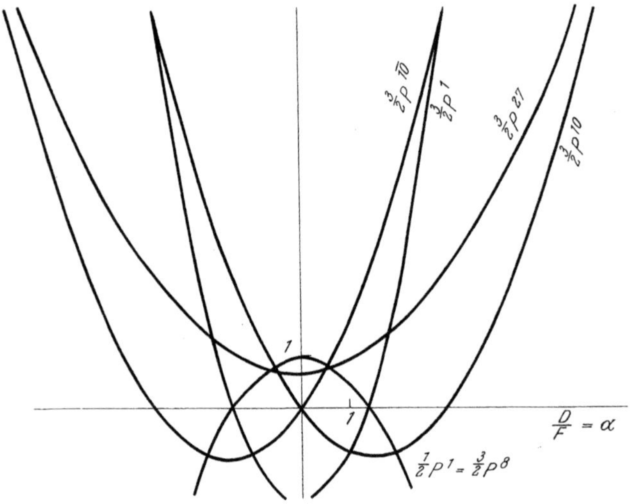

Fig. 1

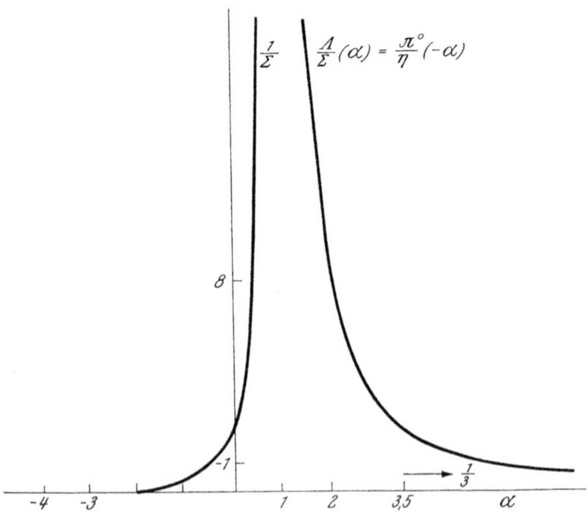

Fig. 2

Problems on Conservation and Lack of Conservation of Weak Interaction Currents

By

R. Vinh Mau, Orsay

With 2 Figures

Introduction

In these lectures, we wish to focus our attention on some aspects of the renormalization problems in weak interactions. We shall suppose the reader to be familiar with the symmetry principles and the empirical facts which have led to the current status of weak interactions. At the present time, we have at our disposal a phenomenological theory of weak interactions which turns out to be very successful in processes such as: μ-meson decay, nuclear β-decay and μ-meson capture by nuclei.

Let us consider the two following classes of phenomena:

1. μ-meson decay:

$$\mu^- \to e^- + \bar{\nu}_e + \nu_\mu.$$

This process can be described by the following interaction:

$$H = -\frac{G}{\sqrt{2}} j_\lambda^{(\mu)}(x)\, j_\lambda^{(e)+}(x) + \text{h. c.} \tag{1}$$

where

$$j_\lambda^{(l)}(x) = i \overline{\Psi}_l \gamma_\lambda (1 + \gamma_5) \Psi_{\nu l}. \tag{2}$$

2. Weak interactions involving both leptons and strongly interacting particles.

For these processes

$$H_w = -J_\lambda(x)\, j_\lambda^{(l)+}(x) + \text{h. c.} \tag{3}$$

where J_λ are operators acting only on the strongly interacting particles (nucleons, pions, K-mesons, hyperons). The nature of $J_\lambda(x)$ is known so far, at least for *low momentum transfer*. For example, in β-decay, the nucleonic "current" matrix element is

$$\langle p \mid J_\lambda \mid n \rangle = \frac{i}{\sqrt{2}} \bar{u}_p \gamma_\lambda (G_V - G_A \gamma_5)\, u_n \tag{4}$$

where u_p and u_n are the spinor solution of the free Dirac equations with the same 4-momenta as the physical proton and neutron. G_V and G_A are respectively the Fermi and Gamow-Teller coupling constants.

A great deal of work has been devoted to the study of μ-meson decay and nuclear β-decay. From these investigations, one of the most remarkable and surprising results is the equality of the μ-decay coupling constant G and the Fermi (or vector) coupling constant G_V of β-decay. A priori, the values of G and G_V might be expected to be very different because of the renormalization effects due to strong interactions in β-decay. But the result $G = G_V$ implies an absence of the renormalization effects.

The conserved vector currents [1]

One knows already another example where renormalization effects are absent, namely, for the electric charge. Leptons have no strong interactions while mesons and baryons have. In spite of this, e^+, μ^+ and π^+, K^+, p, Σ^+ have the same electric charge. The universality of electric charge is a consequence of the existence of a *conserved electromagnetic current.*

Once this feature is noticed, it is natural to look for a similar explanation for the Fermi (or vector) part of the weak interaction current. The idea is that the vector part of the current $J_\lambda(x)$ is conserved, as expressed by

$$\frac{\partial}{\partial x_\lambda} J_\lambda(x) = 0. \tag{5}$$

Let us now make full use of the analogy between weak interactions and electromagnetic theory to find many useful results and consider first the electromagnetic case:
The electromagnetic interaction is:

$$H_{E.M.} = -e J_\lambda^{el}(x) A_\lambda(x) + \text{h. c.} \tag{6}$$

where

$$J_\lambda^{el}(x) = i\left[\overline{\Psi}_N \gamma_\lambda \frac{1+\tau_3}{2} \Psi_N + \Phi_\pi \varrho_3 \partial_\lambda \Phi_\pi + \dots\right]. \tag{7}$$

The dots in this equation represent the contributions from K-mesons and hyperons. For simplicity, we shall consider only nucleons and pions to be strongly interacting particles and drop out the contribution from strange particles. $\vec{\tau}(\tau_1, \tau_2, \tau_3)$ and $\vec{\varrho}\,(\varrho_1, \varrho_2, \varrho_3)$ are respectively 2×2 and 3×3 isospin matrices. We neglect the effect of electromagnetic coupling and the mass difference of $p-n, \pi^\pm-\pi^0$ etc.... in the expression (7).

We can split $J_\lambda^{el}(x)$ into two parts corresponding to the two different isospin characters:

$$J_\lambda^s = i\frac{\overline{\Psi}_N \gamma_\lambda \Psi_N}{2}$$

and

$$J_\lambda^3 = i\left[\overline{\Psi}_N \gamma_\lambda \frac{\tau_3}{2} \Psi_N + \Phi_\pi^+ \varrho_3 \partial_\lambda \Phi_\pi\right]. \tag{8}$$

The conservation law is expressed now by

$$\frac{\partial}{\partial x_\lambda} J_\lambda^s = 0, \quad \frac{\partial}{\partial x_\lambda} J_\lambda^3 = 0. \tag{9}$$

Now, *charge independence* implies not only (9), but also

$$\frac{\partial}{\partial x_\lambda} \vec{J}_\lambda = 0$$

where

$$\vec{J}_\lambda = i \left[\overline{\Psi}_{\mathcal{N}} \gamma_\lambda \frac{\vec{\tau}}{2} \Psi_{\mathcal{N}} + \Phi_\pi^+ \vec{\varrho}\, \partial_\lambda \Phi_\pi \right]. \tag{10}$$

Inspection shows that the vector part of the nucleonic weak interaction current of equ. (4) is the first term of $J_\lambda^1 + i J_\lambda^2$ of equ. (10) (up to the coupling constant $G_V/\sqrt{2}$).

If we now define the weak interaction vector current by:

$$V_\lambda = (J_\lambda^1 + i J_\lambda^2) \frac{G}{\sqrt{2}} = \frac{i\,G}{\sqrt{2}} \left[\overline{\Psi}_p \gamma_\lambda \Psi_n + \sqrt{2} \left(\Phi \frac{\partial \Phi_0}{\partial x_\lambda} - \Phi_0 \frac{\partial \Phi}{\partial x_\lambda} \right) \right] \tag{11}$$

where G is the coupling constant of μ-meson decay, we have of course the conservation law (valid in the charge independence approximation)

$$\frac{\partial}{\partial x_\lambda} V_\lambda = 0. \tag{12}$$

Let us compare the isovector part of the electromagnetic interaction

$$-e J_\lambda^3 A_\lambda$$

and the weak interaction (Fermi part)

$$V_\lambda\, j_\lambda^{(l)+}.$$

There are perfect analogues between them:

$$-e \to \frac{G}{\sqrt{2}}\,; \quad J_\lambda^3 \to J_\lambda^1 + i J_\lambda^2; \quad A_\lambda \to j_\lambda^{(l)+}.$$

We are now in a position to connect the weak interaction vector amplitude to the electromagnetic form factors.

The isovector part of the electromagnetic current $J_\lambda^{el}(x)$ has the following matrix element between the one nucleon states with momenta P and Q

$$\langle \text{nucleon}, Q \mid e J_\lambda^3 (x) \mid \text{nucleon}, P \rangle =$$

$$= \frac{ie}{2} \bar{u}(Q) \left[F_1^V (q^2) \gamma_\lambda + \frac{\mu_p - \mu_n}{2m} F_2^V (q^2) \sigma_{\lambda\mu} q^\mu \right] \tau_3 u(P) \tag{13}$$

where

$$\sigma_{\lambda\mu} = \frac{1}{2i} (\gamma_\lambda \gamma_\mu - \gamma_\mu \gamma_\lambda)$$

$q = P - Q =$ 4-momentum transfer

$1.8 = \mu_p$ static anomalous magnetic moment of proton

$-1.9 = \mu_n$ static anomalous magnetic moment of neutron

F_i^V's are the electromagnetic form factors of the nucleon, normalized to 1 at $q^2 = 0$:

$$F_i^V(0) = 1.$$

With the conserved vector current of equ. (11), the weak interaction vector matrix element can be immediately found to be:

$$\langle \text{nucleon}, Q | V_\lambda | \text{nucleon}, P \rangle =$$
$$= \frac{iG}{\sqrt{2}} \bar{u}_p(Q) \left[F_1^V(q^2)\gamma_\lambda + \frac{\mu_p - \mu_n}{2m} F_2^V(q^2)\sigma_{\lambda\mu} q^\mu \right] \cdot u_n(P). \tag{14}$$

Beta decay involves momentum transfers $q \simeq$ MeV, so that we only need the values $F_1^V(0) = F_2^V(0) = 1$ and equ. (14) yields in this case a *vector* matrix element

$$= \frac{iG}{\sqrt{2}} \bar{u}_p \gamma_\lambda u_n.$$

Comparing it to equ. (4), we get a natural understanding of the equality between the coupling constants in μ-meson decay and β-decay for the vector part of the interaction.

Weak magnetism [2]

The structure of the weak interaction matrix element, as given by equ. (14) is *definitely obtained* from electromagnetic form factors, this is a consequence of the conserved weak vector current. The definite form of the second term in equ. (14) gives rise to the so-called "weak magnetism".

Consider a $(\Delta J = 1, n_0)$ β-decay transition, the conserved vector current theory predicts a correction term to the usual allowed Gamow-Teller matrix element and the *usual* first forbidden Fermi term, this is actually:

$$\frac{iG}{\sqrt{2}} \langle f | \frac{\mu_p - \mu_n}{2m} \vec{q} \times \vec{\sigma}_N | i \rangle \, [e^+ \vec{\sigma}\, (1 + \gamma_5)\, \nu]. \tag{15}$$

This correction term can be obtained directly from the electromagnetic interaction term

$$\frac{-e\, \mu_p - \mu_n}{2m} \left(\vec{V} \times \vec{A} \right) \tag{16}$$

which is responsible of a radiative magnetic dipole transition.

The effects of weak magnetism for $\beta\pm$ energy spectra of the allowed Gamow-Teller transition is reflected in the correction factor

$$1 \mp \tfrac{8}{3}\, aE \tag{17}$$

where E is the electron (positron) energy and

$$a = \left| \frac{G_V}{G_A} \right| \frac{1 + \mu_p - \mu_n}{2m} \; (\mp \text{ for } \beta\pm). \tag{18}$$

In the usual β-decay interaction, the value of a would be:

$$a = \left| \frac{G_V}{G_A} \right| \frac{1}{2m}$$

(this is just the first forbidden interference term between axial and vector).

The "$\mu_p - \mu_n$" term is due to weak magnetism.

Experimental test of weak magnetism

The triad of nuclei B^{12}, C^{12}, and N^{12} provides the possibility of a test of the correction factor in equ. (17).

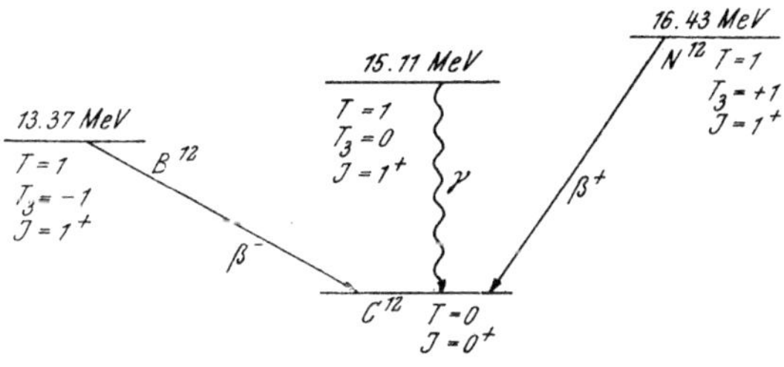

Fig. 1

The ground states of B^{12} and N^{12} and the excited state in C^{12} at 15.11 MeV form an isotopic spin triplet. The interference term between axial and conserved vector has the opposite sign in B^{12} and N^{12} because one involves β^+ and the other β^- and in the ratio of β^- spectrum to the β^+ spectrum, one gets a net effect of the form

$$R(E) \simeq 1 + \frac{16}{3} a E . \tag{19}$$

The experimental measurements [3, 4] of $R(E)$ give a linear function of energy with a slope which is in good agreement with $a \simeq \frac{2.0}{m}$ predicted by the conserved vector current.

Other experimental tests of the conserved vector current

Beta-decay of charged pions

Another experimental test of the conserved vector current is afforded by the β-decay of charged pions

$$\pi^\pm \to \pi^0 + e^\pm \pm \nu .$$

Without assumptions about a conserved vector current this process can only occur because of the pion-nucleon coupling through virtual loops like

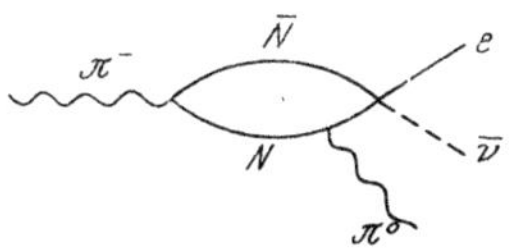

Fig. 2

The contribution due to this diagram cannot be calculated reliably in the present status of strong interactions. The beautiful result of the conserved vector current is that it predicts an unambiguous value for the rate of the process

$$\pi^\pm \to \pi^0 + e^\pm \pm \nu.$$

The decay matrix element is

$$\mathfrak{M} = \langle \pi^0 \mid V_\lambda \mid \pi^+ \rangle \, \bar{u}_e \gamma_\lambda \, (1 + \gamma_5) \, v_\nu.$$

(The axial vector cannot contribute to above matrix element in any kind of theory: it is impossible to construct axial vector bilinear in pion fields because they are spinless.)

The conserved vector current equ. (11) involves the pion fields explicitely and has a known matrix element:

$$\mathfrak{M} = -i \sqrt{2} \frac{G}{\sqrt{2}} (P_{\pi^+} + P_{\pi^0}) \, \bar{u}_e \gamma_\lambda \, (1 + \gamma_5) \, v_\nu.$$

Since the energy release is small ($Q = 4.1$ MeV)

$$\mathfrak{M} \simeq G \, (m_{\pi^+} + m_{\pi^0}) \, \bar{u}_e \gamma_4 \, (1 + \gamma_5) \, v_\nu.$$

The decay rate is then given by:

$$\Gamma = \frac{G^2}{\pi^3} \int_{m_e}^{E_{\max}} dE \, p \, E \, (E_{\max} - E)^2 = \frac{G^2}{30 \, \pi^3} (E_{\max})^5 \, R_0 \left(\frac{m_e}{E_{\max}} \right)$$

where $E_{\max} = 4.52$ MeV is the maximum electron energy and

$$R_0 \, (x) = \sqrt{1 - x^2} \, (1 - {}^9/_2 x^2 + 4 x^4) + {}^{15}/_2 x^4 \log \left(\frac{1 + \sqrt{1 - x^2}}{x} \right).$$

With the value of G given by the μ-mesonlifetime:

$$G = (1.023 \pm 0.002) \cdot 10^{-5} \, m_{\text{proton}}^{-2}$$

one gets:

$$\Gamma = 0.39 \pm 0.04 \text{ sec}^{-1}$$

and this means a branching ratio

$$\Gamma \tau_\pi = (1.0 \pm 0.1) \, 10^{-8}.$$

Experimental works indicate

$$\Gamma \tau_\pi = \begin{cases} (1.7 \pm 0.5) \cdot 10^{-8} & [5] \\ \left(1.1 \pm {}^{1.0}_{0.5}\right) \cdot 10^{-8} & [6] \end{cases}$$

High energy neutrino absorption

The neutrino absorption reactions

$$\bar{\nu} + p \to \bar{l} + n \quad \text{(a)}$$
$$\nu + n \to l + p \quad \text{(b)}$$

at high energies are the most sensitive to weak magnetism assumption because of the large momentum transfers involved. Thus the weak magnetism effects will be large. Furthermore, as shown in [15] (see also T. D. LEE, Cern Seminars 1961), the weak magnetism amplitudes interfere in different ways in (a) and (b) and one has [15]

$$d\sigma_\nu - d\sigma_{\bar{\nu}} = -\frac{\nu + l}{\pi\, m\, \nu^2} G_V G_A (1 + \mu_p - \mu_n) F_2^V f_A q^2 dq^2$$

where ν: incoming lab. energy of neutrino
l: outgoing lab. energy of lepton
F_2^V: is known from the Hofstadter analyses
and f_A: axial vector form factor which will be discussed below

This difference $d\sigma_\nu - d\sigma_{\bar{\nu}}$ becomes appreciable when $\frac{m}{\nu} \sim 1$.

$\beta - \alpha$ angular correlation in Li^8 and B^8

For completeness, let us mention another possible experimental test of conserved vector current [7, 8]. This involves the (β, α) angular correlation in the decays of Li^8 and B^8 to an excited state of Be^8 which breaks up into two alphas.

Again this case, the experimental data [9] confirmed the conservation of vector current.

The above discussion shows that the vector part of the weak interaction current (at least for $\Delta S = 0$) can be well understood in framework of the conserved vector current theory.

Let us turn on problems concerning the axial vector of the interaction.

Lack of conservation of axial vector current

In analogy to the conserved vector current interaction, some speculations have been made about a possible conserved axial vector current [10]. However, such attempts have to overcome many great difficulties:

1) First of all, the existence of the ratio $\left|\frac{G_A}{G_V}\right| = 1.2$ implies that there are some renormalization effects to the axial vector coupling constant.

2) Secondly, the possible forbidding of the electronic decay of the pion $\pi \to l + \bar{\nu}_l$ which was an appealing point of a conserved axial vector current [10] (one had, however, to exempt the μ-meson from some postulates of the scheme) turns out to be a strong argument against the conservation of the axial vector current:

In fact, consider the decay of the pion [10] $\pi \to l + \bar{\nu}_l$. The strong coupling of the pion to baryons means that the decay is caused by the strong interacting current and the Feynman amplitude is

$$\mathfrak{M} = \langle 0 | A_\lambda | \pi \rangle \left[\bar{u}_l (p_1) \gamma_\lambda (1 + \gamma_5) v_\nu (p_2)\right]. \tag{20}$$

The vector current V_λ does not contribute to $\mathfrak{M}$ because the pion is pseudoscalar.

Now, if the axial vector current was conserved one would have

$$\langle 0 | \partial_\lambda A_\lambda | \pi \rangle = 0. \tag{21}$$

But

$$\langle 0 | \partial_\lambda A_\lambda | \pi \rangle = i Q_\lambda \langle 0 | A_\lambda | \pi \rangle. \qquad (22)$$

where Q is the pion 4-momentum and equation (21) implies that

$$\langle 0 | A_\lambda | \pi \rangle = 0. \qquad (23)$$

Therefore the $\pi \to l + \bar{\nu}_l$ process is forbidden.

3) The third kind of difficulties concerns the induced pseudoscalar coupling between nucleons and leptons [11].

On invariance grounds, the axial vector current matrix element has the following form:

$$\langle p | A_\lambda | n \rangle = \frac{i G}{\sqrt{2}} \bar{u}_p (Q) \left[i f_A (q^2) \gamma_\lambda + i f_T \sigma_{\lambda\mu} q^\mu + f_P q_\lambda \right] \gamma_5 u_n (P). \qquad (24)$$

If we require that the matrix element of the divergence of A_λ vanishes, we find

$$\langle p | \partial_\lambda A_\lambda | n \rangle = \frac{i G}{\sqrt{2}} \bar{u}_p (Q) \left[i f_A (q^2) \gamma_\lambda q_\lambda + q^2 f_P (q^2) \right] \gamma_5 u_n (P) = 0. \qquad (25)$$

Now

$$i \bar{u}_p (Q) \gamma_\lambda \gamma_5 q_\lambda u_n (P) = 2m \bar{u}_p (Q) \gamma_5 u_n (P).$$

So that

$$\langle p | \partial_\lambda A_\lambda | n \rangle = \frac{i G}{\sqrt{2}} \bar{u}_p (Q) \left[2m f_A (q^2) + f_P (q^2) q^2 \right] \gamma_5 u_n (P) = 0. \qquad (26)$$

This means

$$f_P (q^2) = - \frac{2m}{q^2} f_A (q^2) \qquad (27)$$

and equ. (24) reads now

$$\begin{aligned} \langle p | A_\lambda | n \rangle = & \frac{i G}{\sqrt{2}} f_A (q^2) \bar{u}_p (Q) \left(i \gamma_\lambda - \frac{2m}{q^2} q_\lambda \right) \gamma_5 u_n (P) + \\ & + \frac{i G}{\sqrt{2}} f_T (q^2) \bar{u}_p \, i \sigma_{\lambda\mu} q^\mu u_n. \end{aligned} \qquad (28)$$

The induced pseudoscalar term when coupled to the leptonic current gives rise to the contribution

$$\mathfrak{M}_P = - \frac{i G}{\sqrt{2}} \frac{2m}{q^2} f_A (q^2) \bar{u}_p (Q) \gamma_5 u_n (P) \bar{u}_l q_\lambda \gamma_\lambda (1 + \gamma_5) v_{\nu_l}. \qquad (29)$$

There results, when use is made of leptonic equations of motion

$$\mathfrak{M}_P = \frac{G}{\sqrt{2}} \frac{2 m m_l}{q^2} f_A (q^2) \bar{u}_p \gamma_5 u_n \bar{u}_l (1 + \gamma_5) v_{\nu l} \qquad (30)$$

(m_l: the lepton mass).

For small q^2, the effective coupling is

$$G_P = \frac{2 m m_l}{q^2} G_A.$$

This is energy dependent but $\sim 10^3 G_A$ in β-decay which surely be ruled out experimentally, even when the relativistic nature of pseudoscalar coupling is allowed for.

Asymptotic conservation of the axial vector current

This notion was introduced by NAMBU [12] who observed that in equ. (28) for the conserved axial current, the pseudoscalar term has a $1/q^2$ factor which could be interpreted as implying a massless, pseudoscalar and charged quantum bridging the nucleon and lepton currents. He suggests that there may not be a strict pseudovector current conservation, but that one may have an approximate conservation which becomes rigorous in the limit $q^2 \gg m^2_\pi$ (m_π: the pion mass). He proposes that the axial vector part of the nucleonic current matrix element has the following form and properties (assuming $G^{-1} A_\lambda G = -A_\lambda$)

$$\langle p | A_\lambda | n \rangle = \frac{G}{\sqrt{2}} \bar{u}_p \left[i \gamma_\lambda F_A (q^2) - \frac{2\, m\, q_\lambda}{q^2 + m_\pi^2} F_P (q^2) \right] \gamma_5 u_n \tag{31}$$

with

$$F_A (0) \simeq F_P (0) \simeq 1.2 \tag{32}$$

$$F_A (q^2) = F_P (q^2) \quad \text{for} \quad q^2 \gg m_\pi^2. \tag{33}$$

The pion is then the analogous of the massless particle mentioned above and F_P and F_A should have in general the form

$$F_P (q^2) = F_P (-m_\pi^2) - (q^2 + m_\pi^2) \int_{m_0^2}^{\infty} \frac{\varrho_P (m^2)\, dm^2}{(q^2 + m^2)(m^2 - m_\pi^2)} \tag{34}$$

$$F_A (q^2) = \int_{m_0^2}^{\infty} \frac{\varrho_A (m^2)\, dm^2}{q^2 + m^2} \tag{35}$$

where $m_0 = 3 m_\pi$ unless there are new particles of low mass [13].

Consequences of the pion decay rate for the μ-meson capture matrix element

In equ. (34), $F_P (-m_\pi^2)$ is just the contribution due to the one pion pole and can be related to $p n \pi$ vertex and $\pi \to l + \bar{\nu}_l$ decay amplitude by

$$-2 m F_P (-m_\pi^2) = i^2 \sqrt{2}\, g_\pi\, m_\pi\, f_\pi (-m_\pi^2) \tag{36}$$

where g_π is the renormalized $\pi - N$ coupling constant, $f_\pi (-m_\pi^2)$ is the $\pi \to l + \bar{\nu}_l$ decay amplitude.

Now, according to equ. (34), $F_P (q^2)$ and $F_A (q^2)$ are slowly varying functions for $|q^2| \ll m_0^2$ so that

$$F_P (0) \simeq F_P (-m_\pi^2). \tag{37}$$

Combining equ. (32), (36) and (37), one obtains

$$f_\pi (-m_\pi^2) = \sqrt{2}\, \frac{m}{m_\pi}\, \frac{F_A (0)}{g_\pi} \tag{38}$$

which was derived by GOLDBERGER and TREIMAN [14] from a somewhat different approach. Numerically the r. h. s. of equ. (38) is 0.873 and experimental value of $f_\pi\ (-m_\pi^2)$ is 0.935. The agreement is rather remarkable if one takes into account the uncertainties for g_π and F_A.

In μ-meson capture, $q^2 \sim 0.9\, m_\mu^2$ (m_μ: the mass of μ-meson), the nucleonic current matrix element given by equ. (31) calculated at $q^2 \sim 0.9\, m_\mu^2$ is

$$\langle p|\, A_\lambda \,|n\rangle = \frac{G}{\sqrt{2}}\, \bar{u}_p \left[i\,\gamma_\lambda\, F_A\,(0.9\, m_\mu^2) - \frac{2\,m\,q_\lambda}{0.9\, m_\mu^2 + m_\pi^2}\, F_P\,(0.9\, m_\mu^2)\right] \gamma_5\, u_n.$$

Again the slow variation of F_A and F_P permits us to set

$$F_A\,(0.9\, m_\mu^2) \simeq F_A\,(0)$$
$$F_P\,(0.9\, m_\mu^2) \simeq F_P\,(0)$$

and

$$\langle p|\, A_\lambda \,|n\rangle = \frac{G}{\sqrt{2}}\, \bar{u}_p\, F_A\,(0) \left[i\,\gamma_\lambda - \lambda\, \frac{q_\lambda}{m_\mu}\right] \gamma_5\, u_n$$

with

$$\lambda = \frac{2\,m\,m_\mu}{0.9\, m_\mu^2 + m_\pi^2} \simeq 7.8.$$

The evidence for the presence of an induced pseudoscalar coupling of about such a magnitude comes from the directional asymmetry of the product neutrons from capture of polarized muons.

In nuclear β-decay, on the contrary, the contribution from the induced pseudoscalar is

$$\mathfrak{M}_P\,(\beta\text{-decay}) = \frac{i\,G}{\sqrt{2}}\, \lambda'\, F_A\,(0)\, \bar{u}_p\, \gamma_5\, u_n\, \bar{u}_l\, (1 + \gamma_5)\, v_\nu$$

with

$$\lambda' = \frac{2\,m\,m_l}{m_\pi^2} \simeq \frac{1}{20}$$

which is consistent with the smallness of observed result.

References

0. For an extensive survey of the weak interactions field, the reader is referred to:
   — TREIMAN, Dispersion Relations and Elementary Particles (Les Houches 1960) P. 515—58.
   — BERMAN, Lectures Notes on Weak Interaction. CERN Report **1961**.
   — JACKSON, Brandeis Lectures, P. 266 (1962).
   — CERN Seminars (1961).
1. GERSHTEIN and ZELDOVITCH, JETP **8**, 570 (1959).
   — FEYNMAN and GELL-MANN. Phys. Rev. **109**, 193 (1958).
2. GELL-MANN, Phys. Rev. **111**, 362 (1958).
3. MAYER-KUCKUK and MICHEL, Phys. Rev. **127**, 545 (1962).
4. LEE, MO, and WU, Phys. Rev. Lett. **10**, 253 (1963).
5. DEPOMMIER et al., Phys. Lett. **2**, 23 (1962).
6. DUNAITSEV et al., Phys. Lett. **1**, 138 (1962).
7. BERNSTEIN and LEWIS, Phys. Rev. **112**, 232 (1958).

8. MORITA, Phys. Rev. **113,** 1584 (1959).
9. NORDBERG et al., Phys. Rev. **125,** 321 (1962).
10. TAYLOR, Phys. Rev. **110,** 1216 (1958).
11. GOLDBERGER and TREIMAN, Phys. Rev. **110,** 1478 (1958).
12. NAMBU, Phys. Rev. Lett. **4,** 380 (1960).
13. DENNERY and PRIMAKOFF, Phys. Rev. Lett. **8,** 350 (1962).
14. GOLDBERGER and TREIMAN, Phys. Rev. **110,** 1178 (1958).
15. LEE and YANG, Phys. Rev. Lett. **4,** 307 (1960). See also T. D. LEE. CERN Seminars (1961).

Dynamical Consequences of Symmetries among Elementary Particles

By

H. Pietschmann, Vienna

With 2 Figures

In the proceeding series of lectures we have heard much about symmetries of elementary particles and the more or less successful classification of them into the various representations of SU_3 [1]. We also heard about the information one obtains by combining group properties with certain dynamical models [2]. The question now arises whether it is possible to obtain additional information on dynamical entities (such as amplitudes or form factors) purely from the symmetry properties of elementary particles without further dynamical assumptions. That this is indeed possible has been shown by investigations carried out in Vienna by several authors (refs. [3] to [6]).

Although this program can be applied to any invariance group we want to restrict ourselves in a first attempt to the well-known Isospin-Group. This has two reasons:

A. The problem is algebraically very complex, so that simplicity gained by this restriction is welcome and

B. we know that Isospin invariance is broken only by the relatively weak electromagnetic (and weak) interactions so that our results can be directly checked against experiments. This helps in estimating the error introduced by several approximations.

Let us now start by writing down the commutation relations of the underlying group (or rather algebra). In an obvious notation

$$[I_l, I_j] = i\, \varepsilon_{ljk}\, I_k \tag{1}$$

or by multiplication with ε_{ijl},

$$\varepsilon_{ljk}\, I_l\, I_j = i\, I_k. \tag{1'}$$

The connection to dynamical entities enters through the assumption of a local density (or current)

$$I_l = \int d^3 x\, J_{l,0}(\vec{x}, t) \tag{2}$$

such that

$$\varepsilon_{ljk}\, J_{l,0}(\vec{x}, t)\, J_{j,0}(\vec{y}, t) = i\, J_{k,0}(\vec{x}, t)\, \delta^{(3)}(\vec{x} - \vec{y}) \tag{3}$$

where we have essentially assumed the canonical form of these commutation relations.

Eq. (3) is the basic equation of our investigation. We specify it to a particular physical process by taking it between states $\langle A |$ and $| B \rangle$. The connection of $\langle A | J_{k,0} | B \rangle$ to the isovector part of the matrix element for scattering of an electron (or muon) on the state A whereby the latter is converted into a state B is given by

$$\mathfrak{M}^V = \langle A | J_{3,\mu} | B \rangle A_\mu^M \tag{4}$$

where A_μ^M is the Möller potential of the scattered lepton

$$A_\mu^M = \frac{e^2}{q^2} \bar{u}_l (p_2) \gamma_\mu u_l (p_1) \tag{5}$$

$q^2 = (p_2 - p_1)^2$ being the squared momentum transfer. (See Fig. 1.) Alternatively we can relate $\langle A | J_{k,0} | B \rangle$ to the neutrino induced process shown in Fig. 2.

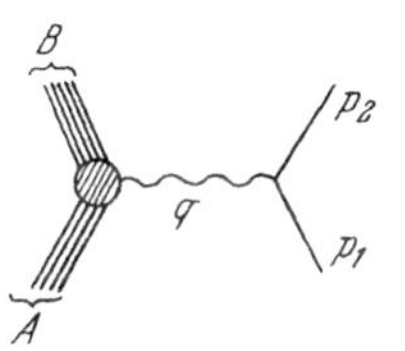

Fig. 1

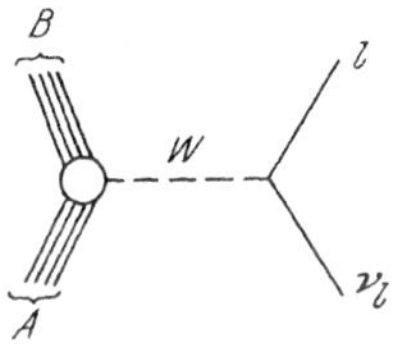

Fig. 2

$$\mathfrak{M}^{WV} = \langle A | J_{+,\mu} | B \rangle B_\mu^W \tag{6}$$

$$B_\mu^W = \frac{G^2}{q^2 - M_W^2} \cdot \bar{u}_l (p_2) \gamma_\mu (1 + \gamma_5) u_{\nu_l} (p_1) \tag{7}$$

Let us turn now to the simple case where A and B are both one nucleon states.

Eq. (3) then reads

$$\varepsilon_{lkj} \langle p | J_{l,0} (x) J_{k,0} (y) | p' \rangle = i \langle p | J_{j,0} (x) | p' \rangle \delta^{(3)} (x - y) \tag{8}$$

with

$$\langle p | J_{k,0} | p' \rangle \propto \bar{u} (p) \tau_k [\gamma_0 F_1^V (q^2) + \sigma_{0\nu} q_\nu F_2^V (q^2)] u (p') \tag{9}$$

where $F_i^V (q^2)$ are the two isovector nucleon form factors. One realizes immediately that the structure of Eq. (8) is similar to that of the well-known unitarity relation. There is, however, a certain drawback in Eq. (8): By introducing a complete set of intermediate states in the unitary relation, one may neglect higher states via the "near by singularity argument". No such arguments can be given here. We have, however, shown in ref. [3] that (at least in the case of the form factors, considered above) the resulting equations are exact in the limit of zero momentum transfer, i. e. $q^2 = 0$ and therefore may be a good approximation for small q^2. An investigation of this technique in the static model (ref. [5]) further encouraged this assumption. It is, however, not possible to rely on the first approximation, the one-nucleon intermediate state, since this yields a point structure for the charge distribution (ref. [4]).

After integration over $\vec{x}$ and $\vec{y}$ Eq. (8) will take the following structure, written out symbolically:

$$\varepsilon \mathop{\mathrm{S}}_{n} \delta^{(3)}\left(n - \frac{p+p'}{2}\right) \langle p \mid I \mid n \rangle \langle n \mid I \mid p' \rangle = i \langle p \mid I \mid p' \rangle. \tag{10}$$

In order to get equations for the form factors we take advantage of the linear independence of γ_0 and $\sigma_{0\nu}\, q_\nu$ and equate their coefficients on both sides of (10). The resulting equations are, of course, invariant; by specifying a particular frame they can also be cast into a covariant form. Taking the first two intermediate states leads to (again in symbolic notation)

$$F = \alpha\, F \cdot F + \beta\, A \cdot A + \ldots \tag{11}$$

where A stands for the 6 amplitudes for electroproduction of a pion on a nucleon. Taking Eq. (3) between a one nucleon and a pion-nucleon state will similarly lead to

$$A = \gamma\, A \cdot F + \ldots \tag{12}$$

so that we have the same number of equations and unknown functions. The question thus arises, what are the experimental parameters to be fed in for a computation of F and A; it turns out that these are the static values of form factors and amplitudes, e. g. the static magnetic moments. These then determine, at least in principle, the functional behaviour.

After this short outline of our method let us report on what has been achieved until now. Unfortunately even the results are too lengthy to be given here, but again in symbolic form we can indicate our conclusions. We can represent the mean square radii of the nucleon isovector charge and magnetic moment distribution in the form (see ref. [4])

$$\langle r_i^2 \rangle = \text{const} + \int_{M^2}^{\infty} du \sum_{\alpha\beta}^{6} \Big[R_{\alpha\beta}\left(\sqrt{u}\right) A_\alpha(u)\, A_\beta(u) + + R'_{\alpha\beta}\left(\sqrt{u}\right) \frac{d}{du} A_\alpha(u) \frac{d}{du} A_\beta(u) \Big] \tag{13}$$

where $A_\alpha(u)$ are again the 6 production amplitudes and $R_{\alpha\beta}$ and $R'_{\alpha\beta}$ are certain rational functions. The static values of the amplitudes are interrelated such that they can be expressed by only two of them (ref. [6]).

$$A_\alpha(0) = \sum_{k=1}^{2} c_{\alpha k}\, A_k(0). \tag{14}$$

It seems to be desirable to extend this program to the full SU_3 group. In this case we obtain much more information; e. g. also the isoscalar formfactors can be investigated but we then lack the full validity of our results because SU_3 is broken. It seems that the experimental agreement of the electromagnetic mass splitting formula is a good hint for the validity of such investigations.

References

1. MAYER, M. E., these lecture notes.
2. WESS, J., these lecture notes.
3. BALACHANDRAN, A. P., and H. PIETSCHMANN, Nucl. Phys. **43**, 321 (1963) and Acta Phys. Austr. **16**, 362 (1963).
4. KUMMER, W., H. PIETSCHMANN, and A. P. BALACHANDRAN, Ann. Phys. N. Y. in press.
5. BIRITZ, H., and H. PIETSCHMANN, Proc. of the Siena Conf. on Elementary Particles, 1963 and Acta Phys. Austr. to be published.
6. STREMNITZER, H., Ph. D. thesis, Inst. f. theor. Phys. University of Vienna (1964).

Unitary Symmetry and the Leptonic Decays of Hadrons

By

D. Horn

Department of Physics, Tel Aviv University, Tel Aviv, Israel

With 1 Figure

Abstract

The structure of the weak currents that appear in the leptonic decays is discussed in the light of unitary symmetry. The known leptonic interaction modes imply the inclusion of more currents than the ones derived from gauge invariance of the octet model. It is therefore suggested that the weak currents emerge from a higher symmetry scheme. The vector part of the interaction is obtained by CVC arguments and the axial-vector part by a Goldberger-Treiman approach. The fit with the experimental situation is discussed.

1. Phenomenology

In recent years some doubts were raised about the validity of the $\Delta S = \Delta Q$ rule. The experimental situation is still not clear but it seems that the rule is not absolutely strict and one must take into account at least small corrections. Furthermore, one would like to explain the smallness of the hyperon-leptonic decay coupling constants.

One can divide the known leptonic decays according to the following groups:

Table 1. Leptonic hadron decays

Group	$\lvert\Delta Y\rvert$	$\lvert\Delta T_z\rvert$	$\lvert\Delta \mathbf{T}\rvert$	Reactions
A	0	1	1	$n \to p + e + \nu$; $\pi \to \mu + \nu$; $\Sigma \to \Lambda + e + \nu$
B	0	1	1, 2	$\pi \to \pi + e + \nu$
C	1	$\frac{1}{2}$	$\frac{1}{2}$	$\Lambda \to p + e + \nu$; $K \to \mu + \nu$
D	1	$\frac{1}{2}$	$\frac{1}{2}, \frac{3}{2}$	$\overset{(-)}{K^0} \to \overset{(+)}{\pi^-} + e + \nu$ $K^+ \to \pi^0 + e + \nu$; $\Sigma^- \to n + e^- + \nu$
E	1	$\frac{3}{2}$	$\frac{3}{2}$	$\Sigma^+ \to n + e^+ + \nu$ $\overset{(-)}{K^0} \to \overset{(-)}{\pi^+} + e + \nu$

(No distinction is made here between the various neutrinos and antineutrinos.)

Group A contains the well known β decay. FEYNMAN and GELL-MANN [1] suggested that the $\Delta S = 0$ decays stem from a conserved vector current which has the character $|\Delta \mathbf{T}| = 1$. If their theory is right then the group B decays should reduce to type A. Groups C and D contain the $\Delta S = \Delta Q$ ($\Delta T_z = \frac{1}{2}$) reactions. Assuming a pure $V-A$ theory one gets for the hyperon-leptonic decays a decay rate that is larger by a factor of 20 than the one observed in nature. Observation of reactions of type E [2] was an indication of the existence of $\Delta S \neq \Delta Q$ weak currents. Our theory must therefore be able to deal with them and predict their decay rates.

2. The Weak Current Representations in Unitary Symmetry

Let us now discuss the weak currents that occur in the processes listed. We can write them schematically as $\overline{B}_2 B_1$ which represents the current that converts baryon 1 into baryon 2. Thus we see that from a group theoretical point of view we have to deal here with a product of the baryon representation with itself. We limit ourselves to the octet model of GELL-MANN and NE'EMAN [3] where all baryons appear in an eightfold representation. The Kronecker product of two octets can be decomposed as follows

$$8 \otimes 8 = 1^+ \oplus 8^+ \oplus 8^- \oplus 10^- \oplus \overline{10}^- \oplus 27^+ \qquad (1)$$

where we have designated the representations by their dimensionality and the symmetric$^{(+)}$ or antisymmetric$^{(-)}$ properties of the matrices.

Figure 1 shows the elements that appear in the different representations. Their explicit form is given in the Appendix.

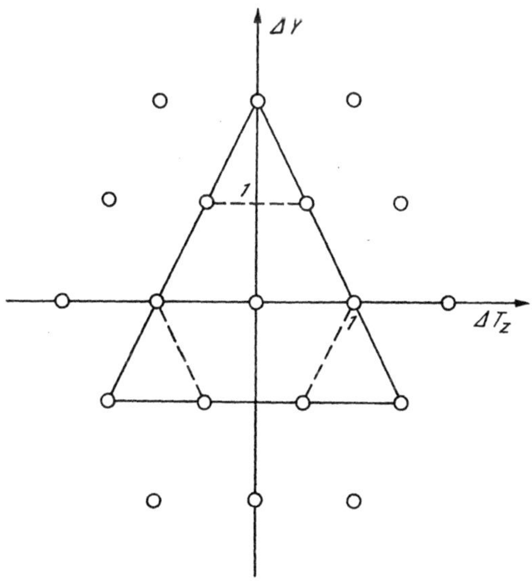

Fig. 1

1$^+$ has only one element, $\Delta T_z = \Delta Y = 0$. The two octets have the six elements on the dashed hexagon and two elements at the centre. The triangle represents the ten elements in the $\overline{\mathbf{10}}$ representation (the **10** is its reflection through the centre). The **27** has elements at all the points of the diagram with doublets on the hexagon and a triplet at the centre.

We are essentially interested in the elements on the right side of the triangle which represents $\Delta Q = 1$ (or their reflection through the origin which has $\Delta Q = -1$). We see from the diagram that it contains two elements which are common to the $\overline{\mathbf{10}}$ and **27** only and these are ($\Delta Y = -1$, $\Delta T_z = \frac{3}{2}$) and ($\Delta Y = 2$, $\Delta T_z = 0$). The first case includes the following terms (here we treat the reflection through the origin in order to get physically feasible processes)

$$\sqrt{\tfrac{1}{2}}\,(\overline{\Sigma^-}\,\Xi^0 - \bar{n}\,\Sigma^+) \text{ in } \mathbf{10}^-$$
$$\sqrt{\tfrac{1}{2}}\,(\overline{\Sigma^-}\,\Xi^0 + \bar{n}\,\Sigma^+) \text{ in } \mathbf{27}^+.$$

This corresponds to group E according to our classification.

The second case has

$$\sqrt{\tfrac{1}{2}}\,(\bar{n}\,\Xi^- - \bar{p}\,\Xi^0) \text{ in } \overline{\mathbf{10}}^-$$
$$\sqrt{\tfrac{1}{2}}\,(\bar{n}\,\Xi^- + \bar{p}\,\Xi^0) \text{ in } \mathbf{27}^+.$$

These are $\Delta S = 2$ reactions which have not yet been observed in nature. The consideration of $\mathbf{10} \oplus \overline{\mathbf{10}}$ and/or **27** is necessary in order to explain class E, but this implies the existence of $\Delta S = 2$ decays, although possibly at a very low rate. Moreover, one should find the same ratio of matrix elements of $(\bar{p}\,\Xi^0)/(\bar{n}\,\Xi^-)$ as $(\bar{n}\,\Sigma^+)/(\overline{\Sigma^-}\,\Xi^0)$. This ratio can tell us the relative amount of the two representations $\mathbf{10} \oplus \overline{\mathbf{10}}$ and **27**. (We consider $\mathbf{10} \oplus \overline{\mathbf{10}}$ together because they are each other's conjugate. All other representations are self conjugate.)

There remain two elements in the diagram which can belong to any representation apart from the singlet. They are ($\Delta Y = 0$, $\Delta T_z = 1$), which includes the groups A and B, and ($\Delta Y = 1$, $\Delta T_z = \frac{1}{2}$), which includes the hyperon and K meson decays C and D and should therefore have a smaller coupling associated with it. On the basis of the above considerations one can compare different processes under the assumption of one important current. For instance, the ratio of the $(\bar{p}\,\Lambda)$ to $(\bar{p}\,n)$ matrix element will be $-\sqrt{\frac{3}{2}}$, $-\sqrt{\frac{1}{6}}$, $\frac{3}{2}$, $\sqrt{\frac{3}{2}}$, 0 according to whether the important representation is $\mathbf{8}^-$, $\mathbf{8}^+$, $\mathbf{27}^+$, $\overline{\mathbf{10}}^-$ or $\mathbf{10}^-$ respectively. In this calculation we assume that the different currents that appear in the same representation are coupled with the same strength to the leptonic current. One can also look upon the points of the diagram in Fig. 1 as representing the intermediate vector mesons W. Then our assumption is that all the W mesons that belong to the same representation have the same "weak charge".

Summing up, we see that the $8 \otimes 8$ product can carry all the necessary representations. It is clear that $10 \oplus \overline{10}$ or 27 are involved in order to explain $|\Delta T| = \frac{3}{2}$ currents. Other representations may be responsible for the stronger processes.

3. Conserved Vector Currents

Feynman and Gell-Mann [1] suggested that the vector part of the $|\Delta T| = 1$ current which is responsible for processes A, is conserved under strong interactions. This explains the lack of renormalization of the vector coupling constant of β decay which is equal to that of muon decay. Let us use this concept of the CVC (conserved vector current) as a guide to find the structure of the vector part of the leptonic decays.

Starting from a Lagrangian theory which is invariant under a certain symmetry transformation, there exists a well known way to find the CVC, namely,

$$j_\mu{}^k = \sum_\alpha \frac{\partial \mathcal{L}}{\partial (\partial^\mu \varphi_\alpha)} \delta^k \varphi_\alpha \tag{2}$$

where $\delta^k \varphi_\alpha$ is the kth symmetry operation on φ_α. Moreover, the operators defined by

$$Q^k = \int j_0{}^k \, d^3 x \tag{3}$$

are constants of motion and serve as the generators of the gauge group.

If we start with a theory of 8 free equal-mass baryons, then all combination of the form

$$\overline{\Psi}_\alpha \gamma_\mu A_{\alpha\beta} \Psi_\beta \tag{4}$$

where A is any 8×8 matrix, will serve as CVC's. This theory is invariant under SU(8). If we add 8 free equal-mass (real) bosons, then only antisymmetric combinations of the form

$$\Phi_\alpha A_{\alpha\beta} \partial_\mu \Phi_\beta, \; A_{\alpha\beta} = -A_{\beta\alpha} \tag{5}$$

will be CVC's. Thus we get a theory which is invariant under R(8).

Introducing an interaction between the particles the symmetry may be reduced. Let us assume that the resultant is SU(3). Then it turns out that the conserved vector currents can only belong to the 8^- representation out of all the representations of $8 \otimes 8$ which are present in (4). If we restrict ourselves further and consider only the isotopic spin invariance then we are left with SU(2) symmetry and the $|\Delta T| = 1$ CVC of Feynman and Gell-Mann.

Let us suppose that all the vector currents which appear in the leptonic decays stem from currents which are conserved at some stage of the theory before all the symmetry breaking takes place. We start with R(8) which provides us with 28 currents, out of which only 8 carry $\Delta Q = 1$. Let us look upon the symmetry breaking effect as a large perturbation. We keep our R(8) currents, labelling them according to SU(3) quantum numbers, and restricting invariance requirements to currents with the same SU(3) representation only. The classification is given in Table 2.

Table 2. Anti-symmetric baryon bilinears with $\Delta Q = 1$

Quantum numbers	R(8)	SU(3)
$\Delta Y = 0 \quad \Delta T_z = 1$	$\overline{\Xi^0}\,\Xi^- - \bar{p}\,n =$	$-\sqrt{\frac{1}{3}}\,8_1^- + \sqrt{\frac{1}{3}}\,10_1 + \sqrt{\frac{1}{3}}\,\overline{10}_1$
	$\overline{\Sigma^0}\,\Sigma^- - \overline{\Sigma^+}\,\Sigma^0 =$	$\sqrt{\frac{2}{3}}\,8_1^- + \sqrt{\frac{1}{6}}\,10_1 + \sqrt{\frac{1}{6}}\,\overline{10}_1$
	$\bar{\Lambda}\,\Sigma^- - \overline{\Sigma^+}\,\Lambda =$	$\frac{\sqrt{2}}{2}\,10_1 - \frac{\sqrt{2}}{2}\,\overline{10}_1$
$\Delta Y = 1 \quad \Delta T_z = \frac{1}{2}$	$\bar{\Lambda}\,\Xi^- - \bar{p}\Lambda =$	$\frac{\sqrt{2}}{2}\,8_{\frac{1}{2}}^- + \frac{\sqrt{2}}{2}\,\overline{10}_{\frac{1}{2}}$
	$\overline{\Sigma^0}\,\Xi^- - \bar{p}\,\Sigma^0 =$	$\sqrt{\frac{1}{6}}\,8_{\frac{1}{2}}^- + \sqrt{\frac{2}{3}}\,10_{\frac{3}{2}} - \sqrt{\frac{1}{6}}\,\overline{10}_{\frac{1}{2}}$
	$\overline{\Sigma^+}\,\Xi^0 - \bar{n}\,\Sigma^- =$	$\sqrt{\frac{1}{3}}\,8_{\frac{1}{2}}^- - \sqrt{\frac{1}{3}}\,10_{\frac{3}{2}} - \sqrt{\frac{1}{3}}\,\overline{10}_{\frac{1}{2}}$
$\Delta Y = 2 \quad \Delta T_z = 0$	$\bar{n}\,\Xi^- - \bar{p}\,\Xi^0 =$	$\overline{10}_0$
$\Delta Y = -1 \quad \Delta T_z = \frac{3}{2}$	$\overline{\Xi^0}\,\Sigma^- - \overline{\Sigma^+}\,n =$	$\overline{10}_{\frac{3}{2}}$

The 8^- representation provides the weak currents. This includes both $|\Delta \mathbf{T}| = 1$ and $|\Delta \mathbf{T}| = \frac{1}{2}$ parts. A priori there is no difference between these two directions in the 8^- manifold and therefore we expect both of them to have the same coupling strength with the leptonic current. In order to explain the remarkably low decay rate of the hyperons as compared with $V - A$ predictions, we must consider the axial vector currents.

4. The Axial Vector Currents

Axial vector currents can be conserved only in a γ_5 — invariant theory which is not the case we are considering here. Therefore pure symmetry cannot provide us with information about them. One systematic approach to the evaluation of the axial vector currents is given by the GOLDBERGER-TREIMAN [4] type relations.

Their original relation was derived for the nucleons in the $|\Delta \mathbf{T}| = 1$ current

$$2 m_N \left(- \frac{G_A}{G_V} \right) \approx \frac{g_{NN\pi}}{f_\pi} . \tag{6}$$

A similar relation can be derived from polological reasoning for the $|\Delta \mathbf{T}| = \frac{1}{2}$ vertex

$$(m_N + m_Y) \left(- \frac{G_A^{NY}}{G_V^{NY}} \right) \approx \frac{g_{NYK}}{f_K} \tag{7}$$

where the K meson plays the same role as the pion in (6). Then it turns out that $(G_A^{NY})^2$ is smaller than $(G_V^{NY})^2$ by two orders of magnitude. Therefore from this standpoint the hyperon decays are mostly vectorial.

The decay probability is approximately proportional to $(|G_V|^2 + 3 |G_A|^2)$. If the decay is mostly vectorial then we get only a quarter of the value predicted by the $V - A$ theory. A further reduction of G_V^{NY} by a factor of 2.6 brings us to the experimental value. Here it should be noted that $8_{\frac{1}{2}}{}^-$ is conserved only in the limit of unitary symmetry. Therefore we should not be astonished to find such a renormalization effect.

5. Discussion of Results

The $\mathbf{10} \oplus \overline{\mathbf{10}}$ currents may be responsible for the small difference between the coupling constants of neutron and muon decays. If we assign the same effective coupling constant $-0.1 \sqrt{3}\, G$ to these two representations we get a correction of the right magnitude. A similar calculation for $\Sigma^- \to \Sigma^0 + e^- + \nu$ will lead to a smaller correction. The transition $\Sigma \to \Lambda$ is a "forbidden" one. It may arise because of $\mathbf{10} \oplus \overline{\mathbf{10}}$ or because of "magnetic moment" terms.

We assumed the hyperon decays to be mostly vectorial. The amplitude of the $\Lambda \to p$ transition is larger than that of beta decay by a factor of $\sqrt{\frac{3}{2}}$. Reducing the $8_{\frac{1}{2}}{}^-$ amplitude by a factor of 2.6 corresponds to a reduction of the branching ratio from 0.5% to the observed value of 0.075% [5]. The predicted value of the branching ratio of $\Xi^- \to \Lambda + e^- + \bar{\nu}$ is then 0.14%, and that of $\Sigma^- \to n + e^- + \bar{\nu}$ is 0.2%.

The amplitude for the $\Delta Q \neq \Delta S$ decay $\Sigma^+ \to n + e^+ + \nu$ is composed of $\mathbf{10}_{\frac{3}{2}}$ alone. Assuming the coupling constant to be $0.1 \sqrt{3}\, G$ we expect a branching ratio of 0.07% in a $V - A$ theory.

The unobserved $\Delta S = 2$ Ξ decays can be brought about by the $\overline{\mathbf{10}}_0$ current if it is coupled to the leptonic current. In a $V - A$ theory with a $0.1 \sqrt{3}\, G$ coupling constant we get the high branching ratio of 1.2%.

As a final remark one should notice that we consider coupling constants and not form factors. In other words we choose our form factors to be equal to their value at $q^2=0$. It may well happen that they vary strongly and thus affect the value of the decay amplitudes.

This lecture follows closely a series of papers on this subject [6] by Dr. Y. Ne'eman and the author.

References

1. Feynman, R. P., and M. Gell-Mann, Phys. Rev. **109,** 193 (1958).
2. e. g. Ely, R. P., et al., Phys. Rev. Lett. **8,** 132 (1962).
   Alexander, G., et al., Phys. Rev. Lett. **9,** 69 (1962).
   Humphrey, W. E., et al., Phys. Rev. Lett. **6,** 487 (1961).
3. Ne'eman, Y., Nucl. Phys. **26,** 222 (1961).
   Gell-Mann, M., report CTSL-20; Phys. Rev. **125,** 1067 (1962).
4. Goldberger, M. L., and S. B. Treiman, Phys. Rev. **110,** 1178 (1958).
   See also M. Gell-Mann in ref. 3.
5. Ely, R. P., et al., Phys. Rev. **131,** 868 (1963).
6. Horn, D., and Y. Ne'eman, Nuovo Cim. **29,** 760 (1963); **31,** 879 (1964) Israel AEC report IA-840.

Appendix: Reduction of $8 \otimes 8$

8⁻			
$Y=1$	$T=1/2$	$T_z=1/2$	$\sqrt{\frac{1}{6}}\{\sqrt{\frac{3}{2}}(\overline{\Xi^-}\Lambda-\overline{\Lambda}p)+\sqrt{\frac{1}{2}}(\overline{\Xi^-}\Sigma^0-\overline{\Sigma^0}p)$ $+(\overline{\Xi^0}\Sigma^+-\overline{\Sigma^-}n)\}$
		$-1/2$	$\sqrt{\frac{1}{6}}\{(\overline{\Xi^-}\Sigma^- -\overline{\Sigma^+}p)+\sqrt{\frac{3}{2}}(\overline{\Xi^0}\Lambda-\overline{\Lambda}n)$ $-\sqrt{\frac{1}{2}}(\overline{\Xi^0}\Sigma^0-\overline{\Sigma^0}n)\}$
0	1	1	$\sqrt{\frac{1}{6}}\{-(\overline{\Xi^-}\Xi^0-\overline{n}p)+\sqrt{2}(\overline{\Sigma^-}\Sigma^0-\overline{\Sigma^0}\Sigma^+)\}$
		0	$-\frac{1}{2}\sqrt{\frac{1}{3}}\{(\overline{\Xi^-}\Xi^- -\overline{p}p)-(\overline{\Xi^0}\Xi^0-\overline{n}n)$ $+2(\overline{\Sigma^-}\Sigma^- -\overline{\Sigma^+}\Sigma^+)\}$
		-1	$\sqrt{\frac{1}{6}}\{-(\overline{\Xi^0}\Xi^- -\overline{p}n)+\sqrt{2}(\overline{\Sigma^0}\Sigma^- -\overline{\Sigma^+}\Sigma^0)\}$
-1	$1/2$	$1/2$	$\sqrt{\frac{1}{6}}\{(\overline{\Sigma^-}\Xi^- -\overline{p}\Sigma^+)+\sqrt{\frac{3}{2}}(\overline{\Lambda}\Xi^0-\overline{n}\Lambda)$ $-\sqrt{\frac{1}{2}}(\overline{\Sigma^0}\Xi^0-\overline{n}\Sigma^0)\}$
		$-1/2$	$\sqrt{\frac{1}{6}}\{\sqrt{\frac{3}{2}}(\overline{\Lambda}\Xi^- -\overline{p}\Lambda)+\sqrt{\frac{1}{2}}(\overline{\Sigma^0}\Xi^- -p\Sigma^0)$ $+(\overline{\Sigma^+}\Xi^0-\overline{n}\Sigma^-)\}$
0	0	0	$-\frac{1}{2}\{(\overline{\Xi^-}\Xi^- -\overline{p}p)+(\overline{\Xi^0}\Xi^0-\overline{n}n)\}$
10⁻			
$Y=2$	$T=0$	$T_z=0$	$\sqrt{\frac{1}{2}}\{\overline{\Xi^-}n-\overline{\Xi^0}p\}$

1	$^1/_2$	$^1/_2$	$\sqrt{\frac{1}{6}}\{\sqrt{\frac{3}{2}}(\overline{\Xi^-}\Lambda-\bar{\Lambda}p)-\sqrt{\frac{1}{2}}(\overline{\Xi^-}\Sigma^0-\overline{\Sigma^0}p)-(\overline{\Xi^0}\Sigma^+-\overline{\Sigma^-}n)\}$
		$-^1/_2$	$\sqrt{\frac{1}{6}}\{\sqrt{\frac{3}{2}}(\overline{\Xi^0}\Lambda-\bar{\Lambda}n)+\sqrt{\frac{1}{2}}(\overline{\Xi^0}\Sigma^0-\overline{\Sigma^0}n)-(\overline{\Xi^0}\Sigma^--\overline{\Sigma^+}p)\}$
0	1	1	$\sqrt{\frac{1}{6}}\{(\overline{\Xi^-}\Xi^0-\bar{n}p)-\sqrt{\frac{3}{2}}(\overline{\Sigma^-}\Lambda-\bar{\Lambda}\Sigma^+)+\sqrt{\frac{1}{2}}(\overline{\Sigma^-}\Sigma^0-\overline{\Sigma^0}\Sigma^+)\}$
0	1	0	$\frac{1}{2}\sqrt{\frac{1}{3}}\{-(\overline{\Xi^-}\Xi^--\bar{p}p)+(\overline{\Xi^0}\Xi^0-\bar{n}n)+(\overline{\Sigma^-}\Sigma^--\overline{\Sigma^+}\Sigma^+)-\sqrt{3}(\bar{\Lambda}\Sigma^0-\overline{\Sigma^0}\Lambda)\}$
		-1	$\sqrt{\frac{1}{6}}\{(\overline{\Xi^0}\Xi^--\bar{p}n)+\sqrt{\frac{3}{2}}(\bar{\Lambda}\Sigma^--\overline{\Sigma^+}\Lambda)+\sqrt{\frac{1}{2}}(\overline{\Sigma^0}\Sigma^--\overline{\Sigma^+}\Sigma^0)\}$
-1	$^3/_2$	$^3/_2$	$\sqrt{\frac{1}{2}}(\overline{\Sigma^-}\Xi^0-\bar{n}\Sigma^+)$
		$^1/_2$	$\sqrt{\frac{1}{6}}\{(\overline{\Sigma^-}\Xi^--\bar{p}\Sigma^+)+\sqrt{2}(\overline{\Sigma^0}\Xi^0-\bar{n}\Sigma^0)\}$
		$-^1/_2$	$\sqrt{\frac{1}{6}}\{\sqrt{2}(\overline{\Sigma^0}\Xi^--\bar{p}\Sigma^0)-(\overline{\Sigma^+}\Xi^0-\bar{n}\Sigma^-)\}$
		$-^3/_2$	$\sqrt{\frac{1}{2}}(\overline{\Sigma^+}\Xi^--\bar{p}\Sigma^-)$

$\overline{\mathbf{10}}$ can be derived from **10** by changing every $\bar{A}B$ of **10** into $\bar{B}A$ thus reversing the sign of Y and T_z.

8^+			
$Y=1$	$T={}^1/_2$	$T_z={}^1/_2$	$\sqrt{\frac{3}{10}}\{-\sqrt{\frac{1}{6}}(\overline{\Xi^-}\Lambda+\bar{\Lambda}p)+\sqrt{\frac{1}{2}}(\overline{\Xi^-}\Sigma^0+\overline{\Sigma^0}p)+(\overline{\Xi^0}\Sigma^++\overline{\Sigma^-}n)\}$
		$-^1/_2$	$\sqrt{\frac{3}{10}}\{(\overline{\Xi^-}\Sigma^-+\overline{\Sigma^+}p)-\sqrt{\frac{1}{6}}(\overline{\Xi^0}\Lambda+\bar{\Lambda}n)-\sqrt{\frac{1}{2}}(\overline{\Xi^0}\Sigma^0+\overline{\Sigma^0}n)\}$
0	1	1	$\sqrt{\frac{1}{5}}\{\sqrt{\frac{3}{2}}(\overline{\Xi^-}\Xi^0+\bar{n}p)+(\overline{\Sigma^-}\Lambda+\bar{\Lambda}\Sigma^+)\}$
		0	$\sqrt{\frac{3}{20}}\{(\overline{\Xi^-}\Xi^-+\bar{p}p)-(\overline{\Xi^0}\Xi^0+\bar{n}n)+\frac{2}{\sqrt{3}}(\bar{\Lambda}\Sigma^0+\overline{\Sigma^0}\Lambda)\}$
		-1	$\sqrt{\frac{1}{5}}\{\sqrt{\frac{3}{2}}(\overline{\Xi^0}\Xi^-+\bar{p}n)+(\bar{\Lambda}\Sigma^-+\overline{\Sigma^+}\Lambda)\}$

-1	$^1/_2$	$^1/_2$	$\sqrt{\frac{3}{10}}\{(\bar{\Sigma}^-\Xi^- + \bar{p}\Sigma^+) - \sqrt{\frac{1}{6}}(\bar{\Lambda}\Xi^0 + \bar{n}\Lambda) - \sqrt{\frac{1}{2}}(\bar{\Sigma}^0\Xi^0 + \bar{n}\Sigma^0)\}$
		$-^1/_2$	$\sqrt{\frac{3}{10}}\{(\bar{\Sigma}^+\Xi^0 + \bar{n}\Sigma^-) - \sqrt{\frac{1}{6}}(\bar{\Lambda}\Xi^- + \bar{p}\Lambda) + \sqrt{\frac{1}{2}}(\bar{\Sigma}^0\Xi^- + \bar{p}\Sigma^0)\}$
0	0	0	$-\sqrt{\frac{1}{20}}\{(\bar{\Xi}^-\Xi^- + \bar{p}p) + (\bar{\Xi}^0\Xi^0 + \bar{n}n) - 2(\bar{\Sigma}^-\Sigma^- + \bar{\Sigma}^+\Sigma^+) + 2(\bar{\Lambda}\Lambda - \bar{\Sigma}^0\Sigma^0)\}$
27^+			
$Y=2$	$T=1$	$T_z=1$	$\bar{\Xi}^- p$
		0	$\sqrt{\frac{1}{2}}(\bar{\Xi}^- n + \bar{\Xi}^0 p)$
		-1	$\bar{\Xi}^0 n$
1	$\frac{3}{2}$	$\frac{3}{2}$	$\sqrt{\frac{1}{2}}(\bar{\Xi}^-\Sigma^+ + \bar{\Sigma}^- p)$
		$\frac{1}{2}$	$\sqrt{\frac{1}{6}}\{-(\bar{\Xi}^0\Sigma^+ + \bar{\Sigma}^- n) + \sqrt{2}(\bar{\Xi}^-\Sigma^0 + \bar{\Sigma}^0 p)\}$
		$-\frac{1}{2}$	$\sqrt{\frac{1}{6}}\{(\bar{\Xi}^-\Sigma^- + \bar{\Sigma}^+ p) + \sqrt{2}(\bar{\Xi}^0\Sigma^0 + \bar{\Sigma} n^0)\}$
		$-\frac{3}{2}$	$\sqrt{\frac{1}{2}}(\bar{\Xi}^0\Sigma^- + \bar{\Sigma}^+ n)$
1	$\frac{1}{2}$	$\frac{1}{2}$	$\sqrt{\frac{1}{30}}\{3\sqrt{\frac{3}{2}}(\bar{\Xi}^-\Lambda + \bar{\Lambda}p) + \sqrt{\frac{1}{2}}(\bar{\Xi}^-\Sigma^0 + \bar{\Sigma}^0 p) + (\bar{\Xi}^0\Sigma^+ + \bar{\Sigma}^- n)\}$
		$-\frac{1}{2}$	$\sqrt{\frac{1}{30}}\{(\bar{\Xi}^-\Sigma^- + \bar{\Sigma}^+ p) + 3\sqrt{\frac{3}{2}}(\bar{\Xi}^0\Lambda + \bar{\Lambda}n) - \sqrt{\frac{1}{2}}(\bar{\Xi}^0\Sigma^0 + \bar{\Sigma}^0 n)\}$
0	2	2	$\bar{\Sigma}^-\Sigma^+$
		1	$\sqrt{\frac{1}{2}}(\bar{\Sigma}^-\Sigma^0 + \bar{\Sigma}^0\Sigma^+)$
		0	$\sqrt{\frac{1}{6}}\{(\bar{\Sigma}^-\Sigma^- + \bar{\Sigma}^+\Sigma^+) - 2\bar{\Sigma}^0\Sigma^0\}$
		-1	$\sqrt{\frac{1}{2}}(\bar{\Sigma}^0\Sigma^- + \bar{\Sigma}^+\Sigma^0)$
		-2	$\bar{\Sigma}^+\Sigma^-$
0	1	1	$\sqrt{\frac{1}{5}}\{(\bar{\Xi}^-\Xi^0 + \bar{n}p) - \sqrt{\frac{3}{2}}(\bar{\Sigma}^-\Lambda + \bar{\Lambda}\Sigma^+)\}$
		0	$\sqrt{\frac{1}{10}}\{(\bar{\Xi}^-\Xi^- + \bar{p}p) - (\bar{\Xi}^0\Xi^0 + \bar{n}n) - \sqrt{3}(\bar{\Lambda}\Sigma^0 + \bar{\Sigma}^0\Lambda)$
		-1	$\sqrt{\frac{1}{5}}\{(\bar{\Xi}^0\Xi^- + \bar{p}n) - \sqrt{\frac{3}{2}}(\bar{\Lambda}\Sigma^- + \bar{\Sigma}^+\Lambda)\}$

0	0	0	$\frac{1}{2}\sqrt{\frac{1}{30}}\{3(\overline{\Xi^-}\Xi^- + \bar{p}p) + 3(\overline{\Xi^0}\Xi^0 + \bar{n}n)$ $-(\overline{\Sigma^-}\Sigma^- + \overline{\Sigma^+}\Sigma^+) - 9(\bar{\Lambda}\Lambda) - \overline{\Sigma^0}\Sigma^0\}$
-1	$\frac{3}{2}$	$\frac{3}{2}$	$\sqrt{\frac{1}{2}}(\overline{\Sigma^-}\Xi^0 + \bar{n}\Sigma^+)$
		$\frac{1}{2}$	$\sqrt{\frac{1}{6}}\{(\overline{\Sigma^-}\Xi^- + \bar{p}\Sigma^+) + \sqrt{2}(\overline{\Sigma^0}\Xi^0 + \bar{n}\Sigma^0)\}$
		$-\frac{1}{2}$	$\sqrt{\frac{1}{6}}\{\sqrt{2}(\overline{\Sigma^0}\Xi^- + \bar{p}\Sigma^-) - (\overline{\Sigma^+}\Xi^0 + \bar{n}\Sigma^-)\}$
		$-\frac{3}{2}$	$\sqrt{\frac{1}{2}}(\overline{\Sigma^+}\Xi^- + \bar{p}\Sigma^-)$
-1	$\frac{1}{2}$	$\frac{1}{2}$	$\sqrt{\frac{1}{30}}\{(\overline{\Sigma^-}\Xi^- + \bar{p}\Sigma^+) + 3\sqrt{\frac{3}{2}}(\bar{\Lambda}\Xi^0 + \bar{n}\Lambda)$ $-\sqrt{\frac{1}{2}}(\overline{\Sigma^0}\Xi^0 + \bar{n}\Sigma^0)\}$
		$-\frac{1}{2}$	$\sqrt{\frac{1}{30}}\{3\sqrt{\frac{3}{2}}(\bar{\Lambda}\Xi^- + \bar{p}\Lambda) + \sqrt{\frac{1}{2}}(\overline{\Sigma^0}\Xi^- + \bar{p}\Sigma^0)$ $+(\overline{\Sigma^+}\Xi^0 + \bar{n}\Sigma^-)\}$
-2	1	1	$\bar{n}\Xi^0$
		0	$\sqrt{\frac{1}{2}}(\bar{n}\Xi^- + \bar{p}\Xi^0)$
		-1	$\bar{p}\Xi^-$
1^+			
$Y=0$	$T=0$	$T_z=0$	$\frac{1}{2}\sqrt{\frac{1}{2}}\{\overline{\Xi^-}\Xi^- + \bar{p}p + \overline{\Xi^0}\Xi^0 + \bar{n}n$ $+\overline{\Sigma^-}\Sigma^- + \overline{\Sigma^+}\Sigma^+ + \overline{\Sigma^0}\Sigma^0 + \bar{\Lambda}\Lambda\}$

This table is taken from H. Goldberg: The Crossing Matrix in the Gell-Mann-Ne'eman Model, Israel AEC report IA-834, 1963.

Leptonic Decay of Hadrons

Part II: Cabibbo's Theory

By

D. Horn

Departement of Physics, Tel Aviv University, Tel Aviv, Israel

In this theory one considers only the currents which transform according to the octet representation. Thus only currents of the form $|\Delta \mathbf{T}| = 1$ and $|\Delta \mathbf{T}| = \frac{1}{2}$ may occur. As usually one assumes the vector currents to be CVC's. Then we get two types of currents

$$j_\mu^{(0)} + g_\mu^{(0)} \text{ and } j_\mu^{(1)} + g_\mu^{(1)}$$

where the brackets denote the ΔS-value. j_μ and g_μ are the vector and the axial vector parts respectively.

Cabibbo [1] now assumes that the total current can be written in the form

$$J_\mu = \cos\theta\,(j_\mu^{(0)} + g_\mu^{(0)}) + \sin\theta\,(j_\mu^{(1)} + g_\mu^{(1)}). \tag{1}$$

This means that J_μ has a unit length in the 1—0 plane and one should find the angle it forms with the axes. One can compare this with the theory presented before where we assumed that both $j_\mu^{(1)}$ and $j_\mu^{(0)}$ are coupled initially with the same strength to the lepton currents and attributed the reduction afterwards to the break down of the symmetry. Here the currents remain conserved and the symmetry is not broken.

One can obtain the value of θ from the ratios of

$$\frac{K^+ \to \mu^+ + \nu}{\pi^+ \to \mu^+ + \nu} \quad \text{and} \quad \frac{K^+ \to \pi^0 + e^+ + \nu}{\pi^+ \to \pi^0 + e^+ + \nu}$$

and it turns out to be

$$\theta = 0.26 \tag{2}$$

in both cases.

The general matrix element for the transition $A \to B$ can be written in the form

$$\langle A \,|\, j_\mu^{(i)} + g_\mu^{(i)} \,|\, B \rangle = i\, f_{ABi}\, O_\mu + d_{ABi}\, E_\mu \tag{3}$$

where f and d are the coefficients defined by Gell-Mann [2] for the commutators and anticommutators of the λ matrices. The O and E are 4-dimensional vectors of the form

$$O_\mu,\ E_\mu = F^{0,E}\gamma_\mu + H^{0,E}\gamma_\mu\gamma_5.$$

Here one considers only allowed transitions and one chooses the form factors to be equal to their value at $q^2=0$.

From CVC arguments the $j_\mu^{(i)}$ must belong to the 8⁻representation and this means that $F^0=1$, $F^E=0$. The $g_\mu^{(i)}$ may belong both to the 8^- and 8^+ representations. From the value of G_A of beta decay one deduces that $H^0+H^E=1.25$. Furthermore choosing the branching ratio for the $\Sigma^- \to \Lambda + e^- + \bar{\nu}$ decay to be 0.9×10^{-4} one arrives at $H^E=\pm 0.95$. The negative solution is discarded because it produces a large branching ratio for $\Sigma^- \to n + e^- + \bar{\nu}$. Thus one arrives at the following values:

$$F^0=1 \qquad F^E=0 \qquad H^0=0.30 \qquad H^E=0.95. \tag{4}$$

The following table gives then the branching ratios as presented in CABIBBO's paper:

Decay	branching ratio $\times 10^{-3}$	type of interaction
$\Lambda \to p + e^- + \bar{\nu}$	0.75	$V-0.72\,A$
$\Sigma^- \to n + e^- + \bar{\nu}$	1.9	$V+0.65\,A$
$\Xi^- \to \Lambda + e^- + \bar{\nu}$	0.35	$V+0.02\,A$
$\Xi^- \to \Sigma^0 + e^- + \bar{\nu}$	0.07	$V-1.25\,A$
$\Xi^0 \to \Sigma^+ + e^- + \bar{\nu}$	0.26	$V-1.25\,A$

One should notice that the vector coupling constant for β decay is now $G\cdot\cos\theta$ instead of G. This gives a correction of 6.6% which may explain the discrepancy between 0^{14} and μ lifetimes. This correction is by 2% too large.

References

1. CABIBBO, N., Phys. Rev. Letters **10**, 531 (1963).
2. GELL-MANN, M., report CTSL – 20; Phys. Rev. **125**, 1067 (1962).

Possible U- and V-Spin Selection Rules in Weak Interactions

By

D. Horn

Department of Physics, Tel Aviv University, Tel Aviv, Israel

As shown in the lectures by Profs. Joos and Mayer one can define three sorts of isotopic spins in SU(3). The first one is the usual isotopic spin **T**, the other ones are designated by **U** and **V**. The classification of the particles according to the various unitary spins is given in Table 1.

Table 1. Classification of Baryons

T	(Λ)	(Ξ^0, Ξ^-)	(p, n)	$(\Sigma^+, \Sigma^0, \Sigma^-)$
U	$\left(\frac{1}{2}\Lambda + \frac{\sqrt{3}}{2}\Sigma^0\right)$	(p, Σ^+)	(Σ^-, Ξ^-)	$\left(n, \frac{\sqrt{3}}{2}\Lambda - \frac{1}{2}\Sigma^0, \Xi^0\right)$
V	$\left(\frac{1}{2}\Lambda - \frac{\sqrt{3}}{2}\Sigma^0\right)$	(Σ^+, Ξ^0)	(n, Σ^-)	$\left(p, \frac{\sqrt{3}}{2}\Lambda + \frac{1}{2}\Sigma^0, \Xi^-\right)$

The meson multiplets can be formed in an analogous manner. According to the choice of the explicit representation there may appear a minus sign in front of several particles in this table. Such a sign plays no significant role in our theory.

There are two known selection rules of **T**-spin, namely $|\Delta \mathbf{T}| = \frac{1}{2}$ for the non-leptonic decays of hyperons and $|\Delta \mathbf{T}| = 1$ for the strangeness conserving leptonic decays. Let us see if similar rules using **U** and **V** spin may hold.

The non-leptonic decays of hyperons obey $\Delta U_z = 1$ and $\Delta V_z = \frac{1}{2}$. There are thus four possible rules, namely $|\Delta \mathbf{U}| = 1$ or 2 and $|\Delta \mathbf{V}| = \frac{1}{2}$ or $\frac{3}{2}$. The only one which gives a definite result is

$$|\Delta \mathbf{U}| = 2$$

from which it follows that

$$\langle \Lambda \pi^- | \Xi^- \rangle = -\sqrt{\frac{3}{2}} \langle n \pi^- | \Sigma^- \rangle .$$

This relation stands both for S and P waves and present experimental data cannot exclude it.

The situation is better in the leptonic decays. Let us consider first the $\Delta S = \Delta Q$ decays. All transitions obey $\Delta U_Z = \frac{1}{2}$. A rule of the form $|\Delta \mathbf{U}| = \frac{1}{2}$ predicts

$$\sqrt{3}\,(\bar{\Lambda}\,\Xi^-) - \sqrt{2}\,(\bar{n}\,\Sigma^-) = (\bar{\Sigma}^0\,\Xi^-)$$

whereas $|\Delta \mathbf{U}| = \frac{3}{2}$ would give

$$(\bar{n}\,\Sigma^-) = -\frac{\sqrt{6}}{3}(\bar{\Lambda}\,\Xi^-) = \sqrt{2}\,(\bar{\Sigma}^0\,\Xi^-)$$

$$(\bar{p}\,\Lambda) = -\frac{\sqrt{6}}{2}(\bar{\Sigma}^+\,\Xi^0).$$

If one combines the last rule together with $|\Delta \underset{\sim}{T}| = \frac{1}{2}$ then a connection is obtained between all amplitudes:

$$(\bar{p}\,\Lambda) = -\frac{\sqrt{6}}{2}(\bar{n}\,\Sigma^-) = (\bar{\Lambda}\,\Xi^-) = -\sqrt{3}\,(\bar{\Sigma}^0\,\Xi^-) = -\frac{\sqrt{6}}{2}(\bar{\Sigma}^+\,\Xi^0).$$

Assuming a V-A theory one can deduce from the known branching ratio of the Λ

$$R\,(\bar{p}\,\Lambda) = 0.75 \cdot 10^{-3}$$

the following branching ratios

$$R\,(\bar{n}\,\Sigma^-) = 2\,^0/_{00}\quad R(\bar{\Lambda}\,\Xi^-) = 1.4\,^0/_{00}\quad R\,(\bar{\Sigma}^0\,\Xi^-) = 0.05\,^0/_{00}$$

$$R\,(\bar{\Sigma}^+\,\Xi^0) = 0.1\,^0/_{00}.$$

The analogous approach for $\mathbf{V}$ does not give any result because the particles in one multiplet are characterized by the same value of $Q - S$ and therefore $\Delta Q = \Delta S$ transitions will leave us in the same multiplet except when Σ^0 or Λ are considered. If we demand that in the case of Σ^0 and Λ the transition occurs also only within the same multiplet then we obtain:

$$(\bar{\Lambda}\,\Xi^-) = \sqrt{3}\,(\bar{\Sigma}^0\,\Xi^-) = (\bar{p}\,\Lambda)$$

which is equivalent to some of the results obtained before.

The strangeness conserving decays obey $\Delta U_Z = -\frac{1}{2}$, $\Delta V_Z = \frac{1}{2}$. There are once again four rules possible: $|\Delta \mathbf{U}| = \frac{1}{2}$ or $\frac{3}{2}$ and $|\Delta \mathbf{V}| = \frac{1}{2}$ or $\frac{3}{2}$. The possibilities of $\frac{3}{2}$ are in contradiction with results of the CVC approach. Thus we are left with $|\Delta \mathbf{U}| = \frac{1}{2}$ which predicts

$$\sqrt{2}\,(\bar{\Xi}^0\,\Xi^-) = \sqrt{3}\,(\bar{\Lambda}\,\Sigma^-) - (\bar{\Sigma}^0\,\Sigma^-)$$

or $|\Delta \mathbf{V}| = \frac{1}{2}$ which predicts

$$\sqrt{2}\,(\bar{p}\,n) = \sqrt{3}\,(\bar{\Lambda}\,\Sigma^-) + (\bar{\Sigma}^0\,\Sigma^-).$$

It should be noted that this approach cannot give a connection between the $\Delta S = 0$ and $\Delta S = 1$ decays. The experimental verification of most of the results is still far from being possible.

This lecture is an abbreviated form of a paper on this subject which was submitted to Nuovo Cimento.

Symmetries of Leptons

By

D. Horn

Department of Physics, Tel-Aviv University, Tel-Aviv, Israel

1. Properties of the Leptons

There are three particles denoted by the name lepton — the electron (e), the muon (μ), and the neutrino (ν). All of them are spin $\frac{1}{2}$ particles and their common property is the leptonic number. A leptonic number $+1$ can be assigned to e^-, μ^-, ν, and -1 to their antiparticles e^+, μ^+, $\bar{\nu}$. The conservation of this number can be shown by the following examples of weak interactions:

$n \to p + e^- + \bar{\nu}$; $\pi^- \to \mu^- + \bar{\nu}$; $\pi^+ \to e^+ + \nu$; $K^+ \to \mu^+ + \nu$; $\mu^\pm \to e^\pm + \nu + \bar{\nu}$.

When several particles are characterized by a certain quantity common to them it is interesting to try to introduce symmetry arguments to classify them, the more so when they have several features that are practically identical.

The muon and the electron are two such particles. Apart from their different masses ($m_\mu = 206.77\, m_e$; $m_e = 0.511$ MeV) they seem to be practically identical. This identity is expressed in their electromagnetic interactions as so well established in the g-2 experiment in CERN [1]. It was shown there that the giromagnetic ratios of these two particles are equal, to an accuracy of 10^{-5}. The muon and the electron also show similar behaviour in weak interactions. This fact is included in the concept of "universality" of the weak interactions. The explicit form of the interaction Hamiltonian can be written, according to the theory of FEYNMAN and GELL-MANN [2] as a multiplication of a current by its hermitian conjugate

$$H_I = \frac{G}{\sqrt{2}} J_\mu J_\mu^*. \tag{1}$$

The current contains the interacting particles in the form

$$J_\mu = \overline{\Psi}_A \gamma_\mu (1 + \gamma_5) \Psi_B \tag{2}$$

which we will denote simply by $(\bar{A}B)$. This way of writing corresponds to the V-A theory. The leptons appear in the pairs

$$J = (\bar{e}\,\nu) + (\bar{\mu}\,\nu) \tag{3}$$

and the interaction Hamiltonian will then contain four terms

$$(\bar{e}\,\nu)\,(\bar{\nu}\,e)+(\bar{\mu}\,\nu)\,(\bar{\nu}\,e)+(\bar{e}\,\nu)\,(\bar{\nu}\,\mu)+(\bar{\mu}\,\nu)\,(\bar{\nu}\,\mu). \tag{4}$$

The neutrino seems to have zero mass (for experimental limits see the lecture by Prof. FAISSNER). In the interaction (4) the neutrino always appears as $\frac{1}{2}(1+\gamma_5)\,\Psi_\nu$, which corresponds, for a massless particle, to choosing only that part of it which has negative helicity. Thus we arrive at the two component theory of the neutrino which states that the neutrino has only negative helicity (only two out of the four spinor components), and its antiparticle has pocitive helicity.

2. The Muonic Number

The form (3) shows explicitly the similarity between the muon and the electron. It looks as if the muon is some excited state of the electron. Therefore it was only natural to look for an electromagnetic transition by which the muon decays into an electron. Some other processes were also expected. The interesting ones are

$$\mu\to 3\,e,\qquad \mu+N\to e+N,\qquad \mu\to e+\gamma. \tag{5}$$

None of these processes has been found experimentally. It is not easy to say what amplitude these processes should have because the calculations involve arbitrary cut-offs. Nevertheless it is a striking fact that these processes do not occur. This is already a good reason to suspect that there is some difference between the two charged leptons. The difference implies a new quantum number. Let us call this new quantum number the muonic number and assign the value of $+1$ to μ^- and -1 to e^-, and the opposite values to their anti-particles.

As the current (3) is also coupled to non-leptonic currents it should have a total muonic number zero in order for this number to be conserved. It is then impossible to have the same ν coupled to e and μ and one must introduce two neutrinos, namely, ν_μ and ν_e, with muonic numbers $+1$ and -1 respectively. The weak current will then have the form

$$J=(\bar{e}\,\nu_e)+(\bar{\mu}\,\nu_\mu). \tag{6}$$

As is now well known these two neutrinos have been found experimentally in the Brookhaven experiment [3].

Let us describe briefly a simple symmetry scheme which takes the above-mentioned quantum numbers into account [4]. We designate a wave function of two leptons by $\Psi(a_1,\,a_2)\equiv\begin{pmatrix}a_1\\a_2\end{pmatrix}$. Then the free Lagrangian of the leptons can be written in the form

$$\mathfrak{L}_f=-\overline{\Psi}(a_1\,a_2)\,A\,\Psi(a_1\,a_2)-\overline{\Psi}(b_1\,b_2)\,B\,\Psi(b_1\,b_2). \tag{7}$$

If we want $\mathfrak{L}_f$ to be invariant under rotation through an angle φ in the two dimensional space spanned by $a_1\,a_2$ (or $b_1\,b_2$) then we get a theory invariant under a one-parameter group. This limits our matrices A and B to the form

$$A = \alpha_1 + \alpha_2 \sigma_2 \qquad B = \beta_1 + \beta_2 \sigma_2. \tag{8}$$

In order to get physical meaning we choose

$$\alpha_1 = \gamma_\mu \partial_\mu + \tfrac{1}{2}(m_\mu + m_e) \qquad \alpha_2 = \tfrac{1}{2}(m_\mu - m_e). \tag{9}$$

Then the a's will stand for the charged leptons. Similarly the b's will denote the neutrinos. After diagonalizing the Lagrangian we get it expressed in terms of the observed particles which are

$$\begin{aligned} \mu &= \frac{1}{\sqrt{2}}(a_1 - i\, a_2) & e &= \frac{-i}{\sqrt{2}}(a_1 + i\, a_2) \\ \nu_\mu &= \frac{1}{\sqrt{2}}(b_1 - i\, b_2) & \nu_e &= \frac{-i}{\sqrt{2}}(b_1 + b\, i_2). \end{aligned} \tag{10}$$

The rotation through the angle φ now corresponds to the phase transformation

$$\begin{aligned} \mu &\to \mu \exp(i\varphi) & \nu_\mu &\to \nu_\mu \exp(i\varphi) \\ e &\to e \exp(-i\varphi) & \nu_e &\to \nu_e \exp(-i\varphi) \end{aligned} \tag{11}$$

which can be written in the form $\mathfrak{l} \to \mathfrak{l} \exp(i\, M \varphi)$. M is the muonic quantum number which is introduced by the gauge transformation that we have chosen. The interaction can of course be written in an invariant form too. The known current is simply

$$J_\mu = \overline{\Psi}(a_1\, a_2)\, \gamma_\mu (1 + \gamma_5)\, \Psi(b_1\, b_2). \tag{12}$$

3. Different Muonic Numbers and Lepton Schemes

Different symmetry requirements can lead to different quantum numbers. In the preceding section we saw how a continuous rotation symmetry led to the additive muonic number. If one demands invariance under reflection instead of rotation then one arrives at a multiplicative quantum number. CABIBBO and GATTO [5] prposed such a theory. In their theory the Lagrangian (7) is invariant under the discrete transformation

$$a_1 \leftrightarrow i\, a_2 \qquad b_1 \leftrightarrow i\, b_2. \tag{13}$$

After diagonalization one arrives at μ, ν_μ and e, ν_e which are eigenstates of σ_3, and they attribute to them the values $+i$ and $-i$ respectively. All other particles in nature will then have the multiplicative muonic number $+1$.

A different approach was suggested by TZOU [6]. He arrives at different multiplicative muonic numbers by requiring

$$a_2(m) = \gamma_5\, a_1(-m). \tag{14}$$

The resulting suggestions are compared in Table 1.

Table 1. Different Muonic Numbers

Particles	Addidative M	C & G [5]	T [6]
$\mu^- \nu_\mu e^+ \bar{\nu}_e$	$+1$	$+i$	$+1$
$\mu^+ \bar{\nu}_\mu e^- \nu_e$	-1	$-i$	-1
the rest	0	1	± 1

All classifications forbid the processes (5). Nevertheless there are other processes in which we get different predictions from the different symmetry schemes. For instance, processes forbidden by the additive number and allowed by the others are:

$$e^+ + \mu^- \leftrightarrow e^- + \mu^+ \quad \mu^+ + \mu^+ \leftrightarrow e^+ + e^+ \quad \mu^+ + \bar{\nu}_\mu \leftrightarrow e^+ + \bar{\nu}_e. \tag{15}$$

A process allowed only by Tzou is

$$\gamma \to \mu^+ + e^- \tag{16}$$

and a process allowed only by Cabibbo and Gatto is

$$\nu_\mu + N \to \nu_e + e^- + \mu^+ + N. \tag{17}$$

The last one is an interesting possibility which can perhaps be checked by neutrino experiments.

Let us now turn again to the additive muonic number. As already explained, the particles are defined by their lepton number. This classification is given in Table 2a.

Table 2. Classification of Leptons

Particle	L	M	Q	H
μ^-	1	1	-1	
ν_μ	1	1	0	-1
e^-	1	-1	-1	
ν_e	1	-1	0	-1

a

Particle	L	M	Q	H
μ^+	-1	-1	1	
$\bar{\nu}_\mu$	-1	-1	0	1
e^-	1	-1	-1	
ν_e	1	-1	0	-1

b

However, one could define the "particle" by choosing, for example, $M = -1$ as a definition of it. Such a classification is given in Table 2b. It is interesting to notice that $L = -Q$ for the charged particles and $L = -H$ (H = helicity) for the neutrinos. Thus L is a spurious quantum number because the particles are unambiguously defined even without it. If we drop L we can see from Table 2b that we are left with three "leptons" only, namely, μ^+, e^- and a neutrino of which $\bar{\nu}_\mu$ is the $H = +1$

part and ν_e is the $H=-1$ part. This 3-lepton scheme is completely equivalent to the 4-lepton scheme except that the notation has to be changed. Here the current will be given by

$$J = \overline{\Psi}_{e-} \gamma_\mu (1+\gamma_5) \Psi_\nu + \overline{\Psi}_\nu \gamma_\mu (1-\gamma_5) \Psi_{\mu+} \qquad (18)$$

which no longer has a "universal" form.

4. Higher Symmetries

Several higher symmetries have already been suggested for the leptons. We will discuss here three of them. All are characterized by the fact that they are valid only in an approximate form. It may be hoped that their predictions can be checked in high energy experiments where the mass differences of the leptons should play only a small role.

FEINBERG and GÜRSEY [7] group the four leptons in a form similar to (7), namely, they define $\Psi_\nu = \begin{pmatrix} \nu_e \\ \nu_\mu \end{pmatrix}$, $\Psi_l = \begin{pmatrix} e \\ \mu \end{pmatrix}$ and require them to form a spinorial representation of SU(2), invariant under the transformation

$$\Psi_\nu \to e^{\frac{i}{2} \mathbf{c} \cdot \boldsymbol{\alpha}} \Psi_\nu \qquad \Psi_l \to e^{\frac{i}{2} \mathbf{c} \cdot \boldsymbol{\alpha}} \Psi_l. \qquad (19)$$

The Lagrangian is invariant under this transformation as long as the mass difference between μ and e is neglected. When this mass difference is taken into account we are left with the theory described in section 2.

Under the above assumption one can derive relations between the amplitudes of several processes. Let us look at the following four reactions:

(a) $e^- + e^- \to e^- + e^-$
(b) $e^- + \mu^- \to e^- + \mu^-$
(c) $e^- + \mu^- \to \mu^- + e^-$
(d) $\mu^- + \mu^- \to \mu^- + \mu^-$.

Then the above theory predicts

$$\sigma_a = \sigma_d; \qquad \sqrt{\sigma_a} + \sqrt{\sigma_b} \geq \sqrt{\sigma_c} \geq \sqrt{\sigma_a} - \sqrt{\sigma_b}. \qquad (20)$$

Similar relations can be found among other processes.

Let us now discuss the 3-lepton scheme. NE'EMAN [8] suggested that the leptons form a vector representation of the three dimensional rotation group. But it turned out that in order to get the right form of the current (18) one has to incorporate into the generators of the group the helicity projection operators $a_L = \frac{1}{2}(1+\gamma_5)$ and $a_R = \frac{1}{2}(1-\gamma_5)$. The three generators are then

$$W_1 = \tfrac{1}{2} \begin{pmatrix} 0 & a_R & 0 \\ a_R & 0 & a_L \\ 0 & a_L & 0 \end{pmatrix} \quad W_2 = \tfrac{1}{2} \begin{pmatrix} 0 & i\,a_R & 0 \\ -i\,a_R & 0 & i\,a_L \\ 0 & -i\,a_L & 0 \end{pmatrix} \quad W_3 = \tfrac{1}{2} \begin{pmatrix} -a_R & 0 & 0 \\ 0 & a_R - a_L & 0 \\ 0 & 0 & a_L \end{pmatrix} \qquad (21)$$

and satisfy the usual commutation relations of generators of R (3). Invariance under the symmetry operations

$$\boldsymbol{\Psi} \to e^{i \boldsymbol{\alpha} \cdot \mathbf{W}} \boldsymbol{\Psi} \tag{22}$$

produces then the right currents, both the known charged ones and also an additional neutral one which seems not to exist in nature.

A different theory was proposed by GATTO [9]. In his theory the 3-leptons form representations of SU(3). The left handed ones are incorporated in the **3** representation while the right handed ones build the $\bar{\mathbf{3}}$. It should of course be noted that both NE'EMAN's and GATTO's theories are valid only in the limit where all masses of the leptons vanish.

A further discussion of GATTO's theory is given in the lecture of Prof. TOYODA. We would only like to mention that all interactions break the symmetry of SU(3) and thus, although SU(3) may serve as a classification, it is difficult to derive physical predictions from it.

References

1. CHARPAK, G., et al., Phys. Rev. Letters **6**, 128 (1961).
2. FEYNMAN, R. P., and M. GELL-MANN, Phys. Rev. **109**, 193 (1958).
3. DANBY, G., et al., Phys. Rev. Letters **9**, 36 (1962).
4. HORN, D., Phys. Letters **2**, 303 (1962).
5. CABIBBO, N., and R. GATTO, Phys. Rev. Letters **5**, 144 (1960); Nuovo Cimento **19**, 612 (1961).
6. TZOU, K. H., Cah. Phys. **128**, 171 (1961); **132**, 367 (1961).
7. FEINBERG, G., and F. GÜRSEY, Phys. Rev. **128**, 378 (1962).
8. NE'EMAN, Y., Nuovo Cimento **27**, 922 (1963).
9. GATTO, R., Nuovo Cimento **27**, 313 (1963).

A Formulation for the Two Neutrino Fields in the SU(3)

By

T. Toyoda*

Recently the rather old problem of neutrino has been put in the new light of theoretical approaches due to the experimental discovery of the two distinct neutrinos. Particularly, being stimulated by the success of the SU(3) symmetry in the baryon family, several authors [1], [2] have attempted to amalgamate the two distinct neutrino fields into somewhat unified schemes.

Just one year ago GATTO published a very interesting paper entitled "Classification of Lepton Currents" [1] in the Nuovo Cimento, and last fall MARSHAK, RYAN, RADHA, and RAMAN [2] proposed a new correspondence between the baryons and the leptons in their note to the Physical Review Letters and distributed the preprints of their full paper entitled "Universal Theory of Semi-Weak Interactions". Therefore, most of you might be familiar with these two theories.

However, since my talk will be closely related with these two papers, it would be useful to summarize very briefly some important points in their papers for our discussions. Let us start from GATTO's paper. —

Although there are several problems as to the applicability of SU(3) even to the baryon family (e. g., a mixing of the φ-meson state and the ϱ-meson [3] state and $\Delta S = -\Delta Q$ [4]), I would like to confine myself to the SU(3) framework in the present lecture.

§ 1. Gatto's Theory

In GATTO's formalism [1] the lepton family is expressed with a set of three Dirac spinors,

$$\psi = \begin{pmatrix} \mu \\ \nu \\ e \end{pmatrix}. \tag{1.1}$$

By making use of the helicity operator, we may split the ψ as

$$\psi = a\psi + \bar{a}\psi \equiv \psi^{(-)} + \psi^{(+)} \tag{1.2}$$

where

$$a = \tfrac{1}{2}(1 + \gamma_5) \text{ and } \bar{a} = \tfrac{1}{2}(1 - \gamma_5).$$

* On leave of absence from Rikkyo University, Tokyo, Japan.

Then GATTO assumed that the lepton currents be invariant under the following unitary transformation:

$$\psi \rightarrow \psi' = \exp\left[i \sum_{j=1}^{8} \varepsilon_j \frac{\Lambda_j}{2} \right] \psi \tag{1.3}$$

where ε_j's are eight independent real parameters, and Λ_j's are given as,

$$\Lambda_j = \Lambda_j^{(-)}\, a + \Lambda_j^{(+)}\, \bar{a}. \tag{1.4}$$

What are the $\Lambda_j^{(-)}$ and $\Lambda_j^{(+)}$? This is one of the essential part of GATTO's theory. Both $\Lambda_j^{(-)}$ and $\Lambda_j^{(+)}$ are elements of $\mathfrak{S}\mathfrak{U}(3)$ which is a Lie algebraassociated with SU(3). However their representations are different each other.

Since we are working in the three dimensional space, we may take two inequivalent representations of $\mathfrak{S}\mathfrak{U}(3)$, $D^3(1,0)$ and $D^3(0,1)$. Let us write those elements as

$$\lambda_j \in D^3(1,0) \text{ and } \tilde{\lambda}_j \in D^3(0,1) \tag{1.5}$$

Then GATTO claimed that

$$\Lambda_j^{(-)} = \lambda_j \tag{1.6}$$

$$\Lambda_j^{(+)} = \omega\, \lambda_j\, \omega^{-1} \text{ or } \omega\, \tilde{\lambda}_j\, \omega^{-1} \tag{1.7}$$

where ω was supposed to be a similarity transformation. Such a similarity transformation is, of course, too general to obtain the meaningful results. Therefore, GATTO imposed a unitary and unimodular condition to ω. This is nothing but the assumption that ω be an element of the SU(3) group.

Now the well-known theorem of the adjoint representation is useful.

Theorem: If $\omega \in$ SU(3) and $\lambda_j \in \mathfrak{S}\mathfrak{U}(3)$

$$(A d \omega)\, \lambda_j = \omega\, \lambda_j\, \omega^{-1} = W_{jk}\, \lambda_k$$

Proof:
$$\begin{aligned} (A d \omega)\,(m\, \lambda_j + n\, \lambda_k) &= m\,(A d \omega)\, \lambda_j + n\,(A d \omega)\, \lambda_k \\ \exp\,[i t\,(A d \omega)\, \lambda_j] &= \exp\,[i t\, \omega\, \lambda_j \omega^{-1}] \\ &= \omega \exp\,(i t\, \lambda_j)\, \omega^{-1}, \end{aligned} \tag{1.8}$$

where t is a real parameter. Since $\lambda_j \in \mathfrak{S}\mathfrak{U}(3)$, $\exp\,(i t \lambda_j) \in$ SU(3). Then

$$\omega \exp\,(i t\, \lambda_j)\, \omega^{-1} \in \mathrm{SU}(3)$$

and

$$(A d \omega)\, \lambda_j \in \mathfrak{S}\mathfrak{U}(3).$$

Therefore, if we restrict ω to a unitary unimodular 3×3 matrix, ambiguity of Eqs. (1.7) may be much reduced.

Furthermore GATTO postulated very stringent conditions on ω, that is, the invariance of charge operator and the reality condition.

Since a positive muon and a negative electron should share the same lepton number, the charge operator of lepton is defined as follows:

$$q=\begin{pmatrix}1 & & \\ & 0 & \\ & & -1\end{pmatrix}=\tfrac{1}{2}(\lambda_3+\sqrt{3}\,\lambda_8)=-\tfrac{1}{2}(\tilde{\lambda}_3+\sqrt{3}\,\tilde{\lambda}_8) \tag{1.9}$$

where λ_3, λ_8, $\tilde{\lambda}_3$ and $\tilde{\lambda}_8$ are diagonalized as usual. Since we are using the mixture of two representations, q can not be defined in general unless we restrict very much the form of ω. Namely, if we take two equivalent representations (I), ω should satisfy the condition

$$\omega q \omega^{-1}=q \tag{1.10}$$

and if we choose two inequivalent representations (II),

$$\omega q \omega^{-1}=-q. \tag{1.11}$$

Then the explicit forms of W_{jk} defined by (1.8) can be easily obtained. In addition the reality condition restrict them strongly and gives us the final results:

$$\begin{aligned}&\lambda_1'=\varrho\,\lambda_1,\ \lambda_2'=\varrho\,\lambda_2,\ \lambda_3'=\lambda_3,\ \lambda_4'=\varrho\sigma\,\lambda_4, && (1.12)\\ &\lambda_5'=\varrho\sigma\,\lambda_5,\ \lambda_6'=\sigma\,\lambda_6,\ \lambda_7'=\sigma\lambda_7,\ \lambda_8'=\lambda_8 && \text{for (I)}\end{aligned}$$

$$\left.\begin{aligned}&\lambda_1'=\varrho\lambda_6,\ \lambda_2'=\varrho\,\lambda_7,\ \lambda_3'=-\tfrac{1}{2}\,\lambda_3+\tfrac{\sqrt{3}}{2}\,\lambda_8,\ \lambda_4'=-\varrho\sigma\,\lambda_4\\ &\lambda_5'=-\,\varrho\sigma\,\lambda_5,\ \lambda_6'=\sigma\,\lambda_1,\ \lambda_7'=\sigma\,\lambda_2,\ \lambda_8'=\tfrac{\sqrt{3}}{2}\,\lambda_3+\tfrac{1}{2}\,\lambda_8\end{aligned}\right\}\text{for (II)} \tag{1.13}$$

where ϱ and σ may take $+1$ or -1 independently of each other.

Here I would like to make a remark on the charge operator. If we apply this formalism to the Sakata model and describe ψ as

$$\psi=\begin{pmatrix}\Lambda\\ n\\ p\end{pmatrix}$$

the charge operator may be defined as

$$q=\begin{pmatrix}0 & & \\ & 0 & \\ & & 1\end{pmatrix}=\tfrac{1}{2}\left(\tfrac{1}{\sqrt{3}}-\,\lambda_8\right)=\tfrac{1}{2}\left(\tfrac{1}{\sqrt{3}}+\tilde{\lambda}_8\right). \tag{1.14}$$

Since there is no similarity transformation such that $\omega\,\tilde{\lambda}_8\,\omega^{-1}=-\tilde{\lambda}_8$, it is not possible to identify $\Lambda^{(-)}$ to $D^3\,(1, 0)$ and $\Lambda^{(+)}$ to $D^3\,(0, 1)$. This point was discussed also by Gatto [4] in the connection of the parity conservation and the strangeness conservation.

Finally Gatto assigned the two distinct neutrinos to the different states of ν as:

$$a\,\nu=\nu_-=\underline{\nu}^{(e)} \tag{1.15}$$

$$a\,\nu^c=\nu_+{}^c=\underline{\nu}^{(\mu)} \tag{1.16}$$

where $\nu^{(e)}$ and $\nu^{(\mu)}$ are an electron-neutrino and a muon-neutrino respectively. In order to see the structure of $\underline{\nu}^{(e)}$ and $\underline{\nu}^{(\mu)}$ we had better write down them in the Majorana representation, on which we shall discuss later in details. For the coment we shall list them below.

$$\underline{\nu}^{(e)} = a\varphi_1 - ia\varphi_2 \tag{1.17}$$

$$\underline{\nu}^{(\mu)} = a\varphi_1 + ia\varphi_2. \tag{1.18}$$

Since φ_1 and φ_2 are two independent Majorana fields, $\underline{\nu}^{(e)}$ and $\underline{\nu}^{(\mu)}$ are certainly independent of each other.

§ 2. Marshak-Ryan-Radha-Raman's Theory

Before the discovery of the two distinct neutrinos GAMBA, MARSHAK, and OKUBO [5] proposed the symmetry (GMO symmetry) between Sakata's three fundamental baryons and three leptons:

$$\begin{matrix} p & n & \Lambda \\ \updownarrow & \updownarrow & \updownarrow \\ \nu & e & \mu \end{matrix} \tag{2.1}$$

On the other hand MARKOV proposed that the eight baryons are fundamental particles, and OIGLANE made the first attempt to set up a correspondence between the baryon and lepton octets [6]:

$$\begin{matrix} p & n & \Lambda & \Sigma^+ & \Sigma^0 & \Sigma^- & \Xi^0 & \Xi^- \\ \nu_e & e^- & \mu^- & \nu_\mu & \mu^+ & \bar{\nu}_\mu & e^+ & \bar{\nu}_e. \end{matrix} \tag{2.2}$$

Recently MARSHAK, RYAN, RADHA, and RAMAN proposed a different correspondence from the above, that is

$$\begin{matrix} p & n & \Lambda & \Sigma^+ & \Sigma^0 & \Sigma^- & \Xi^0 & \Xi^- \\ e^+ & \nu_e & \psi = \frac{1}{\sqrt{2}}(\nu_\mu - \bar{\nu}_\mu) & \mu^+ & \chi = \frac{1}{\sqrt{2}}(\nu_\mu + \bar{\nu}_\mu) & \mu^- & \mu_e & e^- \end{matrix} \tag{2. 3}$$

where ν_e and ν_μ are an electron-neutrino and a muon-neutrino (Dirac 4-component spinors), respectively.

Then several important postulates have been made:

1. the conserved vector current (c.v.c.) hypothesis,
2. the conservation of the "weak hypercharge" X which corresponds to the conventional (strong) hypercharge. The assignment of X of each lepton can be made by the new correspondence mentioned above.
3. four intermediate charged vector bosons, W_0^-, W_0^+, W_1^- and W_1^+. W_0 $(X=0)$, W_1 $(X=1)$.
4. the leptons undergo the following gauge transformation.

$$e \to e' = e^{i\alpha} e$$

$$\mu \to \mu' = e^{-i\alpha} \mu$$

$$\nu_e \to \nu_e' = e^{i\alpha} \nu_e$$

$$\chi \to \chi' = e^{-i\alpha\gamma_5} \chi$$

$$\psi \to \psi' = e^{i\alpha\gamma_5} \psi.$$

In order to construct the charged lepton currents they used the following method instead of the full utilization of SU(3):

$$\begin{pmatrix} \nu_e \\ e \end{pmatrix} \text{ and } \begin{pmatrix} -e^C \\ \nu_e^C \end{pmatrix} \text{ are isospinors,}$$

$$\begin{pmatrix} \mu^C \\ \chi \\ \mu \end{pmatrix} \text{ is an isovector.}$$

In this sense they called their model as the "septet model" instead of the "octet model", because ψ is dropped ad hoc. Then they obtained the following two kinds of charged lepton current:

$$J_0' = [\tfrac{1}{2}(\bar{e}0_+ \nu_e) - \tfrac{1}{2}(\nu_e^C 0_- e^C) + \tfrac{1}{2}(\bar{\mu}0_+\chi) - \tfrac{1}{2}(\bar{\chi}0_- \mu^C)] \tag{2.4}$$

$$J_1' = [\tfrac{1}{2}(\bar{e}0_-\chi) - \tfrac{1}{2}(\bar{\chi}0_+ e^C) + \tfrac{1}{2}(\bar{\mu}0_+ \nu_e^C) - \tfrac{1}{2}(\bar{\nu}_e 0_- \mu^C)] \tag{2.5}$$

where $0_\pm = \gamma_\mu (1 \pm \gamma_5)$. Since we are now interested in the lepton currents, the counter baryon currents are omitted.

The weak interaction Lagrangian is given as

$$\mathfrak{L}_{\text{int}} = ig(J_0 W_0 + J_1 W_1) + \text{h.c.} \tag{2.6}$$

where the counter baryon currents, of course, should be included. Although the Majorana neutrino and the Dirac neutrino are equivalent in the weak interaction as we shall discuss later, it should be kept in mind that in this theory the electron-neutrino is a Dirac spinor and the muon-neutrino appears as a Majorana field.

§ 3. Relation between the Majorana and the Two-component Spinor Fields

In 1959 TOUSCHEK [7] and other authors [7] have clearly shown the relation between the Majorana formalism and the two-component theory of neutrino. The still existing prejudice that the Majorana neutrino can not be compatible with the experimental fact with respect to the double β-decay had in fact to be abandoned after the discovery of the parity violation in the weak interactions, as was pointed out by TOUSCHEK several years ago.

Now, it may be useful to summarize some important points given by Touschek focusing on the relation between the two formalism. Any four-component Dirac spinor field can be expressed by two hermitian spinor fields φ_1 and φ_2

$$\psi = \varphi_1 - i\varphi_2 \tag{3.1}$$

where

$$\varphi_1^+ = \varphi_1, \ \varphi_2^+ = \varphi_2 \tag{3.2}$$

and where φ_1 and φ_2 satisfy the commutation relation

$$\{\varphi_i(x), \varphi_k(y)\} = \tfrac{1}{2}\delta_{ik}\,\delta(x-y). \tag{3.3}$$

The charge conjugate of ψ is then obtained by replacing φ_2 by $-\varphi_2$

$$\psi^C = \varphi_1 + i\,\varphi_2. \tag{3.4}$$

Throughout the present lecture the Majorana gauge will be used. That is, $\gamma_1, \gamma_2, \gamma_3$ are real symmetric and γ_4 is pure imaginary antisymmetric. Since γ_5 is defined as usual $\gamma_5 = \gamma_1\gamma_2\gamma_3\gamma_4$, γ_5 is also pure imaginary antisymmetric and satisfies $\gamma_5{}^2 = 1$.

For the comparison with conventional expression we shall write down the well known relations:

$$\psi^C = C\,\bar{\psi}^T \tag{3.5}$$

$$C^+ C = C\,C^+ = 1 \tag{3.6}$$

$$C^T = -C \tag{3.7}$$

$$-\gamma_\mu^T = C^{-1}\gamma_\mu\,C,\;\; \gamma_5^T = C^{-1}\gamma_5\,C. \tag{3.8}$$

If one puts $C = -\gamma_4$, one can easily see that Eq. (3.4) is identical with Eq. (3.5) [8].

The conventional definition of a two-component spinor field Φ and the Majorana field χ are given as follows:

$$\Phi = a\,\psi = a\,\varphi_1 - i\,a\,\varphi_2 \tag{3.9}$$

$$\chi = \frac{1}{\sqrt{2}}(\psi + \psi^C) = \sqrt{2}\,\varphi_1 \tag{3.10}$$

where

$$a = \tfrac{1}{2}(1 + \gamma_5)$$

and

$$\bar{a} = \tfrac{1}{2}(1 - \gamma_5).$$

The most general canonical transformation (called "extended Pauli group") which keeps the commutation relation (3.3) of Majorana fields has been given by Touschek [9]:

$$\varphi_i{}' = (\exp\,[A + i\,\gamma_5 S])_{ik}\,\varphi_k \tag{3.11}$$

where the φ_i's are the n Majorana fields, A and S are $n \times n$ real antisymmetric and symmetric matrices, respectively. It certainly includes the Pauli transformation

$$\psi' = \frac{1}{\sqrt{2}}(\psi + \gamma_5\,\psi^C) \tag{3.12}$$

as a special case by putting $n = 2$, $A = 0$, $S = \begin{pmatrix} 0 & 1 \\ 1 & 0 \end{pmatrix}\frac{\pi}{4}$. Touschek has shown also the equivalence between the Majorana form of theory and the Weyl formalism, emphasizing that an additional requirement is necessary to obtain the theory of Lee and Yang. This is the conservation of the lepton number.

Let us see the situation by examining the expression for the current density.

$$j_\mu = \bar{\psi}\,\gamma_\mu\, a\,\psi - \psi^T a^T \gamma_\mu^T \bar{\psi}^T. \tag{3.13}$$

Substituting (3.1) in (3.2), one can easily obtain two sets of terms*.

$$\begin{aligned} j_\mu &= (\bar{\varphi}_1 \gamma_\mu\, a\, \varphi_1 - \varphi_1^T a^T \gamma_\mu^T \bar{\varphi}_1^T) + (\bar{\varphi}_2 \gamma_\mu\, a\, \varphi_2 - \varphi_2^T a^T \gamma_\mu^T \bar{\varphi}_2^T) \\ &\quad - i\,(\bar{\varphi}_1 \gamma_\mu\, a\, \varphi_2 + \varphi_1^T a^T \gamma_\mu^T \bar{\varphi}_2^T) + i\,(\bar{\varphi}_2 \gamma_\mu\, a\, \varphi_1 + \varphi_2^T a^T \gamma_\mu^T \bar{\varphi}_1^T), \\ &= (\bar{\varphi}_1 \gamma_\mu \gamma_5 \varphi_1 + \bar{\varphi}_2 \gamma_\mu \gamma_5 \varphi_2) - i\,(\bar{\varphi}_1 \gamma_\mu \varphi_2 - \bar{\varphi}_2 \gamma_\mu \psi_1) \qquad (3.14) \\ &= (\varphi_1 \gamma_4 \gamma_\mu \gamma_5 \varphi_1 + \varphi_2 \gamma_4 \gamma_\mu \gamma_5 \varphi_2) - \\ &\quad - i\,(\varphi_1 \gamma_4 \gamma_\mu \varphi_2 - \varphi_2 \gamma_4 \gamma_\mu \varphi_1). \end{aligned}$$

Due to the commutation relation (3.3) and to the fact that $\gamma_4\,\gamma_\mu\,\gamma_5$ is antisymmetric and $\gamma_4\,\gamma_\mu$ is symmetric in the Majorana gauge, all terms of (3.13) can remain. Then the expression (3.13) is obviously a parity violating current.

However, it should be noted that the lepton number conservation is implicity assumed in the above discussion. Namely, in order to get the usual phase transformation

$$\Phi \to e^{i\alpha}\,\Phi \tag{3.15}$$

one has assumed implicitly the same phase Touschek transformation

$$\begin{aligned} \varphi_1 &\to e^{i\gamma_5 \alpha}\,\varphi_1 \\ \varphi_2 &\to e^{i\gamma_5 \alpha}\,\varphi_2 \end{aligned}\;. \tag{3.16}$$

Under this transformation all terms of (3.14) are certainly invariant.

However, from our definition (3.1), where φ_1 and φ_2 are completely independent solutions of the mass-zero Dirac equation, we may adopt another set of Touschek transformations:

$$\begin{aligned} \varphi_1 &\to e^{i\gamma_5 \alpha}\,\varphi_1 \\ \varphi_2 &\to e^{-i\gamma_5 \alpha}\,\varphi_2 \end{aligned} \tag{3.17}$$

or more generally we assume,

$$\begin{aligned} \varphi_1 &\to e^{i\gamma_5 \alpha}\,\varphi_1 \\ \varphi_2 &\to e^{i\gamma_5 \beta}\,\varphi_2 \end{aligned}\;. \tag{3.18}$$

It is obvious that the conventional phase transformation, which may be interpreted as leptonic number conservation, cannot be defined for φ_1

---

* The calculation can be done in a straightforward way either by the Majorana expression or by the conventional C operator.

For example,

$$\varphi_1{}^T a^T \gamma_\mu^T \bar{\varphi}_1{}^T = -\varphi_1{}^T C^{-1}\, a\, C\, C^{-1}\, \gamma_\mu\, C\, \bar{\varphi}_1{}^T = \bar{\varphi}_1\, \gamma_\mu\, \bar{a}\, \varphi_1.$$

Therefore, one may obtain

$$\bar{\varphi}_1\, \gamma_\mu\, a\, \varphi_1 - \varphi_1{}^T a^T \gamma_\mu^T \bar{\varphi}_1{}^T = \bar{\varphi}_1\, \gamma_\mu\, a\, \gamma_5\, \varphi_1$$

and similarly,

$$\bar{\varphi}_1\, \gamma_\mu\, a\, \varphi_2 + \varphi_1{}^T a^T \gamma_\mu^T \bar{\varphi}_2{}^T = \bar{\varphi}_1\, \gamma_\mu\, \varphi_2.$$

and φ_2, because these spinors cannot represent antiparticles. Unless we introduce one more physical postulate for the neutral spinor field, there is no reason to prefer the transformation (3.16) or (3.17). In the conventional treatment the transformation (3.16) has been implicitly assumed, but that is because a transformation was chosen which was similar to the one used for the charged electron, where two Majorana fields are necessary.

Let us apply transformation (3. 16) to ψ, Φ, and χ

$$\psi \to e^{i\gamma_5 \alpha}\, \psi \tag{3.19}$$

$$\Phi \to e^{i\alpha}\, \Phi \tag{3.15}$$

$$\chi \to e^{i\gamma_5 \alpha}\, \chi. \tag{3.20}$$

These transformations give us the possibility of defining a lepton number for the Φ field but not for the ψ or the χ field.

On the other hand, if one takes case (3.17) or the more general case (3.18), one may lose the concept of leptonic number, but may obtain the parity conservation, because then cross products of φ_1 and φ_2 must be dropped due to their non-invariance under the transformation. Namely, under this assumption one must have

$$j_\mu{}' = \varphi_1\, \gamma_4\, \gamma_5\, \gamma_\mu\, \varphi_1 + \varphi_2\, \gamma_4\, \gamma_5\, \gamma_\mu\, \varphi_2. \tag{3.21}$$

As was clearly stated in TOUSCHEK's paper [7] a usual four-component neutrino field is really a combination of two irreducible representations of the Lorentz group (that is, two Majorana fields) so that one may easily introduce another degree of freedom or quantum number, which may be conserved in weak interactions.

§ 4. Derivation of Gatto's Transformation from Touschek's Extended Pauli Transformation

As is well-known, the most general canonical transformation of the n Majorana fields has been found by TOUSCHEK [7] and called the extended Pauli transformation. This transformation group was first used by THIRRING [10] to determine the minimum number of fundamental fields which are necessary in order to guarantee the conservation of baryon number, isotopic spin and strangeness. Later TONIN [11] investigated the symmetry properties of the strong interaction also by making use of this group. Let us try to derive GATTO's [1] transformation from it as one of the special cases. In general, three Dirac fields can be expressed by six independent Majorana fields:

$$\psi_1 = \varphi_1 - i\,\varphi_2,\ \psi_2 = \varphi_3 - i\,\varphi_4,\ \psi_3 = \varphi_5 - i\,\varphi_6 \tag{4.1}$$

where φ_i's are transformed as follows:

$$\varphi_i{}' = \exp\,(A + i\,\gamma_5\, S)_{ij}\, \varphi_j \quad (i, j \text{ run from 1 to 6}). \tag{4.2}$$

Here A is a 6×6 anti-symmetric real matrix and S is a 6×6 symmetric real matrix. For the sake of writing convenience we shall put

$$T_{ij} = \exp\,(A + i\,\gamma_5\, S)_{ij}. \tag{4.3}$$

The transformation of the three Dirac fields can be obtained under some conditions. First we shall apply the transformation (4.2) to each Majorana component of the Dirac fields.

$$\psi_1' = \varphi_1' - i\,\varphi_2' = (T_{1j} - i\,T_{2j})\,\varphi_j \tag{4.4}$$

$$\psi_2' = \varphi_3' - i\,\varphi_4' = (T_{3j} - i\,T_{4j})\,\varphi_j \tag{4.5}$$

$$\psi_3' = \varphi_5' - i\,\varphi_6' = (T_{5j} - i\,T_{6j})\,\varphi_j. \tag{4.6}$$

From Eqs. (4.1) we can express φ_i's by pairs of two complex-conjugate ψ_i as

$$\begin{aligned}
\varphi_1 &= \tfrac{1}{2}\,(\psi_1 + \psi_1^*)\\
\varphi_2 &= \tfrac{i}{2}\,(\psi_1 - \psi_1^*)\\
\varphi_3 &= \tfrac{1}{2}\,(\psi_2 + \psi_2^*)\\
\varphi_4 &= \tfrac{i}{2}\,(\psi_2 - \psi_2^*)\\
\varphi_5 &= \tfrac{1}{2}\,(\psi_3 + \psi_3^*)\\
\varphi_6 &= \tfrac{i}{2}\,(\psi_3 - \psi_3^*)
\end{aligned}$$

so that we can rewrite Eqs. (4.4), (4.5) and (4.6) as follows:

$$\begin{aligned}
\psi_1' = {} & (T_{11} - i\,T_{21})\,\tfrac{1}{2}\,(\psi_1 + \psi^*) + (T_{12} - i\,T_{22})\,\tfrac{i}{2}\,(\psi_1 - \psi_1^*) + \\
& + (T_{13} - i\,T_{23})\,\tfrac{1}{2}\,(\psi_2 + \psi_2^*) + (T_{14} - i\,T_{24})\,\tfrac{i}{2}\,(\psi_2 - \psi_2^*) + \\
& + (T_{15} - i\,T_{25})\,\tfrac{1}{2}\,(\psi_3 + \psi_3^*) + (T_{16} - i\,T_{26})\,\tfrac{i}{2}\,(\psi_3 - \psi_3^*).
\end{aligned}$$

Therefore, if we impose the following conditions on T:

$$T_{11} - i\,T_{21} = i\,T_{12} + T_{22} \quad T_{31} - i\,T_{41} = i\,T_{32} + T_{42}$$
$$T_{51} - i\,T_{61} = i\,T_{52} + T_{62} \tag{4.7}$$

$$T_{13} - i\,T_{23} = i\,T_{14} + T_{24} \quad T_{33} - i\,T_{43} = i\,T_{34} + T_{44}$$
$$T_{53} - i\,T_{63} = i\,T_{54} + T_{64} \tag{4.8}$$

$$T_{15} - i\,T_{25} = i\,T_{16} + T_{26} \quad T_{35} - i\,T_{45} = i\,T_{36} + T_{46}$$
$$T_{55} - i\,T_{65} = i\,T_{56} + T_{66} \tag{4.9}$$

we have a linear transformation of ψ_i and ψ_i^* as

$$\psi_i' = G_{ij}\,\psi_j$$
$$\psi_i^{*\prime} = G_{ij}^*\,\psi_j^*$$

where G is given explicitly as

$$G = \begin{pmatrix} T_{11} - i\,T_{21} & T_{13} - i\,T_{23} & T_{15} - i\,T_{25} \\ T_{31} - i\,T_{41} & T_{33} - i\,T_{43} & T_{35} - i\,T_{45} \\ T_{51} - i\,T_{61} & T_{53} - i\,T_{63} & T_{55} - i\,T_{65} \end{pmatrix}. \tag{4.10}$$

Furthermore we shall impose a condition on T, that is,

$$\det(G) = 1. \tag{4.11}$$

Recalling that T is a real matrix which can contain $6 \times 6 = 36$ independent real parameters and G is a complex matrix, we can find how many independent real parameters can remain under the conditions (4.7) and (4.11). Namely

$$36 - 3 \times 3 \times 2 - 2 = 16.$$

On the other hand GATTO's transformation contains $8 + 8 = 16$ real independent parameters and $1 + \gamma_5$ and $1 - \gamma_5$ because his similarity transformation ω is nothing else but an element of SU(3):

$$\omega = \exp\left(i \sum_j \mu_j \lambda_j\right) \tag{4.12}$$

where μ_i's are eight independent real parameters. Since GATTO's transformation is a unitary transformation, it is necessary to prove that G is a unitary matrix. The proof is straightforward. Take, for example,

$$\begin{aligned}(G^+ G)_{11} &= G_{11}^* G_{11} + G_{21}^* G_{21} + G_{31}^* G_{31} \\ &= T_{11}^2 + T_{21}^2 + T_{31}^2 + T_{41}^2 + T_{51}^2 + T_{61}^2 \\ &= 1\end{aligned} \tag{4.13}$$

$$\begin{aligned}(G^+ G)_{12} &= G_{11}^* G_{12} + G_{21}^* G_{22} + G_{31}^* G_{32} \\ &= (T_{11} + i T_{21})(T_{13} - i T_{23}) + (T_{31} + i T_{41})(T_{33} - i T_{43}) + \\ &\quad + (T_{51} + i T_{61})(T_{53} - i T_{63}) \\ &= T_{11} T_{13} + T_{21} T_{23} + T_{31} T_{33} + T_{41} T_{43} + T_{51} T_{53} + T_{61} T_{63} \\ &+ i (T_{21} T_{13} - T_{11} T_{23} + T_{41} T_{33} - T_{31} T_{43} + T_{61} T_{53} - T_{51} T_{63}) \\ &= \sum_{i=1}^{6} T_{i1} T_{i3} - i \sum_{i=1}^{6} T_{i2} T_{i3} \\ &= 0,\end{aligned} \tag{4.14}$$

where the unitarity of T and conditions (4.7) are fully utilized, that is,

$$\sum_{i=1}^{6} T_{ij} T_{ik} = \delta_{jk} \tag{4.15}$$

and $T_{21} = -T_{12}$, $T_{11} = T_{22}$, $T_{41} = -T_{32}$, $T_{31} = T_{42}$, $T_{61} = -T_{52}$, and $T_{51} = T_{62}$.

Hence G is a unitary unimodular 3×3 matrix which contains 16 real parameters and each element may involve 4×4 unit matrix and γ_5 matrix. Therefore, by making use of a suitable representation of SU(3) group generators, we may rewrite the G in GATTO's notation.

Remark: If T is a simple 6×6 real matrix, the unitarity condition of T restricts the number of independent real parameters to $6\,(6-1)/2 = 15$. TOUSCHEK's transformation is indeed a real matrix and also unitary, while it contains $6 \times 6 = 36 = 6\,(6+1)/2 + 6\,(6-1)/2$ real parameters. This dilemma may be solved by seeing the fact that the matrix elements of TOUSCHEK's transformation in the 6×6 matrix representation are not real numbers but in general linear combinations of two 4×4 matrices, that is, 1 and γ_5.

§ 5. Lepton Currents

First we shall construct charged lepton currents, borrowing GATTO's idea in a somewhat different way. Let

$$\Psi = \begin{pmatrix} \mu \\ \nu \\ e \end{pmatrix} + \begin{pmatrix} \mu^c \\ \nu' \\ e^c \end{pmatrix} \equiv \Psi_+ + \Psi_- \tag{5.1}$$

where μ and e mean a positive muon (μ^+) and a negative electron (e^-), respectively. The definition of ν is rather different from the usual one, that is,

$$\nu = \bar{a}\, \nu\, (\mu) + a\, \nu\, (e), \quad \nu' = \bar{a}\, \nu^c\, (\mu) + a\, \nu^c\, (e) \tag{5.2}$$

where $\nu\,(\mu)$ and $\nu\,(e)$ represent a muon neutrino and an electron neutrino, respectively. The gauge transformations of these fields are postulated as follows:

$$\mu \quad \to e^{i\alpha}\, \mu \tag{5.3}$$

$$e \quad \to e^{i\alpha}\, e \tag{5.4}$$

$$\nu\,(\mu) \quad \to e^{-i\gamma_5 \alpha}\, \nu\,(\mu) \tag{5.5}$$

$$\nu\,(e) \quad \to e^{i\gamma_5 \beta}\, \nu\,(e). \tag{5.6}$$

If we use the notation (5.1) for the lepton family, we can construct lepton currents in a unified way,

$$j = \bar{\Psi}\, \gamma\, (\Lambda_+\, \bar{a} + \Lambda_-\, a)\, \Psi \tag{5 7}$$

where the γ's are the usual Dirac matrices*, and Λ_+ and Λ_- are current vectors in the "unitary spin space" of the D^3 (1,0) and D^3 (0,1) representations, respectively, given in GATTO's paper by formulae (12), (41) and Table 1 [2].

The ambiguity between the two representations can be reduced by imposing the invariance of the charge operator and the reality conditions on the λ_i matrices which are 3×3 matrix representations of SU(3). (For details, see reference [1]).

One can calculate (5.7) straightforwardly and obtain the following expressions which are invariant under the transformations (5.3)—(5.6). Since we are interested in the charged lepton currents which involve neutrinos, we have to consider only $(1/2)\,(\lambda_1 \pm i\,\lambda_2)$ and $(1/2)\,(\lambda_6 \pm \pm i\,\lambda_7)$ for Λ_+ in our representation of matrices λ_i's. The forms for Λ_- are, then, automatically determined except for a few ambiguities of parameters ϱ and σ which can take $+1$ or -1 independently of each other.

* Although we use the Majorana gauge, the discussions hereafter do not depend on the representation, except for the usual assumption of hermiticity of the γ_μ's.

Case 1

$$\Lambda_+ = \frac{1}{2}(\lambda_1 - i\,\lambda_2),\ \Lambda_- = \frac{1}{2}(\lambda_1' - i\,\lambda_2') = \frac{\varrho}{2}(\lambda_6 - i\,\lambda_7)$$

$$\begin{aligned} j_0 &= \overline{\nu(\mu)}\,\gamma\,\bar{a}\,\mu + \overline{\nu^C(\mu)}\,\gamma\,\bar{a}\,\mu + \varrho\,\bar{e}\,\gamma\,a\,\nu(e) + \varrho\,\bar{e}\,\gamma\,a\,\nu^C(e) \\ &= \overline{\nu(\mu)}\,\gamma\,\bar{a}\,\mu - \overline{\mu^C}\,\gamma\,a\,\nu(\mu) - \varrho\,\overline{\nu^C(e)}\,\gamma\,\bar{a}\,e^C + \varrho\,\bar{e}\,\gamma\,a\,\nu^C(e). \end{aligned} \tag{5.8}$$

Case 2

$$\Lambda_+ = \frac{1}{2}(\lambda_6 - i\,\lambda_7),\ \Lambda_- = \frac{1}{2}(\lambda_6' - i\,\lambda_7') = \frac{\sigma}{2}(\lambda_1 - i\,\lambda_2)$$

$$\begin{aligned} j_1 &= \bar{e}\,\gamma\,\bar{a}\,\nu(\mu) + \bar{e}\,\gamma\,\bar{a}\,\nu^C(\mu) + \sigma\,\overline{\nu(e)}\,\gamma\,a\,\mu + \sigma\,\overline{\nu^C(e)}\,\gamma\,a\,\mu \\ &= \bar{e}\,\gamma\,\bar{a}\,\nu(\mu) - \overline{\nu(\mu)}\,\gamma\,a\;e^C - \sigma\,\mu^C\,\gamma\,\bar{a}\,\nu^C(e) + \sigma\,\overline{\nu^C(e)}\,\gamma\,a\,\mu. \end{aligned} \tag{5.9}^*$$

For the sake of comparison with other authors results [3], we may replace our $\mu, \nu(\mu)$, $\nu^C(e)$ by μ^C, χ, ν_e, respectively, and put $\varrho = -1$ and $\sigma = -1$. Then we obtain

$$-j_0 = \bar{e}\,\gamma\,a\,\nu_e - \bar{\nu}_e\,\gamma\,\bar{a}\,e^C + \bar{\mu}\,\gamma\,a\,\chi - \bar{\chi}\,\gamma\,\bar{a}\,\mu^C \tag{5.10}$$

$$j_1 = \bar{e}\,\gamma\,\bar{a}\,\chi - \bar{\chi}\,\gamma\,a\,e^C + \bar{\mu}\,\gamma\,\bar{a}\,\nu_e - \bar{\nu}_e\,\gamma\,a\,\mu. \tag{5.11}$$

Up to now we have followed GATTO's approach. We want to show now how GATTO's result can be obtained in a completely different way, which in the same time shall also show the connection of GATTO's theory with MARSHAK, RYAN, RADHA, and RAMAN's theory. To this aim let us start with 8×8 matrix representations of SU(3). If one recalls the well-known relation, $D^3(1,0) \times D^3(0,1) = D^8(1,1) + D^1(0,0)$, it does not seem so surprising that one may obtain very similar results, because we have already used both representations $D^3(1,0)$ and $D^3(0,1)$. We shall define ψ and A as follows:

$$\Psi = \begin{bmatrix} e^C \\ \nu_e{}^C \\ \nu_e \\ e \\ \mu^C \\ \chi^C \\ \mu \\ \chi \end{bmatrix} \tag{5.12}$$

* In this derivation the well-known identity

$$\bar{\psi}\,\gamma\,a\,\varphi = -\,\overline{\varphi^C}\,\gamma\,\bar{a}\,\psi^C$$

is used. This identity holds if ψ and φ are anticommutable fields and γ_4 is hermitian as usual.

$$A = \begin{bmatrix} \bar{a} & & & & & & & \\ & 1 & & & & & & \\ & & 1 & & & & & \\ & & & a & & & & \\ & & & & \bar{a} & & & \\ & & & & & -\frac{1}{\sqrt{2}} & & \\ & & & & & & a & \\ & & & & & & & -\frac{1}{\sqrt{6}} \end{bmatrix} \tag{5.13}$$

and also define two charged currents as:

$$j_0' = \overline{\Psi}\,\overline{A}\,\gamma\,(F_1 - i\,F_2)\,A\,\Psi \tag{5.14}$$

$$j_1' = -\,\overline{\Psi}\,A\,\gamma\,(F_4 - i\,F_5)\,\overline{A}\,\Psi \tag{5.15}$$

where F_1, F_2 ..., F_8 are familiar 8×8 matrices of $D^8\,(1,1)$ and we diagonalize F_3 and F_8 as follows:

$$F_3 = \tfrac{1}{2} \begin{bmatrix} 1 & & & & & & & \\ & -1 & & & & & & \\ & & 1 & & & & & \\ & & & -1 & & & & \\ & & & & 2 & & & \\ & & & & & 0 & & \\ & & & & & & -2 & \\ & & & & & & & 0 \end{bmatrix}$$

$$F_8 = \frac{\sqrt{3}}{2} \begin{bmatrix} 1 & & & & & & & \\ & 1 & & & & & & \\ & & -1 & & & & & \\ & & & -1 & & & & \\ & & & & 0 & & & \\ & & & & & 0 & & \\ & & & & & & 0 & \\ & & & & & & & 0 \end{bmatrix}$$

Of course, the correspondence between baryons and leptons in this approach is given by the following table:

Table 1

p	$\longleftrightarrow$	e^C
n	$\longleftrightarrow$	ν_e^C
Ξ^0	$\longleftrightarrow$	ν
Ξ^-	$\longleftrightarrow$	e
Σ^+	$\longleftrightarrow$	μ^C
Σ^0	$\longleftrightarrow$	χ^C
Σ^-	$\longleftrightarrow$	μ
Λ	$\longleftrightarrow$	γ

Representations of $F_1 - i\,F_2$ and $F_4 - i\,F_5$ and also related discussions are given in the Appendix. The reason why we use $F_4 - i\,F_5$ instead of $F_6 - iF_7$ is simply due to our adoption of the conventional representation for the F_i's.

Recalling the following fundamental commutation relations for the F_i's,

$$[F_8,\ F_1 - i\,F_2] = 0 \tag{5.16}$$

$$[F_8,\ F_4 - i\,F_5] = -\tfrac{\sqrt{3}}{2}\,(F_4 - i\,F_5) \neq 0 \tag{5.17}$$

we may find that the lepton current j_0' conserves the "hypercharge X" which is represented by the diagonal elements of F_8, while j_1' does not.

We now impose a first condition,

$$\chi^C = \chi \tag{5.18}$$

and obtain

$$-j_0' = \bar{e}\,\gamma\,a\,\nu_e - \overline{\nu_e^C}\,\gamma\,\bar{a}\,e^C + \bar{\mu}\,\gamma\,a\,\chi - \bar{\chi}\,\gamma\,\bar{a}\,\mu^C \tag{5.19}$$

$$j_1' = \bar{e}\,\gamma\,\bar{a}\,\chi - \bar{\chi}\,\gamma\,a\,e^C + \mu\,\gamma\,a\,\nu_e^C - \nu_e\,\gamma\,a\,\mu^C. \tag{5.20}$$

These are exactly the same as the charged lepton currents given by Marshak, Ryan, Radha and Raman [2].

At this point the following remarks should be done, focusing on obtaining possible relations between Gatto's approach and Marshak, Ryan, Radha and Raman's. As it is evident that j_0' and j_1' are invariant under the gauge transformations (5.3)—(5.6), all terms obtained from (5.14) and (5.15) automatically conserve their gauge charges. On the contrary, in Gatto's approach, the many terms contained in the explicit formulation of definition (5.7) for the current have to be eliminated by the supplementary imposition of the gauge transformations (5.3)—(5.6).

Let us now impose a second requirement

$$\nu_e^C = \nu_e. \tag{5,21}$$

Then Marshak, Ryan, Radha and Raman's results (5.19) and (5.20) coincide with our previous results (5.10) and (5.11) obtained by Gatto's method.

§ 6. Discussion

In § 2 an equivalence between Gatto's method and Marshak, Ryan, Radha and Raman's have been shown by postulating that the two distinct neutrino fields are two independent Majorana fields. Those two Majorana fields obey different Touschek transformations. Namely,

$$\nu\,(e) = \nu_e \rightarrow e^{i\,\gamma_5\,\alpha}\,\nu_e \tag{6.1}$$

$$\nu\,(\mu) = \chi \rightarrow e^{-i\,\gamma_5\,\alpha}\,\chi. \tag{6.2}$$

This is an essential part of our formalism.

Although the results obtained by the two different methods coincide with each other, it should be noted that there is some important difference in these treatments. Both can work on the same footing, that is, the SU(3) symmetry. However, in the 3×3 representation procedure we have to impose explicitly the invariance requirement with respect to the gauge transformation which leads us to the lepton number conservation law (strictly speaking, the helicity conservation law).

Indeed in this approach the lepton number conservation is an independent physical postulate and is quite important for comparing the two-component theory of the neutrino with the Majorana theory.

On the other hand, in the 8×8 representation scheme the correspondence between baryons and leptons is essential, and the lepton number conservation is automatically satisfied.

Since two different Majorana fields for the two distinct neutrinos are proposed in the present paper, it may be worthwhile to see the relation with the MARSHAK, RYAN, RADHA, and RAMAN's treatment, where the electron-neutrino is a usual Dirac field and the muon-neutrino is a Majorana field.

We shall list the correspondence proposed by MARSHAK, RYAN, RADHA, and RAMAN's in Table 2, where the part of the correspondence of charged particles is omitted because we do not change it. In order to see the structure of the neutrino fields, their Majoran representations are also listed in the same table.

Table 2

Baryon		Lepton	Majorana representations
n	$\longleftrightarrow$	$\nu_e C$	$\varphi_1(e) + i\,\varphi_2(e)$
Ξ^0	$\longleftrightarrow$	ν_e	$\varphi_1(e) - i\,\varphi_2(e)$
Σ^0	$\longleftrightarrow$	$\chi = \frac{1}{\sqrt{2}}(\nu_\mu + \nu C_\mu)$	$\sqrt{2}\,\varphi_1(\mu)$
Λ	$\longleftrightarrow$	$\psi = \frac{1}{\sqrt{2}}(\nu_\mu - \nu C_\mu)$	$-\sqrt{2}\,i\,\varphi_2(\mu)$

Here one can see that χ and ψ are independent hermitian and antihermitian fields, respectively. Hence it is certainly possible to impose different Touschek transformations on χ and ψ independently. In fact, MARSHAK, RYAN, RADHA, and RAMAN assumed that

$$\chi \to e^{-i\gamma_5 \alpha}\,\chi \tag{6.3}$$

$$\psi \to e^{i\gamma_5 \alpha}\,\psi \tag{6.4}$$

which correspond to

$$\varphi_1(\mu) \to e^{-i\gamma_5 \alpha}\,\varphi_1(\mu) \tag{6.5}$$

$$\varphi_2(\mu) \to e^{i\gamma_5 \alpha}\,\varphi_2(\mu). \tag{6.6}$$

Then terms like $(e\,\gamma\,\bar{a}\,\psi)$ are exclused by the invariance requirement under the transformations (6.3) and (6.4).

Therefore even in the "septet model" proposed by MARSHAK, RYAN, RADHA, and RAMAN the exclusion of the ψ neutrino instead of being purely postulated as they did, may be on the contrary deduced from the lepton number conservation as in the present formalism.

Moreover we can make the following remarks concerning the gauge transformation properties of both theories. In MARSHAK, RYAN, RADHA, and RAMAN's case the transformations of baryons and leptons appear on a rather different footing owing to the fact that the ψ neutrino is dropped, leaving thus the Λ particle without a corresponding lepton.

Furthermore the transformation properties (6.3) and (6.4) for leptons can not be applied to the corresponding baryons owing to their non-zero masses. As in our present formalism instead there is no cancellation of any leptonic neutrino, the neutral baryons can be grouped into two doublets, (n, Ξ^0) and (Σ^0, Λ) in correspondence to the two neutrinos $\nu_e = \nu_e^C$, $\chi = \chi^C$.

For what concerns the transformation properties the following device can be used for pairs of baryons, neglecting only their mass differences [12].

Let us take two different neutral Dirac fields with the same mass. Using the 8-component representations discussed by DALLAPORTA and TOYODA [13], we may in fact write down the following free field equation:

$$\{\Gamma^\mu \partial_\mu + i m \Gamma_5\} \chi = 0 \tag{6.7}$$

where Γ_5 and Γ^μ are defined by

$$\Gamma^\mu = \begin{pmatrix} & \gamma_\mu \\ \gamma_\mu & \end{pmatrix}, \quad \Gamma_5 = \begin{pmatrix} -\gamma_5 & \\ & \gamma_5 \end{pmatrix}. \tag{6.8}$$

Since Γ_5 commutes with Γ_μ, Eq. (48) is invariant under the generalized gauge transformation

$$\chi \to \{\exp(\pm i \Gamma_5 \alpha)\} \chi, \tag{6.9}$$

which is formally similar to the transformations (6.3) and (6.4) for neutrinos. In such a way, one may get an equivalence between transformation properties of baryons and leptons.

Concerning the physical meaning of our proposal, we may stress that both two neutrino fields are here represented by Majorana fields which have opposite helicities.

Moreover, the use of two independent Majorana fields allows us to introduce another degree of freedom into the structure of the neutrino. Indeed, if we interprete the "extended Pauli group" (3.11) in the case $n = 2$, we can easily construct the "hypercharge spin" in a way similar to that used by TOUSCHEK to construct the usual isotopic spin [7].

If one considers the number of fields which are necessary for describing in a unified way lepton currents in the three approaches here discussed, GATTO's, MARSHAK, RYAN, RADHA, and RAMAN's and the present one, the situation appears rather equivalent because in all three approaches the total number of fields is the same in the same group SU(3) [14].

Appendix

Representations of λ_i and F_i

Although the representations adopted in the present paper are the same as those described in many other papers, we shall list all of them for the reader's convenience.

$$\lambda_1 = \begin{bmatrix} & 1 & \\ 1 & & \\ & & \end{bmatrix}, \quad \lambda_2 = \begin{bmatrix} & -i & \\ i & & \\ & & \end{bmatrix}, \quad \lambda_3 = \begin{bmatrix} 1 & & \\ & -1 & \\ & & \end{bmatrix}, \quad \lambda_4 = \begin{bmatrix} & & 1 \\ & & \\ 1 & & \end{bmatrix}$$

$$\lambda_5 = \begin{bmatrix} & & -i \\ & & \\ i & & \end{bmatrix}, \quad \lambda_6 = \begin{bmatrix} & & \\ & & 1 \\ & 1 & \end{bmatrix}, \quad \lambda_7 = \begin{bmatrix} & & \\ & & -i \\ & i & \end{bmatrix}, \quad \lambda_8 = \begin{bmatrix} 1 & & \\ & 1 & \\ & & -2 \end{bmatrix}$$

$$F_1 = \frac{1}{2} \left[\begin{array}{cccc|cccc} & 1 & & & & & & \\ 1 & & & & & & & \\ & & & -1 & & & & \\ & & -1 & & & & & \\ \hline & & & & & -\sqrt{2} & & \\ & & & & -\sqrt{2} & & \sqrt{2} & \\ & & & & & \sqrt{2} & & \\ & & & & & & & \end{array}\right], \quad F_2 = \frac{i}{2} \left[\begin{array}{cccc|cccc} & -1 & & & & & & \\ 1 & & & & & & & \\ & & & 1 & & & & \\ & & -1 & & & & & \\ \hline & & & & & \sqrt{2} & & \\ & & & & -\sqrt{2} & & -\sqrt{2} & \\ & & & & & \sqrt{2} & & \\ & & & & & & & \end{array}\right]$$

$$F_3 = \frac{1}{2} \left[\begin{array}{cccc|cccc} 1 & & & & & & & \\ & -1 & & & & & & \\ & & 1 & & & & & \\ & & & -1 & & & & \\ \hline & & & & 2 & & & \\ & & & & & 0 & & \\ & & & & & & -2 & \\ & & & & & & & 0 \end{array}\right], \quad F_4 = \frac{1}{2\sqrt{2}} \left[\begin{array}{cccc|cccc} & & & & & -1 & & -\sqrt{3} \\ & & & & & & -\sqrt{2} & \\ & & & & \sqrt{2} & & & \\ & & & & & 1 & & \sqrt{3} \\ \hline & & \sqrt{2} & & & & & \\ -1 & & & 1 & & & & \\ & -\sqrt{2} & & & & & & \\ -\sqrt{3} & & & \sqrt{3} & & & & \end{array}\right]$$

$$F_5 = \frac{i}{2\sqrt{2}} \left[\begin{array}{cccc|cccc} & & & & & 1 & & \sqrt{3} \\ & & & & & & \sqrt{2} & \\ & & & & \sqrt{2} & & & \\ & & & & & 1 & & \sqrt{3} \\ \hline & & -\sqrt{2} & & & & & \\ -1 & & & -1 & & & & \\ & -\sqrt{2} & & & & & & \\ -\sqrt{3} & & & -\sqrt{3} & & & & \end{array}\right], \quad F_6 = \frac{1}{2\sqrt{2}} \left[\begin{array}{cccc|cccc} & & & & -\sqrt{2} & & & \\ & & & & & 1 & & -\sqrt{3} \\ & & & & & -1 & & \sqrt{3} \\ & & & & & & \sqrt{2} & \\ \hline -\sqrt{2} & & & & & & & \\ & 1 & -1 & & & & & \\ & & & \sqrt{2} & & & & \\ & \sqrt{3} & \sqrt{3} & & & & & \end{array}\right]$$

$$F_7 = \frac{i}{2\sqrt{2}} \left[\begin{array}{cccc|cccc} & & & & \sqrt{2} & & & \\ & & & & & -1 & & \sqrt{3} \\ & & & & & -1 & & \sqrt{3} \\ & & & & & & \sqrt{2} & \\ \hline -\sqrt{2} & & & & & & & \\ & 1 & 1 & & & & & \\ & & & -\sqrt{2} & & & & \\ & -\sqrt{3} & -\sqrt{3} & & & & & \end{array}\right], \quad F_8 = \frac{\sqrt{3}}{2} \left[\begin{array}{cccc|cccc} 1 & & & & & & & \\ & 1 & & & & & & \\ & & -1 & & & & & \\ & & & -1 & & & & \\ \hline & & & & & & & \\ & & & & & & & \\ & & & & & & & \\ & & & & & & & \end{array}\right]$$

These matrices satisfy the fundamental commutation relations. For example, for λ_i the following relations are important.

$$[\lambda_3, \lambda_1 \pm i\lambda_2] = \pm 2(\lambda_1 \pm i\lambda_2)$$
$$[\lambda_3, \lambda_4 \pm i\lambda_5] = \pm(\lambda_4 \pm i\lambda_5)$$
$$[\lambda_3, \lambda_6 \pm i\lambda_7] = \mp(\lambda_6 \pm i\lambda_7)$$

$$[\lambda_8, \lambda_4 \pm i\lambda_5] = \pm\sqrt{3}\,(\lambda_4 \pm i\,\lambda_5)$$
$$[\lambda_8, \lambda_6 \pm i\lambda_7] = \pm\sqrt{3}\,(\lambda_6 \pm i\,\lambda_7)$$
$$[\lambda_1 + i\,\lambda_2, \lambda_1 - i\,\lambda_2] = 4\,\lambda_3$$
$$[\lambda_4 + i\,\lambda_5, \lambda_4 - i\,\lambda_5] = 2\,\lambda_3 + 2\sqrt{3}\,\lambda_8$$
$$[\lambda_6 + i\,\lambda_7, \lambda_6 - i\,\lambda_7] = -2\,\lambda_3 + 2\sqrt{3}\,\lambda_8.$$

One can find another set of representation by replacing $\lambda_6 \to \lambda_4$, $\lambda_7 \to \lambda_5$, $\lambda_3 \to -\lambda_3$ and $\lambda_2 \to -\lambda_2$. Since eigenvalues of λ_3 and $-\lambda_3$ are equal, this procedure can be done by a similarity transformation. If we use these representations for λ_i, we can redefine lepton currents, instead of Eq. (5.8):

$$j = \Psi\gamma\,(\Lambda_+ \bar{a} + \Lambda_- a)\,\Psi$$

where

$$\Lambda_+ = \tfrac{1}{2}\,(\lambda_1 + i\,\lambda_2) \qquad \text{for case 1,}$$
$$\Lambda_- = \tfrac{1}{2}\,(\lambda_4 - i\,\lambda_5) \qquad \text{for case 2.}$$

References

1. Gatto, R., Nuovo Cimento **28,** 567 (1963).
2. Marshak, R. E., C. Ryan, T. K. Radha, and K. Raman, Preprint, to be published in Nuovo Cimento.
3. Okubo, S., Preprint, NYO-10246, to be published in Physics Letters.
4. Behrends, R. E., and L. F. Landowitz, Physical Review Letters **11,** 296 (1963).
5. Gamba, A., R. E. Marshak, and S. Okubo, Proc. Nat. Acad. of Sci. **65,** 881 (1959).
6. Oiglane, S., Soviet JETP **13,** 548 (1961).
7. Touschek, B., Rendiconti della Scuola Internazionale di Fisica "Enrico Fermi", Corso XI 40 (1960).
(Proceedings of Varenna Summer School, June 29—July 11, 1959). See also B. Touschek, Nuovo Cimento **5,** 1281 (1957).
The first paper which referred to the possibility of Majorana neutrino for the double β-decay is:
Touschek, B., Zeit. f. Phys. **125,** 108 (1948).
The first paper which referred to the equivalence between the Majorana field and Weyl's two-component spinor field is:
Serpe, J., Physica **18,** 295 (1952).
The proof of the equivalence between the Majorana field and Weyl's two-component spinor has been given independently by:
McLennon J. A., Jr., Phys. Rev. **106,** 821 (1957).
Case, K. M., Phys. Rev. **107,** 307 (1957).
Radicati, L. A., and B. Touschek, Nuovo Cimento **51,** 1693 (1957).
8. Pauli, W., Nuovo Cimento **6,** 204 (1957).
9. Touschek, B., Nuovo Cimento **8,** 181 (1958). The Symmetry Properties of Fermi Dirac Fields.
10. Thirring, W., Nucl. Phys. **10,** 97 (1959); ibid **14,** 569 (1960).
11. Tonin, M., Nuovo Cimento **18,** 304 (1960).
12. Gürsey, F., Nuovo Cimento **7,** 411 (1958).
13. Dallaporta, N., and T. Toyoda, Nuovo Cimento **12,** 539 (1959).
14. Tonin, M., Private communication.

Unitary Symmetry, Weak Interaction and Intermediate Bosons

(An abstract of a paper by **B. d'Espagnat** and **Y. Villachon**)
presented by Miss *Y. Villachon*

With 2 Figures

Considering that, with respect to the strong interactions, the theory of unitary symmetry has met with at least appreciable success, many authors have in recent years tried to discover what sorts of links, *if any*, the weak interactions have with that scheme (see [1], [2], [3], [4], [5], [6], [7]).

By now the idea that the weak interactions are mediated by vector intermediate bosons has gained much credit. Therefore, we have considered the fundamental intermediate boson-current interaction Lagrangian, instead of a Fermi-type interaction Lagrangian, and we have studied its possible insertion into the unitary symmetry group.

As regards the strong interactions we use the octet model. Then, what we do, is the following:

a) First of all we keep as a fundamental postulate the one already stated, that the bulk of the weak currents J_i of the strongly interacting particles is in the regular representation, i. e. "8" + "1", of U_3. We use the group U_3, and not SU_3, because we need one of the "triangular" representations "3", "6" or "15" for the intermediate bosons W_i, which must have integer charges; it is, however equal to zero for all particles except the intermediate bosons, and thus causes no difficulty.

b) Associating the W's with a representation of U_3 only makes sense if the interactions have some group properties. Therefore, we postulate that the fundamental Lagrangian is ascribed to the fundamental representation of the group (and to its contragradient):

$$\mathfrak{L} = g\,(L + L^+); \qquad L \in \text{“3”} \tag{1}$$

and where:

$$L = \sum_i K_i\, J_i^+\, W_i.$$

Then it appears immediately that, if "X" is the representation which the W's belong to, we have:

$$\text{“8”} \otimes \text{“}X\text{”} = \text{“3”}; \tag{2}$$

there are three possible solutions to this "equation" which are:

$$\text{“}X\text{”} = \text{“3”},\ \text{“}\bar{6}\text{”}\ \text{or}\ \text{“15”}. \tag{3}$$

Before describing briefly these three distinct versions, I want to make a few remarks:

1. first it can be shown that, as regards the non-leptonic decays, *the rules* $|\overrightarrow{\Delta I}| = 1/2$ *for* $|\Delta S| = 1$ *and* $|\Delta S| < 2$ follow *from ascribing* L *to* "3": for that purpose let us consider a non-leptonic weak interaction phenomenon with exchange of an intermediate boson W, which after being emitted is reabsorbed:

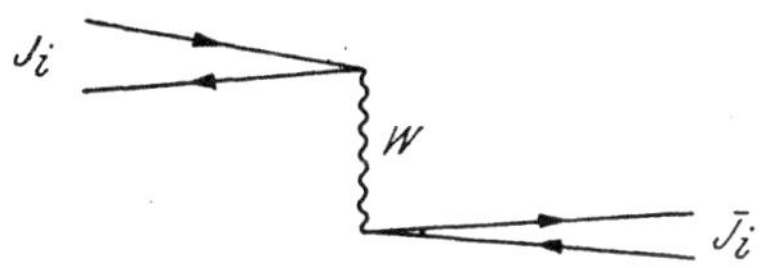

Fig. 1

each of the two vertices transforms under U_3 as a component of "3" or "$\bar{3}$", so that the whole graph transforms as the direct product $(L \cdot L^+)$:

$$\text{“}3\text{”} \otimes \text{“}\bar{3}\text{”} = \text{“}1\text{”} + \text{“}8\text{”} \tag{4}$$

and neither of this two representations contains $|\overrightarrow{\Delta I}| > 1$ and $\Delta S \geq 2$.

2. *proof that* $\mathfrak{L}$ *is invariant under a* SU_2 *subgroup of* U_3.

The general expression of L, is:

$$L = a\, L_0{}^{(0)} + b\, L_{1/2}^{(0)}, \tag{5}$$

where $L_I^{(Q)}$ is a member of "3" with isotopic spin I and charge Q.

The charge conserving transformation R of U_3:

$$R = \exp\left\{\sqrt{6}\,\alpha\,(E_3 - E_{-3})\right\}$$

transforms J_i and W_k into J_i' and W_k' which are linear combinations of the J_i and W_k respectively, and therefore R transforms $L_I^{(Q)}$ into $L'_I{}^{(Q)}$ and we have a new expression for L:

$$L = (a \cos\alpha - b \sin\alpha)\, L_0{}^{(0)\prime} + (a \sin\alpha + b \cos\alpha)\, L_{1/2}^{(0)\prime}. \tag{6}$$

If we choose

$$\begin{cases} a = \cos\alpha \\ b = -\sin\alpha \end{cases} \tag{7}$$

we see that $L = L_0{}^{(0)\prime}$, which is a scalar in a subgroup "G" (SU_2) of U_3 derived from the isotopic spin subgroup "I" of U_3 by means of the transformation R.

3. *Possibility of a* $\Delta S = -\Delta Q$ *admixture.*

If the W's are in the representation "X" = "3", (1) shows that the J_i's belong to:

$$\text{“}\bar{3}\text{”} \otimes \text{“}3\text{”} = \text{“}1\text{”} + \text{“}8\text{”}, \tag{8}$$

which do not contain any $\Delta S = -\Delta Q$ currents.

But for "X" $=6$, the equation:

$$\text{“}\bar{6}\text{”} \otimes \text{“}3\text{”} = \text{“}8\text{”} + \text{“}10\text{”} \tag{9}$$

shows that we can incorporate some admixture of currents in "10" which contains $\Delta S = -\Delta Q$ currents.

The version with "X" $=$ "15" brings no new features and is just more complicated.

It can be shown that the *version of the theory*, "X" $=$ "3", implies in fact only five real interacting intermediate bosons altogether. As regards the leptonic decays, this model turns out to be similar to the one proposed by Cabibbo [4] with the identification:

$$\theta_{\text{CABIBBO}} = \frac{\pi}{2} + \alpha \tag{10}$$

and the comparison with experiment gives:

$$\alpha = -85^0. \tag{11}$$

In the second version, "X" $=$ "6", a small admixture ε of currents belonging to "10" is allowed. We have four distinct charged intermediate boson fields, as shown by Fig. 2.

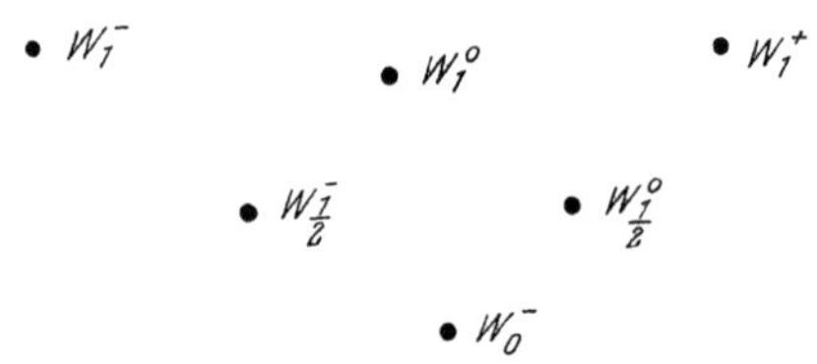

Fig. 2

We have three possibilities to introduce the lepton-boson couplings:

a) only W_1^+ is coupled with the leptons. This possibility hardly differs from the one of the version "X" $=$ "3";

b) leptons coupled to the three $W^{(-)\prime}$ fields with equal strength. If we assume for simplicity that $\varepsilon = 0$ and using the same data as Cabibbo [4], we find for:

$$\alpha \approx 50^0. \tag{12}$$

A renormalisation [8] leads to:

$$\alpha \approx 47^0 \tag{13}$$

and we see that the value

$$\alpha = 45^0 \tag{14}$$

corresponding to the currents of strong interacting particles coupled with equal "strength" to the intermediate bosons, is compatible with present experimental information.

With this choice (14) of α and if we put $\varepsilon = 0$, we find for the ratio of the effective coupling constants G_β and G_μ for β-decay and μ-decay:

$$\left| \frac{G_\beta}{G_\mu} \right| = 0.985 \tag{15}$$

which is to be compared with the experimental value [8]:

$$\left| \frac{G_\beta}{G^\mu} \right| = 0.98. \tag{16}$$

We can calculate the upper limit of ε by requiring that the prediction of the conserved current theory be satisfied with an error less or equal to 10% on the coupling constant [9], and we obtain:

$$\varepsilon \leq \frac{\sqrt{2}}{15}\,. \tag{17}$$

With this value of ε, (14) turns out to be:

$$0.970 \leq \left| \frac{G_\beta}{G_\mu} \right| \leq 1. \tag{18}$$

We can also evaluate an upper limit for the ratio x of $\Delta S = -\Delta Q$ to $\Delta S = \Delta Q$ amplitudes, and it gives:

$$x \leq 0.28 \tag{19}$$

a value which is in good agreement with recent data:

$$x \leq 0.20. \tag{20}$$

c) Leptons coupled to all four charged fields. The results, this possibility leads to, are less interesting.

References

1. Ne'eman, Y., Nucl. Phys. **26**, 222 (1961).
   Gell-Mann, M., Phys. Rev. **125**, 1067 (1962).
2. Cabibbo, N., and R. Gatto, Nuovo Cimento **21**, 872 (1962).
3. Okubo, S., Progr. Theor. Phys. **27**, 949 (1962).
4. Cabibbo, N., Phys. Rev. Letters **10**, 531 (1963).
5. d'Espagnat, B., and J. Prentki, Nuovo Cimento **24**, 497 (1962).
6. Lee, B. W., Phys. Rev. Letters **12**, 83 (1964).
7. Baker, M., and S. L. Glashow, Nuovo Cimento **26**, 803 (1962).
8. Sakurai, J. J., Phys. Rev. Letters **12**, 79 (1964).
9. Depommier, J., J. Heintze, C. Rubbia, and V. Soergel, Phys. Letters **5**, 61 (1963).

Propagator of Vector Bosons

By

Walter Thirring, Vienna

With 1 Figure

In view of the increasing experimental evidence on vector particles we will reexamine the problem of their propagator within Lagrangian fieldtheory. It is known from Quantum Electrodynamic [1] that inconsistencies arise unless the currents are carefully defined by limiting procedures involving the vector field. We shall see that additional difficulties arise if the vector field U_i is coupled to a nonconserved current.

Our starting point will be in the field equations [2]

$$U_k^{i,k} - U_k^{k,i} + \mu_0^2 U^i = j^i; \qquad i, k = 0, 1, 2, 3 \tag{1}$$

and the canonical commutation rules

$$\begin{gathered}[U_\alpha(x), \pi_\beta(x')] = i\,\delta_{\alpha\beta}\,\delta(x-x'),\ \pi_\beta = U_{0,\beta} - U_{0,\beta},\ \alpha, \beta = 1, 2, 3 \\ [U_\alpha(x),\ U_\beta(x')] = [\pi_\alpha(x), \pi_\beta(x)] = 0.\end{gathered} \tag{2}$$

From (1) also follows

$$\mu_0^2\, U^i{}_{,i} = j^i{}_{,i}, \quad U_0 = \mu_0^{-2}(j_0 + \pi_{\alpha,\alpha}). \tag{3}$$

From Lorentz-invariance we know the general form of the two point function

$$\begin{aligned}\langle 0 | U_i(x)\, U_j(x') | 0 \rangle &= \int d^4k\, e^{-ik(x-x')}\,(k_i k_j\, r_2(k^2) - k^2 g_{ij}\, r_1(k^2)) = \\ &= \int dc^2\,[(g_{ij}\,\Box - \partial_i\,\partial_j)\, r_T(c^2) + \partial_i\,\partial_j\, r_L(c^2)]\, \Delta^+(x-x', c^2);\end{aligned} \tag{4}$$

$$r_T = r_1 \qquad r_L = r_1 - r_2.$$

If the Hilbert space is positive definite we have in addition

$$0 \leq r_1 \leq r_2. \tag{5}$$

However since the theory probably exists only if the particles to which the vector field is coupled are properly regularized we shall not rely on (5). The canonical commutation rules (2) imply

$$\int dc^2\, r_1(c^2)\, c^2 = 1. \tag{6}$$

Furthermore if j_0 is supposed to commute with U_α we obtain from (2) and (3)

$$\int dc^2\, r_2\,(c^2) = \mu_0^{-2}. \tag{7}$$

However computing

$$\langle 0 | [U_i{}^{,i}, U_j] | 0 \rangle$$

and

$$\langle 0 | [U_{i,k}{}^{k} - U_{k,i}{}^{k} + \mu_0^2\, U_i, U_j] | 0 \rangle$$

with the aid of the field equation we obtain in addition

$$\int dc^2\, c^2\, r_L = 0, \quad \int dc^2\, r_1\, c^4 = \mu_0^2 \tag{8}$$

provided that the equal time commutators $[\dot{U}_\alpha, j^i]$ is taken to be zero. If (5) holds, (6), (7) and (8) imply free fields. This difficulty is resolved in electrodynamics only by including a factor

$$\lim_{\varepsilon_\mu \to 0} \exp\,(i\, e\, U^\mu\, \varepsilon_\mu)$$

in the current which does not commute with U at equal times. Hence in electrodynamics only (7) and (6) hold [3] and we shall assume the same here.

Turning to the time-ordered product we obtain from (4)

$$T\,\langle 0 | U_i\,(x)\; U_j\,(x') | 0 \rangle = \int d\, c^2 \left\{ (g_{ij}\, c^2 - \partial_i\, \partial_j)\, r_T\,(c^2) + \partial_i\, \partial_j\, r_L\,(c^2) \right\} \cdot$$
$$\cdot\, \Delta_f\,(x - x', c^2) - \delta\,(x - x')\; \delta_{j0}\, \delta_{i0} \int dc^2\, r_2\,(c^2). \tag{9}$$

Defining as the propagator the time-ordered product without the non-covariant term $\delta_{j0}\,\delta_{i0}$ we have in momentum space

$$G_{ij}\,(p) = \int dc^2\, \frac{c^2\, r_T}{c^2 - p^2} \left(g_{ij} - \frac{p_i\, p_j}{c^2} \right) + p_i\, p_j \int \frac{dc^2\, r_L}{c^2 - p^2} =$$
$$= P^T{}_{ij} \int \frac{dc^2\, r_T\, c^2}{c^2 - p^2} + P^L{}_{ij} \left(\frac{1}{\mu_0^2} + \int \frac{dc^2\, c^2\, r_L}{c^2 - p^2} \right), \tag{10}$$
$$P^T{}_{ij} = g_{ij} - \frac{p_i\, p_j}{p^2}\,, \qquad P^L{}_{ij} = \frac{p_i\, p_j}{p^2}$$

assuming that all integrals converge. This expression will now be compared with the result of (unrenormalized) perturbation theory calculated with the free vector propagator

$$G_0 = \left(g_{ij} - \frac{p_i\, p_j}{\mu_0^2} \right) \frac{1}{\mu_0^2 - p^2} = P^L\, \mu_0^{-2} + P^T\,(\mu_0^2 - p^2)^{-1}. \tag{11}$$

If we decompose the proper self-energy part Σ_{ij} into a longitudinal and transversal part

$$\Sigma_{ij}\,(p^2) = P^L{}_{ij}\, \Sigma_L\,(p^2) + P^T{}_{ij}\, \Sigma_T\,(p^2) \tag{12}$$

we have

$$G\,(p) = P^L \frac{1}{\mu_0^2 - \Sigma_L\,(p^2)} + P^T \frac{1}{\mu_0^2 - p^2 - \Sigma_T\,(p^2)}\,. \tag{13}$$

Comparing with (10) we identify

$$\frac{1}{\mu_0{}^2 - \Sigma_L} = \frac{1}{\mu_0{}^2} + \int \frac{dc^2\, c^2\, r_L}{c^2 - p^2}, \qquad \frac{1}{\mu_0{}^2 - p^2 - \Sigma_T} = \int \frac{dc^2\, c^2\, r_T}{c^2 - p^2} \tag{14}$$

which implies

$$\lim_{p^2 \to \infty} \Sigma_L\, (p^2) = 0. \tag{15}$$

The sum-rule (6) tells us

$$\lim \Sigma_T\, (p^2)/p^2 = 0 \tag{16}$$

and from (7) we infer

$$\Sigma_T\, (0) = \Sigma_L\, (0). \tag{17}$$

Finally we want to investigate whether these general requirements are met in an explicit calculation of Σ in perturbation theory. To lowest order Σ is related to $\langle 0 |\, j_i\, (x)\, j_k\, (x')\, | 0 \rangle = \Sigma^+_{ik}\, (x - x')$ which will again be of the structure

$$\Sigma^+_{ik}\, (p^2) = \int d\, c^2\, \delta\, (p^2 - c^2)\, [p_i\, p_k\, \varrho_2\, (c^2) - g_{ik}\, c^2\, \varrho_1\, (c^2)]. \tag{18}$$

For instance, for weak interactions the diagram

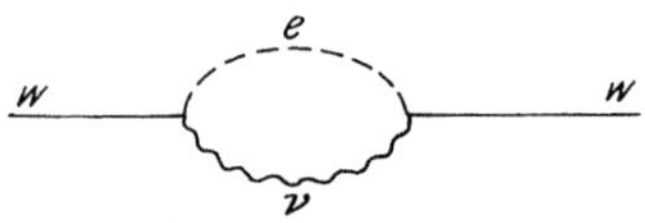

Fig. 1

gives

$$\begin{aligned} \varrho_1 &= \frac{G^2}{6\,\pi}\, \frac{c^2 - m_e{}^2}{c^4} \left(c^2 - \frac{m_e{}^2}{2} - \frac{m_e{}^4}{2\, c^2}\right) \\ \varrho_2 &= \frac{G^2}{6\,\pi}\, \frac{c^2 - m_e{}^2}{c^4} \left(c^2 + m_e{}^2 - \frac{2\, m_e{}^4}{c^2}\right). \end{aligned} \tag{19}$$

Going over to the time-ordered product we firstly assume

A: a noncovariant term $\sim \delta_{\mu 0}\, \delta_{\nu 0}$ is somehow compensated. Then we have

$$\Sigma_T = \int dc^2\, \varrho_1 \frac{c^2}{c^2 - p^2}, \quad \Sigma_L = \int dc^2 \left[(\varrho_1 - \varrho_2)\, \frac{c^2}{c^2 - p^2} - \varrho_2\right]. \tag{20}$$

This implies (16) and (17) but violates $\Sigma_L\, (\infty) \to 0$.

Hence we try

B: a non-covariant term

$$\sim \delta_{\mu 0}\, \delta_{\nu 0} - g_{\mu\nu}$$

is to be dropped. This prescription turns out to be the correct way to handle electrodynamics and gives at least $\Sigma_L \equiv 0$ for $\varrho_1 = \varrho_2$. It yields

$$\Sigma_T = \int dc^2\, \varrho_1\, \frac{p^2}{c^2 - p^2}, \quad \Sigma_L = \int dc^2\, (\varrho_1 - \varrho_2)\, \frac{p^2}{c^2 - p^2} \tag{21}$$

which gives $\Sigma_T(0) = \Sigma_L(0) = 0$ and (16) but violates (15) since we have $\varrho_2 \geq \varrho_1 \geq 0$. Only for $m_e = 0$ where the current is conserved we have like in electrodynamics $\varrho_1 = \varrho_2$ and all conditions are satisfied. Thus it is not easy to see how a consistent vector theory is to be constructed if the current is not conserved.

References

1. SCHWINGER, J., Lecture Trieste Seminar for Theor. Phys. 1962.
2. See e. g. G. WENTZEL, Quantum Theory of Fields.
3. See W. THIRRING, Principles of Quantum Electrodynamics.

The Field Theory of the Intermediate Boson in Weak Interactions

By

W. Kummer, CERN

With 8 Figures

Introduction

This talk shall give a short review of recent trends in the description of weak interactions by a theory which goes beyond the purely phenomenological approach. The great success of the latter in the theoretical investigation of processes at energies which occur in β-decay, μ-decay, μ-capture, and to some extent in strange particle decays, necessarily leads to field theoretical models which contain relatively heavy particles (compared to the energies of the above processes) which mediate the interactions between the four fermions.

We will first give a short introduction into the theory of the intermediate vector boson (W) in general and then speak about methods which allow us to extract information from this theory. In the considerations of higher order corrections these methods must differ from usual perturbation theory because the theory is not renormalizable. We will concentrate on the "peratization" of Feinberg and Pais [1] in the second chapter. A test of their iteration scheme of peratization will be presented in the next lecture.

Theory of the Intermediate Boson

The four-fermion interaction for a weak process

$$p_1 + p_2 \rightarrow p_1' + p_2'$$

(the particles are simply denoted by their momenta) can be phenomenologically described by the effective interaction Lagrangian

$$\mathfrak{L}_{\text{int, eff}} = \frac{G}{\sqrt{2}} j_\mu^{(1)} j^{\mu(2)} + \text{h. c.} \tag{1}$$

as far as leptonic (μ-decay) or semileptonic decays (β-decay, μ-capture, decay of strange particles into leptons) are concerned. Possibly this is true for nonleptonic decays as well.

In (1) for the current e. g. of leptons one has to insert

$$j_\mu^{\text{lept}} = \bar{u}_\nu \gamma_\mu (1 + \gamma_5) u_e. \tag{2}$$

Considering $\mathfrak{L}_{\text{eff}}$ as fundamental we note the impossibility of calculating higher orders within the framework of usual perturbation theory. A redefinition of a finite number of quantities as masses, charges etc. is not enough to obtain a finite answer for the scattering matrix, the theory is nonrenormalizable. You will hear nevertheless how to handle this theory, in the lecture of Dr. Domokos.

One may ask if there exist other theories which lead for decay processes to the same effective interaction (1). Certainly there are many ways some nonlocality can appear. The most obvious one is to introduce an intermediate vector boson W. The probability amplitude is then no longer given by the first order matrix element from (1)

$$\mathfrak{M}_{\text{eff}} = -\frac{iG}{\sqrt{2}}\, j^{(1)}_{\mu}\, j^{\mu(2)} \tag{1'}$$

but it is really a consequence of a second order process with exchange of a W (Fig. 1).

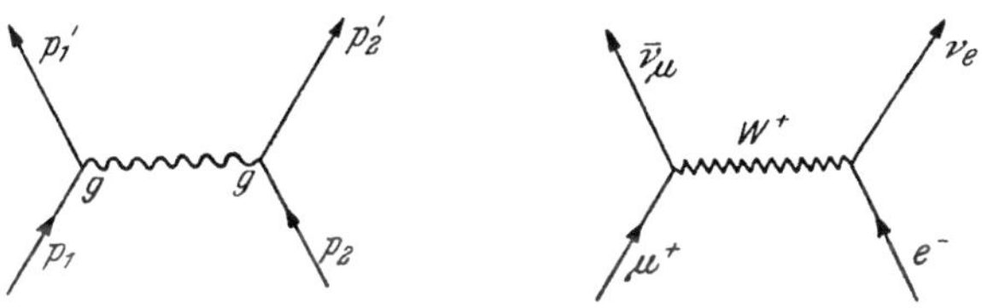

Fig. 1. Exchange of a W

This interaction is described by a Lagrangian

$$\mathfrak{L}^{W}_{\text{int}} = g\,\left(j^{(1)}_{\mu} + j^{(2)}_{\mu}\right)\, W^{\mu} \tag{3}$$

so that

$$\mathfrak{M} = g^2\, j_{\mu}\, P^{\mu}_{\nu}\, j^{\nu} \tag{4}$$

where P^{ν}_{μ} is the propagator of a vector particle

$$P^{\nu}_{\mu} = -\frac{i\left[\delta^{\nu}_{\mu} + \dfrac{q_{\mu}\, q^{\nu}}{m^2}\right]}{q^2 + m^2} \tag{5}$$

$$q = p_1 - p_1{}' = p_2{}' - p_2\,. \tag{6}$$

For processes in which $q^2/m^2 \ll 1$ and if leptons are involved, so that

$$\bar{u}_{\mu}\, \frac{q^{\gamma\mu}_{\nu}}{m}\, (1 + \gamma_5)\, u_e = \bar{u}\, (1 - \gamma_5)\, u \cdot \frac{m_e}{m} \ll 1$$

we may compare (4) and (1'). We obtain for the coupling constant g of the W

$$\frac{g^2}{m^2} = \frac{G}{\sqrt{2}} \tag{7}$$

and from the experimental value $\dfrac{G}{\sqrt{2}} = 10^{-5}/m_N^2$ where m_N is the nucleon mass

$$g \simeq 3.10^{-3} \frac{m}{m_N}. \tag{8}$$

If we try to calculate higher orders in this theory, we find again that it is not renormalizable, the diagrams diverge with increasing degree. In order to avoid ambiguous cutoff procedures LEE and YANG [2] restricted themselves to the *lowest order process*. With two dublets

$$\begin{pmatrix} W^+ \\ W^0 \end{pmatrix} \qquad \qquad \begin{pmatrix} W^- \\ \overline{W^0} \end{pmatrix}$$

and the further assumption that there is *no coupling of* W_0 *to* the leptons (absence of neutral leptonic currents) they were able to describe the $\Delta I = \frac{1}{2}$, $\Delta S = \Delta Q$ — rules in the decay of strange particles as well. The second assumption is necessary to explain the experimental absence (or smallness) of the decay

$$K_0^{(1)} \rightarrow \mu^+ + \mu^-$$

which would go through an W^0 under the first assumption (fig. 2).

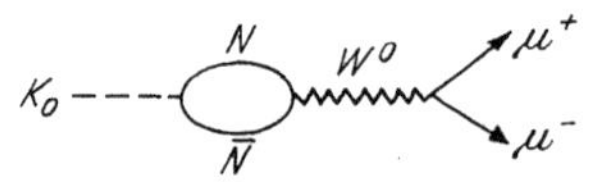

Fig. 2. Decay of $K_0^{(1)}$ into leptons

On the other hand neutral currents are definitely needed for producing $\Delta I = \frac{1}{2}$ in *nonleptonic* decays of strange particles. Charged nucleon currents have $\Delta I = 1$, strangeness changing currents $\Delta I = \frac{1}{2}$. From the composition of angular momenta we know that a state with total $\Delta I = \frac{1}{2}$ is necessarily a mixture of the products $|\Delta I = 1\rangle \, |\Delta I = \frac{1}{2}\rangle$ and $|\Delta I = 0\rangle \, |\Delta I = \frac{1}{2}\rangle$ so that a current with $\Delta I = \frac{1}{2}$ is needed.

Let me emphasize however the unsatisfactory features of such a theory. As we already mentioned we need for explaining $\Delta I = 0$ in the reaction (fig. 3)

$$\Lambda^+ \rightarrow n + \pi^0$$

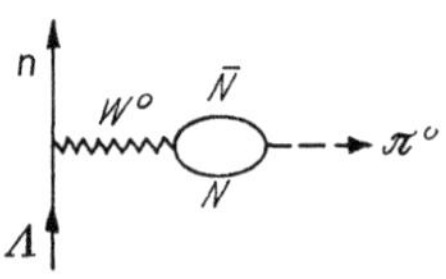

Fig. 3. Λ decay through W^0

within the Lee-Yang assumptions a neutral W^0.

Of the same order is certainly the correction from strong interactions (fig. 4).

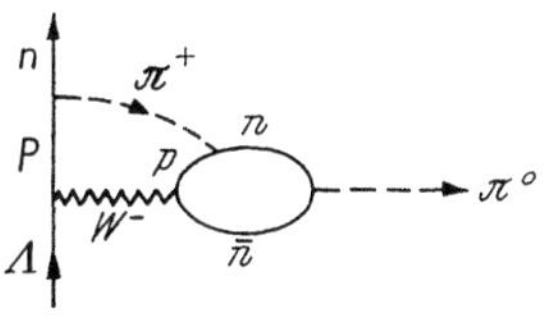

Fig. 4. Possible correction by strong interactions

We may conclude that the intermediate state $\pi^+ W^-$ simulates a W^0 and that there is really no need for introducing explicitely a W^0. Moreover this can considered to be then the explanation of the absence of neutral currents in semileptonic or leptonic processes: There is no such additional exchange of strongly interacting particles between the two fermion lines possible and the W^+ and W^- have to play alone.

One expects to find W^+ and W^- in high energy neutrino experiments

$$\nu_\mu + n \to \mu^- + p$$

where no longer $q^2 \ll m^2$. But then a strong influence of the formfactors $F_V^{1,2}(q^2)$ and $F_A^{1,2}(q^2)$ in the nucleon currents shows up. Even if $F_V^{1,2}(q^2)$ are taken from electromagnetic experiments according to the conserved vector hypothesis the axial vector formfactors $F_A^{(1)(2)}(q^2)$ remain unknown. The latter may e.g. simulate a "propagator" $(q^2 + m_0{}^2)^{-1}$ in the matrixelement as well.

The production of $W^\pm$ itself on the other hand

$$\nu_\mu + \{Z\} = \{Z\} + \mu^- + W^+ \searrow \begin{cases} \mu^+ + \nu_\mu \\ e^+ + \nu_e \end{cases}$$

is a very difficult experiment, so that the existence of the W has not yet been proved conclusively.

Higher Order Field Theory of the W in Weak Interactions, "Peratization".

We already mentioned that the usual perturbation theory is divergent in an nonrenormalizable way (excepted the lowest order only) and has to be made finite somehow. This is exactly what "peratization" means (the world is derived from the Greek $\alpha \pi \varepsilon \iota \varrho o \nu$).

The possibility for a finite result certainly exists if it is the perturbation approach alone which causes the trouble in nonrenormalizable theories.

Let us consider a simple example. After introduction of a cutoff Λ the usual perturbation theory may lead for a fictious process to the matrix element

$$M(\Lambda) = g^2 [M_0 + M_1 (g^2 \Lambda + g^4 \Lambda^4 + g^6 \Lambda^6 + \ldots)] \qquad (9)$$

For $\Lambda \to \infty$ each term diverges with an increasing power of Λ but if we sum the geometric series first

$$M(\Lambda) = \frac{g^2 M_1}{1 - g^2 \Lambda^2} + g^2 (M_0 - M_1) \tag{10}$$

then

$$M = \lim_{\Lambda \to \infty} M(\Lambda) \tag{11}$$

gives a finite result which is different from the lowest order even if g^2 is a very small quantity. In our example the higher order terms change $g^2 M_0$ into $g^2 (M_0 - M_1)$.

The fact that M as a whole exists whereas the single terms of perturbation theory diverge, has been put as an *assumption* in the LEE-YANG [3] theory of electromagnetic interactions of the W (which is nonrenormalizable too).

In the peratization of weak interactions it is *shown* that this is really true for a certain infinite sum of diagrams. This, of course, is the weakest point of the peratization since a priori we cannot expect the contributions of other diagrams to be smaller than the ones of the class under consideration. This class is the infinite sum of all ladder graphs in leptonic reactions, e. q.

$$\nu_\mu + e^- \to \nu_e + \mu^- \tag{12}$$

$$\nu_\mu + e^- \to \nu_\mu + e^-. \tag{13}$$

In the first one on odd number of W is exchanged (fig. 5)

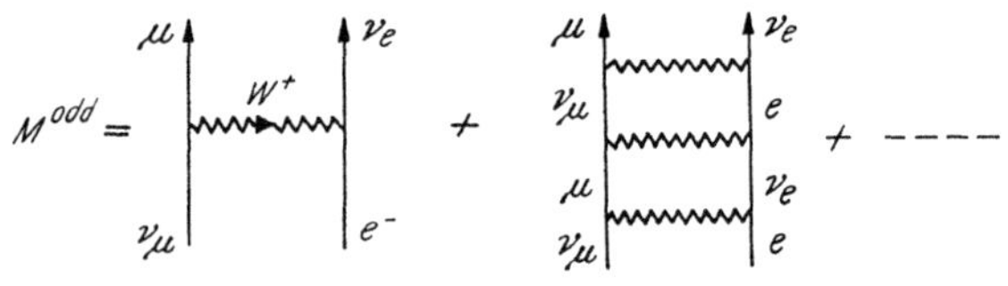

Fig 5. Sum of odd diagrams

in the second one only the ladders with an even number of rungs is contained (fig. 6)

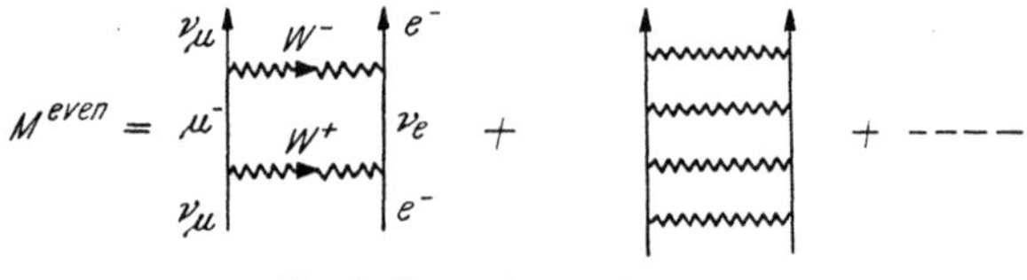

Fig. 6. Sum of even diagrams

We have to treat both infinite sums together because they are related by the following simple prescription:

Take both amplitudes with the outgoing particles virtual (off the mass shell), then

$M^{\text{even}} = M^{\text{odd}}$ with one more W-propagator and two more fermion propagators,

$M^{\text{odd}} = M^{\text{Born approx.}} + M^{\text{even}}$ with one more W-propagator and two more fermion propagators.

This means graphically $(p_1 + p_2 = p_1'' + p_2'' = p_1' + p_2')$, (fig. 7)

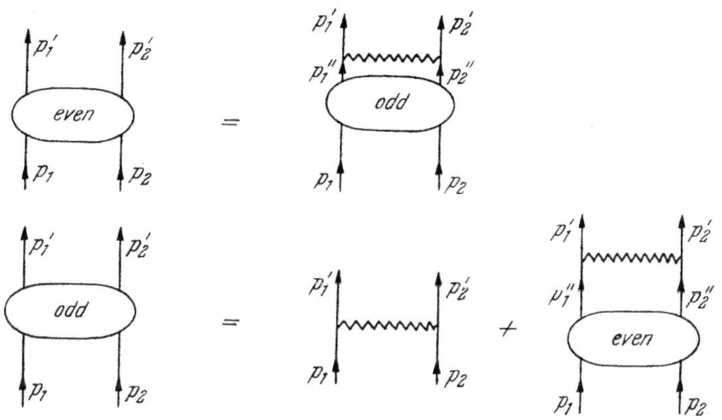

Fig. 7. Graphical representation of the integral equations

Using ordinary Feynman rules and taking the masses of the leptons equal zero (which is certainly allowed for a mass of the W of the order of the nucleon mass or bigger) we get the coupled integral equations

$$M^{\text{odd}}(p_1', p_2', p_1, p_2) = M^{(1)} + \frac{i g^2}{(2\pi)^4} \int \gamma\, \mu^{(1)} (1 + \gamma_5^{(1)}) \frac{1}{\not p_1''} \times$$
$$\times \gamma_\nu^{(2)} (1 + \gamma_5^{(2)}) \frac{1}{\not p_2''} \frac{\left[\delta_{\mu\nu} - \dfrac{(p_1'' - p_2')_\mu (p_1'' - p_2')_\nu}{m^2}\right]}{(p_1'' - p_1')^2 + m^2} \times \tag{14}$$
$$\times M^{\text{even}}(p_1'', p_2'', p_1, p_2)\, d^4 p_1''$$

$$M^{\text{even}}(p', p_2', p_1, p_2) = \int \{\text{same kernel}\}\, M^{\text{odd}}(p_1'', p_2'', p_1, p_2)\, d^4 p_1'' \tag{15}$$

$$M^{(1)} = -i g^2 \gamma_\mu^{(1)} (1 + \gamma_5^{(1)}) \gamma_\nu^{(2)} (1 + \gamma_5^{(2)}) \frac{[\delta_{\mu\nu} + (p_1' - p_1)_\mu (p_1' - p_1)_\nu / m^2]}{(p_1' - p_1)^2 + m^2}. \tag{16}$$

Equations of this type are called Bethe-Salpeter-equations in the ladder approximation. They can be decoupled by introducing

$$M^{\pm} = M^{\text{odd}} \pm M^{\text{even}} \tag{17}$$

In terms of this new quantity (14) and (15) read in an abbreviated but obvious notation

$$M^{\pm} = M^{(1)} \pm \int K M^{\pm}\,. \tag{18}$$

The equations are too difficult for a complete solution. Therefore in their first investigation of this problem Feinberg and Pais concentrate on the most divergent part of the kernel. Then of course some questions remain open: Does this give really the most important part of the solution' ? May the lesser singular parts not add up to something equally important? This is the reason why we will give later another method which seems to have a better foundation and was developed by the same authors.

The most diverging part of the kernel is the one with $-p_{1\mu}'' p_{2\nu}''$ in the numerator. Together with the γ-matrices it just cancels the lepton propagators

$$M^{\pm} = M^{(1)} \pm \frac{i g^2}{m^2 (2\pi)^4} \int \left(1 - \gamma_5^{(1)}\right) \left(1 - \gamma_5^{(2)}\right) \frac{M_{\pm}' d^4 p_1''}{(p_1'' - p_1')^2 + m^2}. \tag{19}$$

Because of the cancellation of the lepton propagators this equation corresponds to the sum of the "shrunk" diagrams (fig. 8).

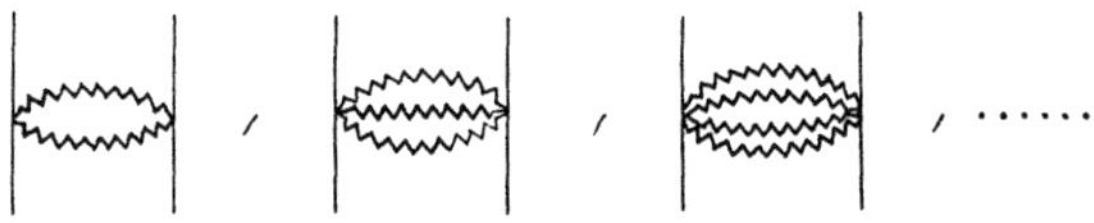

Fig. 8. Sum of "shrunk" diagrams

$M^{(1)}$ contains $\gamma_\mu (1 + \gamma_5)$ for each fermion line and there are projection operators $1 - \gamma_5$ in the kernel, therefore

$$M^{\pm} = \gamma_\mu^{(1)} \left(1 + \gamma_5^{(1)}\right) \gamma_\nu^{(2)} \left(1 + \gamma_5^{(2)}\right) M_{\mu\nu}^{\pm}. \tag{20}$$

Moreover the integral equation depends on the incoming momenta p_1, p_2 only as parameters. In terms of

$$\begin{aligned} p &= p_1' - p_1 \\ p' - p &= p_1'' - p_1' \end{aligned} \tag{21}$$

we must have from invariance

$$M_{\mu\nu}^{\pm} = \alpha (p^2) \, \delta_{\mu\nu} + \beta (p^2) \, p_\mu p_\nu \tag{22}$$

and

$$M_{\mu\nu}^{\pm} (p) = \frac{-i g^2}{p^2 + m^2} \left(\delta_{\mu\nu} + \frac{p_\mu p_\nu}{m^2}\right) \mp \frac{4 i g^2}{(2\pi)^4 m^2} \int M_{\mu\nu}^{\pm} (p') \frac{d^4 p'}{(p' - p)^2 + m^2}. \tag{23}$$

This integral equation is of the convolution type. Hence it can be reduced to an algebraic one by a Fourier transformation

$$M_{\mu\nu}^{\pm} (y) = \frac{1}{(2\pi)^4} \int M_{\mu\nu}^{\pm} (p) \, e^{-ipy} \, d^4 p. \tag{24}$$

In this way α and β of (22) can be obtained. We will show how this works for

$$T^{\pm} = M_{\mu\nu}^{\pm} = 4 \alpha (p^2) + p^2 \beta (p^2). \tag{25}$$

With

$$\Delta_F(y) = \frac{1}{(2\pi)^4} \int \frac{e^{-ipy}}{p^2 + m^2} d^4 p \tag{26}$$

we arrive at

$$T^{\pm}(p^2) = -i g^2 \int e^{ipy} \frac{\left(4 - \frac{\Box}{m^2}\right) \Delta_F(y)}{1 \pm \frac{4 i g^2}{m^2} \Delta_F(y)} . \tag{27}$$

As

$$(\Box - m^2) \Delta_F(y) = -\delta^4(y)$$

we have

$$T = -\frac{i g^2}{m^2} \frac{1}{1 \mp \frac{4 i g^2}{m_2} \Delta_F(0)} - 3 i g^2 \int \frac{e^{ipy} \Delta_F(y)}{1 \mp \frac{4 i g^2}{m^2} \Delta_F(y)} . \tag{28}$$

We have manipulated in this calculation formally with the function Δ_F which has a strong singularity in the light cone. Therefore one should in fact regularize first with a cutoff Λ, e. g. by replacing $[(p'-p)^2 + m^2]^{-1}$ in (23) by

$$\Lambda^4 [\Lambda_2 + (p'-p)^2]^{-2} [(p'-p)^2 + m^2]^{-1}$$

and let Λ go to infinity *after* the integral equation has been solved. We will keep this in mind for the other approach which will be discussed afterwards. In the last equations we really have not made a mistake by forgetting about this regularization.

Since $\Delta_F(0) = \infty$ the first term in (28) vanishes and for small g^2 and $p^2 = 0$ from (25) [we assume $\beta(0)$ to be finite]

$$T(0) = 4\alpha = -\frac{3 i g^2}{m^2} \tag{29}$$

and [c. f. (22)]

$$M^{\pm}_{\mu\nu}(0) = -\frac{3}{4} \frac{i g^2}{m^2} = \frac{3}{4} M^{(1)}_{\mu\nu}(0) . \tag{30}$$

We see that the inclusion of higher order ladders has the effect of reducing the Born approximation by a factor 3/4. The result is finite. An important consequence is that

$$\begin{aligned} M^{\text{odd}} &= \frac{1}{2}(M^+ + M^-) = \frac{3}{4} M^{(1)} \\ M^{\text{even}} &= \frac{1}{2}(M^+ - M^-) = O(g^4) \end{aligned} \tag{31}$$

and therefore the process with $\Delta Q = 0$

$$\nu_\mu + e^- \to \nu_\mu + e^-$$

which corresponds to the presence of a "neutral current" in the sense of the phenomenological analysis is of higher order in the weak coupling constant. This is in agreement with the experimental facts.

Similar conclusions do not seem to be justified for semileptonic processes because the nucleonic formfactors of the W together with other strong corrections play an important role. The formfactors may even make the higher order diagrams convergent if they converge rapidly enough to zero for increasing momenta of the virtual particles. As a consequence the equality of the β- and μ-decay constant (as explained by the conserved vector current hypothesis) is difficult to understand within this theory.

The procedure of taking the most divergent part is not free from objections. Therefore later on FEINBERG and PAIS ([1], II) proposed another approach which goes as follows. In (14), (15), and (16) they consider the special case

$$p_1 = p_2 = 0, \tag{32}$$

so that

$$\begin{aligned} p_1' &= -p_2' = p, \\ p_1'' &= -p_2'' = p'. \end{aligned} \tag{33}$$

There are several arguments in favour of this special simplified case. We already neglected the masses of the leptons so that $p_1^2 = p_2^2 = 0$. If the lepton energies are small compared to m, as it is the case in leptonic reactions, (32) is a justified approximation. Moreover a power counting argument shows that the most singular part very probably is contained in the treatment of this case. In each graph the most diverging part for $p_1 = p_2 = p_1' = p_2' = 0$ comes from the $p_{1\mu}'' \, p_{2\nu}''$ in the numerator of the W-propagator. The next term in a series in an external momentum stems either from one of the denominators, then e. g.

$$\frac{1}{(p_1'' - p_1')^2 + m^2} \simeq \frac{1}{p_1''^2 + m^2} + \frac{2\, p_1'' \, p_1'}{(p_1''^2 + m^2)^2} + \cdots$$

and it is clearly less divergent, because it goes as $|\, p_1'' \,|^{-3}$. If it originates from the numerator then it cannot be the most divergent part $p_{1\mu}'' \, p_{2\nu}''$ but has to be the consequence of a term $p_{1\mu}' \, p_{2\nu}''$ which induces a lesser divergence.

With (32) and (33) we get from (14), (15), and (16) after algebraic manipulations for $T\,(p^2)$ as given by (25) and for $\beta\,(p^2)$ of (22) two uncoupled integral equations

$$\begin{aligned} T^{\pm}\,(p, \Lambda) = &- \frac{i\, g^2 \left(4 + \dfrac{p^2}{m^2}\right)}{p^2 + m^2} \pm \\ &\pm \frac{4\, i\, g^2 \Lambda^4}{(2\,\pi)^4} \int \frac{d^4 p' \left(4 + \dfrac{(p' - p)^2}{m^2}\right) T^{\pm}\,(p'\, \Lambda)}{p'^2\,(p'^2 + \Lambda^2)^2\,[(p' - p)^2 + m^2]} \end{aligned} \tag{34}$$

$$\begin{aligned} \beta^{\pm}\,(p, \Lambda) = &- \frac{i\, g^2}{m^2\,(p^2 + m^2)} \pm \\ \pm \frac{16\, i\, g^2 \Lambda^4}{3\,(2\,\pi)^4\, m^2} &\frac{(p_\mu\, p_\nu - \frac{1}{4}\,\delta_{\mu\nu}\, p^2)}{p^4} \int \frac{(p - p')_\mu\,(p - p')_\nu\, \beta^{\pm}\,(p')\, d^4 p'}{(p'^2 + \Lambda^2)^2\,[(p' - p)^2 + m^2]} \end{aligned} \tag{35}$$

which already contain the regularization factor $\Lambda^4\,(\Lambda^2 + p'^2)^{-2}$. This factor can be understood as a regularization of the lepton propagators.

Feinberg and Pais propose an iteration scheme for solving these equations which we will scetch for equation (34): They divide

$$T = T_1 + T_2 \tag{36}$$

where

$$T_1 = -\frac{i g^2}{m^2} \pm \frac{4 i g^2 \Lambda^4}{m^2 (2\pi)^4} \int \frac{d^4 p' \,[T_1(p') + T_2(p')]}{p'^2 (p'^2 + \Lambda^2)^2} \tag{37}$$

$$T_2 = -\frac{3 i g^2}{p^2 + m^2} \pm \frac{12 i g^2 \Lambda^4}{(2\pi)^4} \int \frac{d^4 p' \,[T_1(p') + T_2(p')]}{p'^2 (p'^2 + \Lambda^2)^2 \,[(p' - p)^2 + m^2]}\,. \tag{38}$$

From (37) T_1 is a constant, independent of p. One solves *exactly* for T_1 in each iteration step for T_2 which is considered to be small. Hence to lowest order in g^2 we get for T_2 from (38)

$$(T_2)_{p^2=0} = -\frac{3 i g^2}{m^2} \pm \frac{3 g^2}{2\pi^2} \log \frac{\Lambda}{m}\, T_1(\Lambda)\,. \tag{39}$$

We have to insert this now (for $p^2 = 0$) into (37) and solve *exactly* for T_1

$$T_1 = -\frac{i g^2}{m^2} \pm \frac{4 i g^2 \Lambda^4}{(2\pi)^4 m^2}\, T_1 \int \frac{d^4 p'}{(p'^2 + \Lambda^2)^2 p'^2} + 0\left(\log \frac{\Lambda}{m}\right) + T_1 0(\Lambda^2)\,. \tag{40}$$

The last two terms stem from T_2, they are of the indicated order in Λ. We note that

$$T_1 = 0\left(\frac{\log \Lambda}{\Lambda^2}\right)$$

and hence

$$\lim_{\Lambda \to \infty} T_1(\Lambda) \cdot \log \frac{\Lambda}{m} = 0\,. \tag{41}$$

Thus to lowest order

$$T - T_2 = \frac{3 i g^2}{m^2}$$

as in (29). The next step is a simple iteration of T_2 in (38). Up to this order one finds

$$\alpha^{\pm}(0) = \frac{T^{\pm}(0)}{4} = -\frac{3}{4} \frac{i g^2}{m^2} \left(1 \mp \frac{3 g^2}{4\pi^2}\right) + 0(g^6)\,. \tag{42}$$

References

1. FEINBERG, G., and A. PAIS, I. Phys. Rev. **131**, 2724 (1963); II. Phys. Rev. in print.
2. LEE, T. D., and C. N. YANG, Phys. Rev. **119**, 1410 (1960).
3. LEE, T. D., and C. N. YANG, Phys. Rev. **128**, 885 (1962).

On the Iteration Procedure in the Peratization of Weak Leptonic Interactions*

By

W. Kummer, CERN

Abstract

The field theory of weak interactions as formulated by Feinberg and Pais in terms of an intermediate vector boson leads for the sum of uncrossed ladder graphs in weak leptonic scattering to the solution of two singular integral equations, namely (34) and (35) of the last lecture. We investigate the iteration procedure of the trace-equation. Having shown its soundness in this case we can believe its validity for the second equation too. We test the Feinberg-Pais iteration scheme by looking for an exact solution of a model equation which is closely related to the original one because it differs by a slight modification in the low frequency region only. The iteration scheme applied to this theory gives therefore the same result in the second and fourth order in the coupling constant g because the result is determined essentially by the high frequency region. By this change of the original equation the integral equation in momentum space is equivalent to a boundary value problem of a second order differential equation of the hypergeometric type.

It is shown that a solution really exists quite independent of the iteration scheme. This is connected with the fact that the analytic continuation of the hypergeometric function is well known. Exchanging the sequence of the perturbation expansion in g^2 and the limiting procedure of removing the regularizing cutoff in the full solution turns out to be an allowed operation. This is the deeper reason why peratization works in the Feinberg-Pais iteration scheme. A simple example is given which shows that this is not true in general. Hence it seems doubtful if other theories or processes can be treated within a similar iteration scheme.

---

* To be published in Nuovo Cimento.

Higher Approximations to Weak Interactions

By

G. Domokos*, CERN-Geneva

With 10 Figures

Summary

The subject of this lecture is a study of the higher corrections to weak interactions by means of the Bethe-Salpeter equation. We first recapitulate the necessary knowledge of the general theory of the Bethe-Salpeter (BS) equation and turn only afterwards to the weak interactions themselves. The kernel of the BS equation will be always approximated by some simple diagrams; nevertheless it seems to us that this is sufficient in order to gain a qualitative insight into the problem.

The problems we will treat:

(1) The infinities of the BS amplitude and its low energy behaviour.

(2) Higher order corrections to some physical processes.

(3) Scattering at high energies via weak interactions in the multiperipheral approximation.

We shall be concerned mainly with four fermion interaction; a W-meson-theory can be treated along the same lines — actually simpler — as the Fermi-theory.

Apart from the classical papers on the BS equation, which are now-a-days commonly known, the present lecture is based on the following works:

I. Domokos, G. and P. Surányi, "Bound States and Analytic Properties in Angular Momentum" (Preprint; Dubna, 1963; to be published in Nuclear Physics).

II. Domokos, G. P. Surányi, and A. Vančura, "Fermi Interactions at High Energies" (Preprint, Dubna, 1963; to be published in Nuclear Physics).

III. Domokos, G., "On the Theory of Fermi Interactions" (Preprint, CERN 1964; to be published in Nuovo Cimento).

IV. Dennery, P., and G. Domokos, "Higher Order Corrections to Weak Interactions" (Preprint CERN 1964; to be published in Nuovo Cimento).

(Theses papers contain further references, in particular to the classical works and related ones dealing with strong interactions.)

---

* On leave from the Joint Institute for Nuclear Research, Dubna, USSR. Permanent adress: Central Research Institute for Physics, Budapest.

Because of lack of time we cannot attempt complete derivations and full mathematical rigor. The procedure we shall follow is rather to make the results plausible and indicate the proof; sometimes we shall refer to the original papers where the complete proof can be found. Corresponding to the style of a lecture, we quote papers other than mentioned above, immediately in the text.

1. Introduction. The Bethe-Salpeter equation

We wish to study reactions with two particles both in the initial and final states. In field-theory the four-point Green function is the most convenient tool to characterize the process. In an adiabatic theory (cf. N. N. Bogolubov, D. V. Shirkov, Introduction to the Theory of Quantized Fields, Interscience, N. Y. Ch. VI) the latter is defined as follows:

$$G(x_1 \ldots x_n) = \langle 0 | T\left(\varphi(x_1) \ldots \varphi(x_n) S\right) S^+ | 0 \rangle, \tag{1.1}$$

where $\varphi(x)$ is some — "bare" — field quantity (in the interaction picture), S = the adiabatic scattering operator, $S = T \exp i \int \mathfrak{L}_{\text{int}}(x)\, dx$. The Green function as defined by (1.1) has a close connection with the usual Feynman amplitude, T — and therefore with the transition matrix element. The connection can be seen best in terms of diagrams:

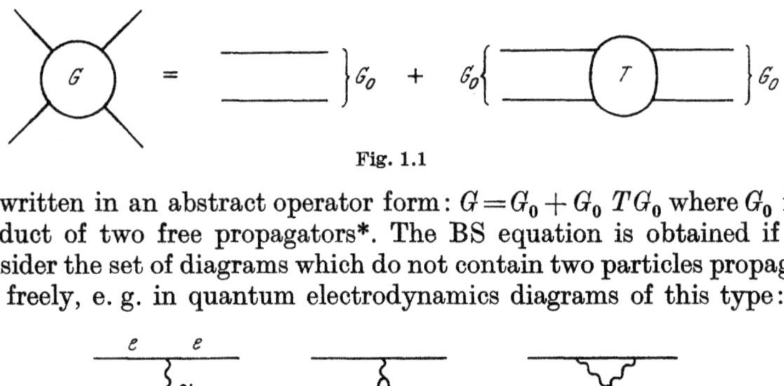

Fig. 1.1

or, written in an abstract operator form: $G = G_0 + G_0 T G_0$ where G_0 is a product of two free propagators*. The BS equation is obtained if we consider the set of diagrams which do not contain two particles propagating freely, e. g. in quantum electrodynamics diagrams of this type:

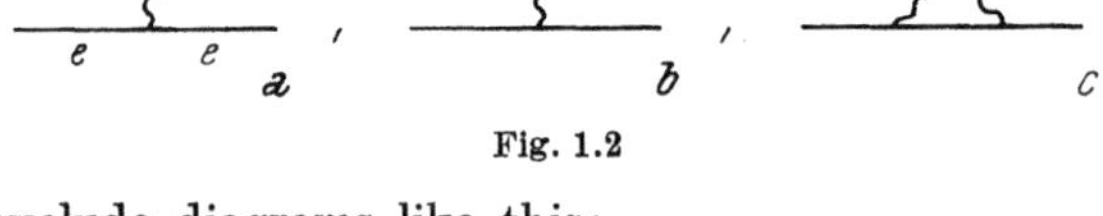

Fig. 1.2

etc., but exclude diagrams like this:

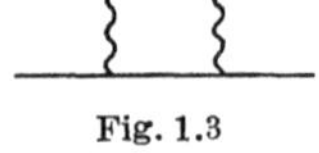

Fig. 1.3

* As a matter of fact, one should consider the full one-particle propagators instead of the bare ones in G_0. We neglect, however, this difference in what follows.

Evidently, the general diagram having four external lines is obtained by iterating those of Fig. 1. 2 (the "doubly connected" diagrams)

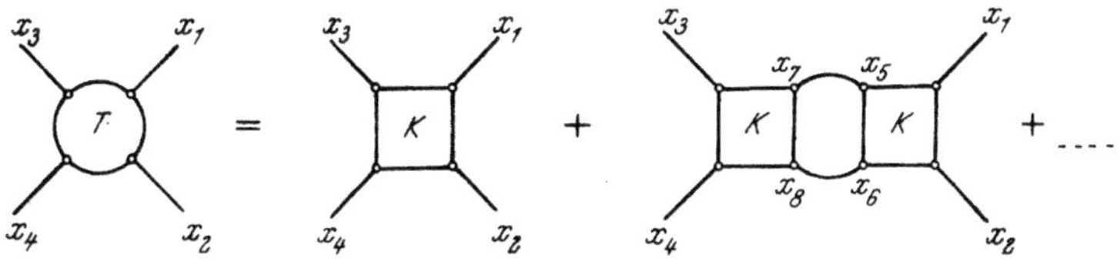

Fig. 1.4

where the square box denotes the sum of all doubly connected diagrams (we shall call it the BS-kernel). From this one easily deduces an equation which T has to satisfy:

$$T = K + K G_0 T \tag{1.2}$$

or, written for the Green function:

$$G = G_0 + G_0 K\, G. \tag{1.3}$$

It is an important point, however, that equs. (1.2) and (1.3) can be given a meaning independently of perturbation theory (K. SYMANZIK, Hercegnovi lectures, 1961). In order to obtain really an equation, one has to write out the abstract relations (1.2) and/or (1.3) in some definite representation. Let us start from the momentum representation. Because of the translation invariance of (1.1) we have three independent momenta. The choice of them (commonly used) is indicated by writing out the Fourier transform of (1.1):

$$G(x_1\, x_2 \,|\, x_3\, x_4) = \frac{1}{(2\pi)^6} \int d^4 p\, d^4 p'\, d^4 E \cdot G(p, p' \,|\, E) \cdot$$
$$\cdot \exp\left[-ip(x_1 - x_2) + ip'(x_3 - x_4) + iE\, \frac{x_3 + x_4 - x_1 - x_2}{2} \right], \tag{1.4}$$

which shows that E is the total four-momentum, p, p' are the relative momenta in the "initial" and "final" states, respectively. Then we obtain e. g. for two spin 1/2 particles of mass m:

$$G_0 = \delta(p - p')\, \frac{\hat{p}^{(1)} + \frac{1}{2}\hat{E}^{(1)} + m}{(p - \frac{1}{2}E)^2 - m^2} \cdot \frac{\frac{1}{2}\hat{E}^{(2)} - \hat{p}^{(2)} + m}{(p + \frac{1}{2}E)^2 - m^2} \equiv \frac{\delta(p - p')}{F}.$$

So that if one multiplies (1.3) by the inverse of the free propagators, one obtains

$$[\hat{p}^{(1)} + \tfrac{1}{2}\hat{E}^{(1)} - m]\, [\tfrac{1}{2}\hat{E}^{(2)} - \hat{p}^{(2)} - m]\, G(p, p' \,|\, E) =$$
$$= \delta(p - p') + \frac{1}{(2\pi)^4 i} \int dq\, K(p, q \,|\, E)\, G(q, p' \,|\, E), \tag{1.5}$$

where $\hat{a} = \gamma_0 a_0 - \underline{\gamma}\underline{a}$, the superscriptes "1" and "2" referring to the two fermions. (Obviously, G, K are direct products of 4×4 matrices in spinor space, which should be decomposed according to a basis of the "direct square" of the Dirac algebra.)

In what follows we make two important simplifications:

(1) We investigate the "leading singularity" ("the most divergent part") of the diagrams in Fig. 1.4, as defined e. g. in I or in FEINBERG-PAIS, Phys. Rev. **131,** 2724 (1963).

This means that the masses of the propagating particles can be dropped.

(2) We shall consider kernels, which depend on $(p-q)$ only ("local kernels").

Remarks:

The first simplification is perfectly justified, as far as one is interested in the most singular part of the amplitude. (Intuitively one believes that this gives the largest contribution to the diagram.) As a matter of fact, we could drop E as well. However, we shall see later that it is convenient to retain an energy variable throughout. The second one seems to be absolutely unjustified from the physical or mathematical point of view: diagrams like Fig. 1.2c depend not only on $(p-q)$ but on p, q, E separately. Nobody has ever shown that the "local" diagrams (e. g. Fig. 1.2a, b) are more important. Nevertheless, we have to make the assumption (2) in order to render the problem tractable at all, hoping that the result will reflect some qualitative properties of the full amplitude.

Let us now assume that we have already decomposed the Green matrix and projected out the equations for the corresponding invariant functions. In general we shall be left with a system of coupled equations which, for a suitable linear combination of the invariant functions, can be written in the following form:

$$F_i \, G_i \, (p, p' \mid E) = \delta \, (p-q) + \frac{1}{(2\,\pi)^4\, i} \int d^4 q \sum_{j=1}^{n} P_{ij} K_{ij} \left((p-q)^2, E\right) \cdot \\ \cdot \, G_j \, (q p' \mid E) \tag{1.6}$$

where F_i is a linear combination of p^2, pE, E^2 and P_{ij} is a polynomial in p^2, q^2, pE, qE, E^2. We now go over the Euclidean metric in the space of the relative momenta [cf. G. C. WICK, Phys. Rev. **96,** 1124 (1954)] i. e. rotate the contour of integration in the q^0 plane to the imaginary axis. (A discussion of the problems connected with this procedure in the case of singular interactions will be given in the last section.)

Expand now the functions in terms of hyperspherical harmonics. The latter arise from the separation of the four dimensional Laplacian in polar coordinates (A. KRATZER, W. FRAUTZ, Transzendente Funktionen, Akademie-Verlag, Leipzig, 1960). They constitute an orthonormal system on the surface of the four dimensional unit sphere; their most important properties are listed in the Appendix. [Unlike the widespread custom we define eigenvalues of the square of the four dimensional orbital momentum being *positive* integers only:

$$\tfrac{1}{4}\, L_{\mu\nu}\, L^{\mu\nu}\, \psi_n\,(x) = \tfrac{1}{2}\,(n^2-1)\,\psi_n\,(x)$$
$$n = 1, 2, 3, \ldots, \text{ with } L_{\mu\nu} = i^{-1}\,(x_\mu\, \partial_\nu - x_\nu\, \partial_\mu)].$$

Choosing the 0-axis in the direction of E_μ (CMS), we can see that in general neither F, nor $\mathfrak{K}_{ij} \equiv P_{ij} K_{ij}$ are diagonal in (n, l, m). To see this, let us calculate e. g. the matrix elements of the quantity $(pE)^2 = = p^2 E^2 \cos^2\psi \equiv p^2 E^2 x^2$. Denote:

$$\langle n' \, l' \, m' | \, x^2 \, | n l m \rangle \equiv \delta_{l l'} \, \delta_{m m'} \, f^l_{n n'} .$$

Then we have:

$$f^l_{n n'} = f^l_{n' n} = \frac{1}{N_{n l} \, N_{n' l}} \int\limits_{-1}^{+1} dx \, C^{l+1}_{n'-l-1} (x) \, C^{l+1}_{n-l-1} (x) \cdot x^2 \, (1 - x^2)^{l + 1/2} .$$

As x^2 is a reducible tensor of second rank, the matrix elements $f^l_{n n'}$ will be different from zero for $n' = n$, $n' = n \pm 2$ only. In order to find the explicit expression of $f^l_{n n'}$, one has to apply the recursion formula

$$(n + 2) \, C^\lambda_{n+2} (x) = 2 \, (\lambda + n + 1) \, C^\lambda_{n+2} (x) - (2 \, \lambda + n) \, C^\lambda_n (x)$$

twice. As a result, we find the non-vanishing matrix elements:

$$f^l_{n n} = - \frac{1}{2} \frac{l \, (l + 1)}{(n - 1) \, (n + 1)} + \frac{1}{2} = \frac{(n - 1) \, (n + 1) - l \, (l + 1)}{2 \, (n - 1) \, (n + 1)} ,$$

$$f^l_{n+2, n} = \frac{1}{4 \, (n + 1)} \sqrt{\frac{(n + l + 2) \, (n + l + 1) \, (n - l + 1) \, (n - l)}{n \, (n + 2)}} ,$$

$$f^l_{n-2, n} = \frac{1}{4 \, (n - 1)} \sqrt{\frac{(n + l) \, (n + l - 1) \, (n - l - 1) \, (n - l - 2)}{n \, (n - 2)}} .$$

Nevertheless, exactly as in the case of an expansion in terms of three-dimensional spherical harmonics, we can define a suitable linear combination of the amplitudes, for which the product

$$\mathfrak{K} \, G \equiv \sum_j \sum_{n''} \langle n | \, \mathfrak{K}_{ij} \, | n'' \rangle \, \langle n'' | \, G_j \, | n' \rangle$$

becomes diagonal. However, for the same combination,

$$\sum_{n''} \langle n | \, F \, | n'' \rangle \, \langle n'' | \, G \, | n' \rangle$$

will not be diagonal.

We now *omit* the off-diagonal terms of F in the combination, where $\mathfrak{K} G$ *is* diagonal. (These terms are obviously which contain the scalar product pE.) So finally, for the suitably combined amplitudes, $H^{(i)}$ we get for two Dirac-particles:

$$(p^2 - {}^1/_4 E^2) \, \langle p^2 n | \, H^{(i)} \, (E^2) \, | n' \, p'^2 \rangle = \frac{\delta \, (p - p')}{p^3} +$$

$$+ \frac{1}{(2 \, \pi)^4 \, i} \sum_{n''} \int\limits_0^\infty q^3 \, dq \, \langle p^2 \, n | \, L^{(i)} \, | n'' \, q^2 \rangle \, \langle n'' \, q^2 | \, H^{(i)} \, (E^2) \, | n' \, p \rangle \quad (1.7)$$

where $L^{(i)}$ is the corresponding combination from the matrix elements of the kernel matrix. Let us now notice that actually $\langle p^2 \, n | \, L^{(i)} \, | n' \, p^2 \rangle$ is independent of 1, m (the latter by WIGNER-ECKART) for a local kernel and even diagonal in n. In fact, the contribution of any diagram, giving a local kernel in the sense defined above, can be written in Euclidean metric as follows:

$$L(q, E^2) = \frac{q^{2N}}{\pi} \int\limits_{\mu_0}^{\infty} \frac{d\mu^2 \cdot \sigma(\mu^2, E)}{\mu^{2N}(\mu^2 + q^2)} + \sum_{k=0}^{N-1} C_k(E)\, q^{2k}. \tag{1.8}$$

Let us ignore for the moment the subtraction polynomial in eq. (1.8) and consider $Re\, n > N$.

Then we must have for the denominator:

$$\frac{1}{\mu^2 + p^2} = \int\limits_0^{\infty} r^3\, dr \int\limits_0^{\pi} \sin^2 \psi\, d\psi \int\limits_0^{\pi} \sin \delta\, d\delta \int\limits_0^{2\pi} d\varphi\, e^{ipx}\, G(r).$$

Choosing $p = (p, 0)$ we have $px = pr \cos \psi$ and we can integrate over the three-dimensional solid angle to find:

$$\frac{1}{\mu^2 + p^2} = 4\pi \int\limits_0^{\infty} r^3\, dr\, G(r)\, F(pr)$$

where

$$F(pr) = \int\limits_0^{\pi} d\psi \sin^2 \psi\, e^{ipr\cos\psi} = \frac{\pi}{p} J_1(pr)$$

and $J_1(x)$ is the Bessel function of the first kind of index one. Hence, by a well-known integral formula

$$G(r) = \frac{\mu}{\pi r} K_1(\mu r). \tag{1.9}$$

$K_1(x)$ being MacDonald's function of index one. In a similar way, noting that $(p-q)^2 = p^2 + q^2 - 2pq\cos\omega$ and making use of the addition theorem of Bessel functions:

$$\frac{J_1(r|p-q|)}{r|p-q|} = 2 \sum_{n=1}^{\infty} n \frac{J_n(pr)\, J_n(qr)}{pr \cdot qr} C_{n-1}^1(\cos\omega) =$$

$$= \frac{1}{\sqrt{2}} \sum_{n=1}^{\infty} \frac{J_n(pr)}{pr} \frac{J_n(qr)}{qr}\, n\, Z_{n0}^0(\cos\omega).$$

Expressing $\cos\omega$ in terms of the unit vectors $\hat{p}$ and $\hat{q}$ by means of the four-dimensional cosine theorem and making use of the addition theorem of four-dimensional spherical functions, we arrive at the expansion:

$$L(p, p' \mid E) = \sum_{n,l,m} Z_{nl}^{m*}(\hat{p}) \cdot L_n(p^2, p'^2 \mid E^2)\, Z_{nl}^m(\hat{p}') \tag{1.10}$$

with

$$L_n(p^2, p'^2 \mid E^2) = \int\limits_0^{\infty} r^3\, dr \frac{J_n(pr)}{pr} V(r) \frac{J_n(qr)}{qr}$$

$$V(r) = \int\limits_{\mu_0}^{\infty} d\mu\, \mu^3\, \sigma(\mu) \frac{K_1(\mu r)}{\mu r}. \tag{1.11}$$

(We omitted the variable E^2.)

The form (1.11) suggests to carry out a Hankel transformation on eq. (1.7) with the functions $J_n(p\,r)/p\,r$, which are orthogonal with the weight r^3:

$$\int_0^\infty r^3\,dr\,\frac{J_n(p\,r)\,J_n(p'r)}{p\,r\,p'\,r} = \frac{\delta(p-p')}{(p\,p')^{3/2}}\,.$$

Hence, by the Bessel differential equation we find that p^2 is represented by the following differential operator:

$$p^2 \to \frac{d^2}{dr^2} + \frac{3}{r}\frac{d}{dr} - \frac{n^2-1}{r^2}$$

so in this coordinate-representation we find for the function

$$H_n^{(i)}(r,r') = \int_0^\infty p^3\,dp \int_0^\infty p'^3\,dp'\,\frac{J_n(p\,r)}{p\,r}\,\frac{J_n(p'\,r')}{p'\,r'}\,H_n^{(i)}(p,p')$$

$$\left(\frac{d^2}{dr^2} + \frac{3}{r}\frac{d}{dr} + k^2 - \frac{n^2-1}{r^2}\right) H_n^{(i)}(r,r') = \qquad (1.12)$$

$$= \frac{\delta(r-r')}{r^3} + V(r)\,H_n^{(i)}(r,r')$$

with $k^2 = {}^1/_4\,E^2$.

Obviously, the solutions for $H_n^{(i)}(r,r')$ can be found from those of the homogeneous equation by the usual methods of constructing a Green function. On writing the homogeneous solution in the form: $r^{-3/2}\,\psi_n(r)$ we find:

$$\left[\frac{d^2}{dr^2} + k^2 - \frac{n^2 - {}^1/_4}{r^2} - V(r)\right]\psi_n(r) = 0 \qquad (1.13)$$

where $V(r)$ is given by (1.11).

Let us observe now that (1.13) is exactly of the form of an ordinary nonrelativistic Schrödinger equation, the only difference being that $l + {}^1/_2$ is replaced by n. The analogy goes still further; one can show namely that the partial scattering amplitude can be found by imposing the boundary conditions:

$$\psi_n(0) < \infty \qquad \psi(r) \sim e^{-ikr} + S_n\,e^{ikr}\ (r \to \infty)$$

and then S_n is exactly the scattering matrix element. (This has been observed in II, and in recent works of Lee and Swift and Sawyer.)

One easily checks that if for $\mu \to, \infty$ $\sigma(\mu^2)$ behaves as:

$$\sigma(\mu) \sim \sigma_0\,\mu^\alpha\,(\log\mu)^\beta$$

then the leading term in the "potential" $V(r)$ at $r = 0$ is of the following form:

$$V(r) \sim V_0\,\frac{(\log r)^{\,\beta}}{r^{4+\alpha}}.$$

We have now still to settle the following problems:

(a) Eq. (1.13) has to be continued for $Re\,n < N$, where N was the number of the subtractions in the spectral representation of the kernel.

(b) We have omitted so far the subtraction polynomial in eq. (1.8).

(c) We omitted from the F_i the non-diagonal terms, proportional to powers of pE.

We treat (b) in some detail, and sketch only the answers for (a) and (c). As to (a) we immediately see that eq. (1.13) can be given a meaning for any n; one has to be careful at the points $n=1, 2, \ldots$ only. In fact, if we come to (b) and try to include the subtraction polynomial, we immediately observe that it gives a contribution for a few points only:

$$\sum_{k=0}^{N-1} C_k\, q^{2k}$$

can be rearranged to give

$$\sum_{k=1}^{N} d_k\, C^1_{k-1}\,(\cos\omega)$$

so there are nonvanishing contributions for $n=0, 1, 2, \ldots N$ only. Actually, the subtraction polynomial is a superposition of contact interactions, because it is a polynomial in p, p'.

Now, if we have any linear operator equation of the form:

$$G=G_0+G_0\,(K_1+K_2)G, \tag{1.14}$$

and we can find the solution G_1 of

$$G_1=G_0+G_0K_1G_1,$$

then one can easily check that G obeys the equation:

$$G=G_1+G_1K_2G. \tag{1.15}$$

Let us give the role of K_1 to the spectral integral in the representation of the kernel, that of K_2 to the subtraction polynomial. Then, after expansion in terms of the four-dimensional harmonics, we first of all see that $G=G_1$ except for $n=1, \ldots, N$; at the latter points:

$$G=(1-G_1K_2)^{-1}G_1, \tag{1.16}$$

where G_1 is the Green function, constructed from the solutions of eq. (1.13). In terms of the transition amplitude, $G=G_0+G_0\,TG_0$, we have therefrom:

$$T=T_1+[1-(1+T_1G_0)\,K_2\,(n)G_0]^{-1}\cdot$$
$$\cdot\,(1+T_1\,G_0)\,K_2\,(n)\,(1+G_0\,T_1), \tag{1.17}$$

where T_1 is defined by the equation $G_1=G_0+G_0\,T_1G_0$.

Eq. (1.17) has a simple graphical interpretation:

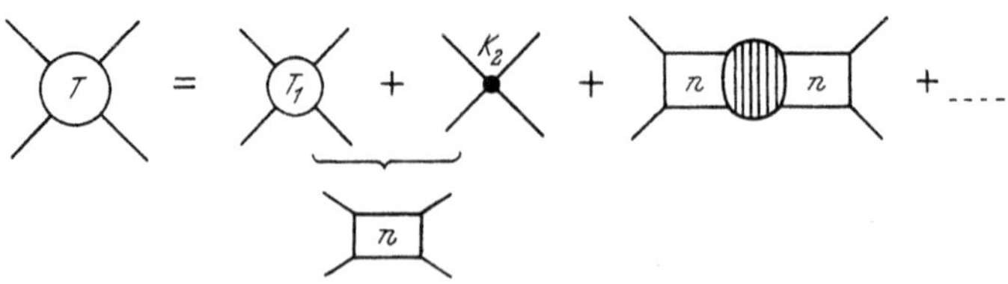

Fig. 1.5

where the symbol $\overline{\underline{|||||}}$ means the two-particle propagator corrected by the "first" part of the interaction (K_1). As to (c) we refer the reader to I—II, where it has been shown that the terms (pE), $(pE)^2$, etc., are bounded operators on the space of the solutions of (1.13). Thus for not too high energies they act as small perturbations to the interaction $V(r)$. After this preparation, we are now ready for attacking our main problem, that of weak interactions.

2. Description of the model

The model we want to study is a theory of weak interactions *without* vector mesons. We assume an elementary four-fermion interaction of the V-A type, with charged currents. In order to simplify matters we will be concerned mainly with purely leptonic processes only. The diagrams we want to consider are the following ladders:

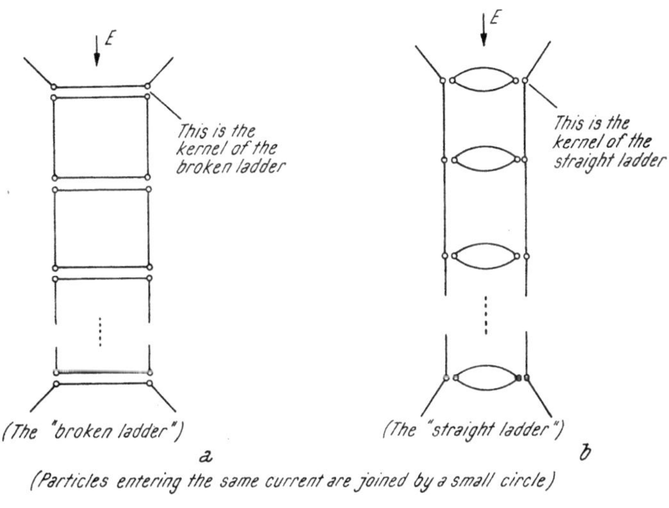

Fig. 2.1

The diagram of Fig. 2.1 a will be used to describe lepton-lepton scattering at high energies in the multiperipheral approximation: It has the property that — in the approximation, where all the fermion masses are neglected — the specification of the internal fermion loops (their orientation and the particles participating in the elementary process) is unnecessary, because it does not change anything (one has closed loops with an even number of vertices). Further it is obvious that lepton-lepton (antilepton-antilepton) and lepton-antilepton scattering differ only in the sign of the amplitude: one can be obtained from the other by reversing the orientation of the last, open fermion lines. Further it turns out that the equation we obtain for the diagram Fig. 2.1 a (the "broken ladder") is a special case of that for Fig. 2.1 b on which we want to study more general questions, like the problem of divergencies, the low energy behaviour etc.

So we turn now to the investigation of the diagram Fig. 2.1 a, and carry over most of our results to the other diagram afterwards.

In order to avoid complications arising from the spinor structure of the equation, we shall pick out just one invariant function of the total Green function, namely the coefficient of $\gamma_\mu^{(1)}(1+i\gamma_5^{(1)})\gamma_\mu^{(2)}(1+i\gamma_5^{(2)})$, where the superscripts 1,2 refer to the "external" leptons. This means that we do not consider the induced interactions but calculate just the correction to the lowest order one. This approximation, although surely not justified *a priori*, simplifies matters considerably and in fact can be shown to give a reasonable approximation, when the total energy of the ladder, E^2 is small. On the other hand, one gains sufficient qualitative insight into the problem even without considering the whole system of coupled equations between the invariant functions.

Denoting the two-lepton propagator by G, the free one by G_0, the interaction kernel by K, the BS equation in operator form reads:

$$G = G_0 + G_0 K G, \tag{2.1}$$

where the quantities involved have the following expressions in momentum representation:

$$\begin{cases} G_0(p, q; E) = \delta(p-q)\left[\tfrac{1}{2}\hat{E}^{(1)} - \hat{p}^{(1)}\right]^{-1}\left[\tfrac{1}{2}\hat{E}^{(2)} + \hat{p}^{(2)}\right]^{-1} \\ K(p, q) = \gamma_\varrho^{(1)}(1+i\gamma_5^{(1)})\gamma^{\varrho(2)}(1+i\gamma_5^{(2)})\cdot K \\ G(p, q; E) = \gamma_\varrho^{(1)}(1-i\gamma_5^{(1)})\gamma^{\varrho(2)}(1-i\gamma_5^{(2)})\, G. \end{cases} \tag{2.2}$$

In eq. (2.2) E is the total CM-four-momentum of the leptons, p, q are their relative momenta. The scalar function K is given by the following expression:

$$K = K_1(t) + K_2(t) = \frac{f^2 t^2}{(2\pi)^4 i}(2\pi)^3 \int_0^\infty \frac{dt'}{t'(t'-t)} + g_1 + g_2 t. \tag{2.3}$$

Here $t=(p-q)^2$, f stands for the Fermi coupling constant, g_1 and g_2 are subtraction constants. The symbols K_1 and K_2 stand for the spectral integral and the subtraction polynomial, respectively. The BS equation for the invariant function G can be derived in a straightforward way. The result is:

$$\left(p^2 - \frac{1}{4}E^2\right)G(p, q; E) = \delta(p-q) + \frac{16 f^2}{(2\pi)^4 i}\int dk\, K(p, k)\, G(k, q). \tag{2.4}$$

We now separate the "analytic part" of G, called G_1 [i. e. that obeying eq. (2.4) with K_1 only], go over to Euclidean metric and expand G_1 according to four-dimensional spherical harmonics. Finally we perform a Hankel transformation on the resulting radial equation. After these operations we are left with the following differential equation for the radial Green function, $G_1(r, r')$:

$$\left[\frac{d^2}{dr^2} + \frac{3}{r}\frac{d}{dr} - \frac{n^2-1}{r^2} + k^2\right] G_1(r, r') = \frac{\lambda^2}{r^6} G_1(r, r') + (r r')^{-3/2}\,\delta(r-r') \tag{2.5}$$

where $k^2 = 1/4\, E^2$, $\lambda^2 = 32\, \pi^{-1} f^2$; n^2 is the square of the four-dimensional orbital momentum. Alternatively, we can write the equation for the wave function, $\psi(r) \equiv r^{-3/2}\, u(r)$

$$\left[\frac{d^2}{dr^2} + k^2 - \frac{n^2 - 1/4}{r^2} - \frac{\lambda^2}{r^6}\right] u(r) = 0. \tag{2.6}$$

Eq. (2.6) has exactly the form of a radial Schrödinger equation with a repulsive potential $\lambda^2 r^{-6}$. Therefore, in order to find the scattering amplitude, we can apply the procedure, familiar in non-relativistic quantum mechanics. Eq. (2.6) cannot be integrated in terms of known transcendental functions, so we are going to develop an approximation procedure, valid for small values of k^2, and potentials more singular than $0(r^{-2})$ at the origin.

3. Approximate solution of the differential equation

We start from the observation that if we drop either the kinetic energy term of the potential from eq. (2.6), it can be integrated exactly in terms of Bessel functions. In fact, the equation:

$$\left(\frac{d^2}{dr^2} + k^2 - \frac{n^2 - 1/4}{r^2}\right) v_0(r) = 0 \tag{3.1}$$

has the following two independent solutions:

$$\begin{aligned} v_0^1(r) &= r^{1/2} H_n^{(1)}(kr) \\ v_0^2(r) &= r^{1/2} H_n^{(2)}(kr) \end{aligned} \tag{3.2}$$

with the boundary conditions:

$$\begin{aligned} v_0^1(r) &\sim \exp i\left\{kr - \frac{\pi}{2}n - \frac{\pi}{4}\right\} \\ v_0^2(r) &\sim \exp\left(-i\left\{kr - \frac{\pi}{2}n - \frac{\pi}{4}\right\}\right). \end{aligned} \qquad (r \to \infty) \tag{3.3}$$

Putting $k^2 = 0$, we arrive at the equation:

$$\left[\frac{d^2}{dr^2} - \frac{n^2 - 1/4}{r^2} - \frac{\lambda^2}{r^6}\right] v_1(r) = 0 \tag{3.4}$$

with the independent solutions:

$$\begin{aligned} v_1^1(r) &= r^{1/2} K_{\frac{n}{2}}\left(\frac{\lambda}{2r^2}\right) \\ v_1^2(r) &= r^{1/2} I_{\frac{n}{2}}\left(\frac{\lambda}{2r^2}\right) \end{aligned} \tag{3.5}$$

and the boundary conditions:

$$\begin{aligned} v_1^1(r) &= 0\left(r^{3/2} \exp\left(-\frac{\lambda}{2r^2}\right)\right) \\ v_1^2(r) &= 0\left(r^{3/2} \exp\frac{\lambda}{2r^2}\right). \end{aligned} \qquad (r \to 0) \tag{3.6}$$

We construct the approximate solution of eq. (2.6) from the functions (3.2) and (3.5), by neglecting the kinetic energy term if r is less than some conveniently chosen value, r_0, and neglecting the potential for $r > r_0$. It is intuitively clear that such an approximation procedure should work well if the potential is very large at small distances while decreases sufficiently rapidly as $r \to \infty$.

We choose for r_0 the distance at which the kinetic and potential energy terms are equal in magnitude, i. e., $r_0 = (\lambda/k)^{1/3}$. [If $k^2 < 0$, which is the domain we are interested in, when going over to the crossed channel, the coefficients of eq. (2.6) are even continuous at $r = r_0$, with $r_0 = (i\,\lambda/k)^{1/3}$].

We look for a solution of eq. (2.6), vanishing at $r=0$ and being a combination of ingoing and outgoing waves at infinity.

Therefore we choose

$$u(r) = \begin{cases} v_1{}^1(r) & (r < r_0) \\ F_1\, v_0{}^1(r) + F_2\, v_0{}^2(r) & (r > r_0). \end{cases} \tag{3.7}$$

Matching the logarithmic derivatives at $r = r_0$, we obtain for the scattering matrix element:

$$S(n, k^2) \equiv e^{2i\delta_n} = \frac{F_1}{F_2} = -\frac{H_n^{(2)}(\varepsilon)\, K'_{n/2}(\varepsilon/2) + H_n^{(2)\prime}(\varepsilon)\, K_{n/2}(\varepsilon/2)}{H_n^{(1)}(\varepsilon)\, K'_{n/2}(\varepsilon/2) + H_n^{(1)\prime}(\varepsilon)\, K_{n/2}(\varepsilon/2)}\,. \tag{3.8}$$

In eq. (3.8) primes mean derivatives with respect to the arguments of the cylindrical functions, and $\varepsilon = (k^2\,\lambda)^{1/3}$.

The approximation procedure described above can be considered as a zeroth order term of a perturbation series to the differential equation:

$$\left[\frac{d^2}{dr^2} + \Theta(r - r_0)\, k^2 - \Theta(r_0 - r)\,\frac{\lambda^2}{r^6} - \frac{n^2 - 1/4}{r^2}\right] u(r) = W(r)\, u(r). \tag{3.9}$$

$\Theta(x)$ being the unit step function and the perturbation operator $W(r)$ has the following expression:

$$W(r) = -k^2\,\Theta(r_0 - r) + \frac{\lambda^2}{r^6}\,\Theta(r - r_0): \tag{3.10}$$

Let us sketch the proof of the convergence of the perturbation expansion for $k^2 < 0$.

In order to simplify matters, let us split off the factor $r^{1/2}$ from the v's in eqs. (3.2) and (3.6): $v = r^{1/2}\, w(r)$ and multiply the equation for $w(r)$ by r^2. Thus we obtain with $k^2 = -\varkappa^2$ $(\varkappa^2 > 0)$:

$$\left[r^2\frac{d^2}{dr^2} + r\frac{d}{dr} - \Theta(r - r_0)\,\varkappa^2 r^2 - \frac{\lambda^2}{r^4}\,\Theta(r_0 - r) - n^2\right] w(r) = $$
$$= V(r)\, w(r) \tag{3.11}$$

with

$$V(r) = \varkappa^2 r^2\,\Theta(r_0 - r) + \frac{\lambda^2}{r^4}\,\Theta(r - r_0).$$

Considering eq. (3.11) as an eigenvalue problem for n^2 with the boundary condition $w(0) = w(\infty) = 0$, the unperturbed solution is:

$$w_\nu^{(0)}(r) = \Theta(r - r_0)\, K_{\nu/2}\left(\frac{\lambda}{2 r_0^2}\right) K_\nu(\varkappa r) + \\ + \Theta(r_0 - r)\, K_{\nu/2}\left(\frac{\lambda}{2 r^2}\right) K_\nu(\varkappa r_0) \qquad (3.12)$$

while the unperturbed eigenvalue, ν satisfies the usual determinental equation:

$$f(\nu) \equiv K_\nu(\varkappa r_0) \frac{d}{dr_0} K_{\nu/2}\left(\frac{\lambda}{2 r_0^2}\right) - K_{\nu/2}\left(\frac{\lambda}{2 r_0^2}\right) \frac{d}{dr_0} K_\nu(\varkappa r_0) = 0. \quad (3.13)$$

In the standard way we find that the solutions of eq. (3.13) are pure imaginary so that the functions $w_\nu^{(0)}$ are real. The functions $w_\nu^{(0)}$ are orthogonal:

$$\int_0^\infty \frac{dr}{r} w_\nu^{(0)}(r)\, w_{\nu'}^{(0)}(r) = N_\nu\, \delta_{\nu\nu'}$$

(N_ν being a normalization coefficient). The perturbed eigenvalue, n, is of course given by:

$$n = \nu + V_{\nu\nu} + \sum_{\nu'} \frac{V_{\nu\nu'} V_{\nu'\nu}}{\nu^2 - \nu'^2} + \dots \qquad (3.14)$$

where $V_{\nu\nu'}$, are the matrix elements of V taken between the *normalized* functions (3.12). It is easy to see that

$$|V_{\nu\nu'}| < (\varkappa r_0)^2 = (\varkappa^2 \lambda)^{2/3}$$

(which is really small for small values of $\varkappa^2$, as we have expected).

Making use of the familiar expression for the number of roots $N(R)$ inside a circle with radius $|n| = R$:

$$\int_0^R \frac{N(x)}{x} dx = \frac{1}{2\pi} \int_0^{2\pi} \log |f(R e^{i\varphi})|\, d\varphi - \log |f(0)|$$

we obtain:

$$N(R) < C R \log R,$$

where C is some constant.

Hence, for the lth eigenvalue, ν_l, we have the lower bound:

$$\nu_l^2 > C' \frac{l^2}{(\log l)^2},$$

where C' is again some constant. So we obtain that the series (3.14) converges. In a similar way one can demonstrate the convergence of the series for the perturbed eigenfunctions as well.

We can check by inspection that our S-matrix element is a meromorphic function of n in the whole complex n-plane; it satisfies the symmetry relation, characteristic of potentials with a hard core*

$$S(-n, k^2) = e^{-2\pi i n} S(n, k^2). \qquad (3.15)$$

* See e. g. PREDAZZI and REGGE, Nuovo Cim. **24**, 518 (1963).

Finally we quote the expression for the transition matrix element, defined by the relation:

$$T(n, k^2) = \frac{1}{i\pi k^2}[S(n, k^2) - 1];$$

$$T(n, k^2) = \frac{-1}{\pi k^2} \frac{J_n(\varepsilon) K'_{n/2}(\varepsilon/2) + J'_n(\varepsilon) K_{n/2}(\varepsilon/2)}{H_n^{(1)}(\varepsilon) K'_{n/2}(\varepsilon/2) + H_n^{(1)\prime}(\varepsilon) K_{n/2}(\varepsilon/2)}. \tag{3.16}$$

One can check by direct calculation that the expression (3.16) gives the correct continuation of the transition matrix element for $k^2 < 0$, $|k^2 \lambda| \ll 1$.

4. Subtraction terms and asymptotic behaviour of the scattering amplitude

All the expressions derived up to now are originally valid for $Re\, n > 2$ [as we had two subtractions in the kernel (2.3)]. However, as explained in Sec. 1. We can continue them even for $Re\, n < 2$, if only we take into account that at the subtraction points, $n = 1, 2$, the expression for the transition amplitude is *not* given by (3.16) but by a suitably modified expression, as written out in Sec. 1. To formulate the content of that equation in an intuitive way, the subtraction polynomial of the kernel can be conceived as a set of new local interaction Lagrangeans. The transition amplitude at the subtraction points is given by the infinite chain diagrams formed with these additional interactions, taking into account the vertex- and propagator corrections caused by the *analytic* part of the transition amplitude, T.

In fact, at the subtraction points the Green function G is given by

$$G = G_1 + G_1 K_2 G. \tag{4.1}$$

Taking into account the relation between the Green function and the transition amplitude off the mass shell:

$$G_1 = G_0 + G_0 T_1 G_0. \tag{4.2}$$

and defining at the subtraction points

$$G = G_0 + G_0 T G_0$$

with the help of eq. (4.1) we obtain the correction to the analytic part of the amplitude:

$$W(n, k^2) = [1 - (1 + T_1 G_0) K_2(n) G_0]^{-1} \times \times (1 + T_1 G_0) K_2(n) (1 + G_0 T_1) \tag{4.3}$$

where $W(n, k^2) \equiv T(n, k^2) - T_1(n, k^2)$.

It is clear that $W(n, k^2)$ can differ from zero for $n = 1, 2$ only, because of the properties of $K_2(n)$ [cf. eq. (2.3)].

If we write out the operator equation (4.3) in momentum representation, we arrive at an integral equation with degenerate kernel; its solution reads e. g. for $n = 1$:

$$\langle p | W(1, k^2) | p' \rangle = g_1 \frac{F(k^2, p) \widetilde{F}(k^2, p')}{1 - g_1 \Phi(k)} \tag{4.4}$$

where

$$F(k^2, p) = 1 + \int_0^\infty \frac{q'^3 \, dq' \langle p \mid T(1, k^2) \mid q' \rangle}{q'^2 - k^2}$$

$$\widetilde{F}(k^2, p) = 1 + \int_0^\infty \frac{q'^3 \, dq' \langle q' \mid T(1, k^2) \mid p \rangle}{q'^2 - k^2} \qquad (4.5)$$

$$\Phi(k) = \int_0^\infty \frac{q'^3 \, dq'}{q'^2 - k^2} + \int_0^\infty q'^3 \, dq' \int_0^\infty q''^3 \, dq'' \frac{\langle q' \mid T(1, k^2) \mid q'' \rangle}{(q'^2 - k^2)(q''^2 - k^2)}.$$

[The first integral in the expression of $\Phi(k)$ is divergent as it stands; it could be defined in the usual way by means of subtractions; however, we shall see in what follows that we can operate formally with such divergent integrals.] Let us now observe that the expressions (4.5) can be expressed with the help of the Green functions in coordinate space. In fact, remembering that

$$\lim_{r \to 0} \frac{J_1(p r)}{p r} = \frac{1}{2}$$

one immediately recognizes [cf. eq. (4.2)] that

$$\Phi(k) = 4 \lim_{\substack{r_1 \to 0 \\ r_2 \to 0}} G_1(r_1, r_2) \equiv G_1(0, 0)$$

$$F(k^2, p) = 2 G_0(p)^{-1} \lim_{r' \to 0} \int_0^\infty r^3 \, dr \frac{J_1(p r)}{p r} G_1(r, r') \quad (n = 1).$$

For small values of r one can immediately write down the expression of the Green function of eq. (2.5)

$$G_1(r, r') = \frac{1}{2} \left[\Theta(r - r') \, I_{\frac{1}{2}}\left(\frac{\lambda}{2 r'^2}\right) K_{\frac{1}{2}}\left(\frac{\lambda}{2 r'^2}\right) + \right.$$
$$\left. + \Theta(r' - r) \, I_{\frac{1}{2}}\left(\frac{\lambda}{2 r'^2}\right) K_{\frac{1}{2}}\left(\frac{\lambda}{2 r'^2}\right) \right].$$

Now, the value of $G_1(0, 0)$ depends on the order of the limits: if $r' > r$ or $r' < r$, then $G_1(0, 0) = 0$, if the limit is taken along the line $r' = r$, $G_1(0, 0)$ is a finite constant. However, as one can easily check, $F(k^2, p) = \widetilde{F}(k^2, p) = 0$. Thus we see that $W(1, k^2) = 0$ and the same result is obtained for $W(2, k^2)$. Hence, *the transition amplitude is determined by its analytic part everywhere*. Let us immediately remark that this property is a consequence of the strong singularity of the Green function at the origin. Had we to do with a "regular" theory (the interaction at small distances is weaker than the centrifugal term), the Green function at small distances would behave as r^{n-1} thus allowing a nonvanishing contribution from the subtraction terms at $n = 1$.

5. Treatment of the "straight ladder"

We consider the interaction of massless, but distinguishable fermions; we call them for the sake of definiteness μ, ν_μ and e, ν_e respectively. They interact via an interaction of the $V-A$ type with charged currents.

We consider both "allowed" transitions, e. g., the reaction

$$\nu_\mu + e \to \nu_e + \mu \tag{5.1}$$

and "forbidden" ones, e. g.,

$$\nu_\mu + e \to \nu_\mu + e. \tag{5.2}$$

The basic process we are going to consider is an allowed reaction of the type (5.1) taking place through the exchange of a fermion loop (Fig. 5.1)

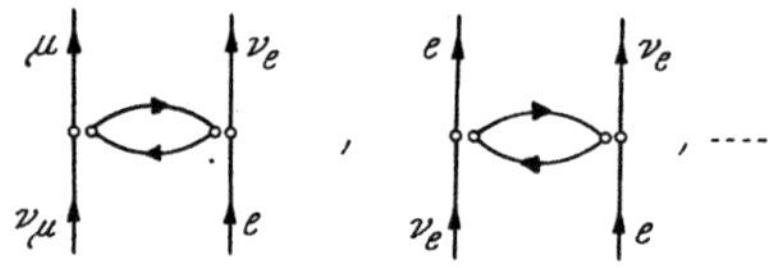

Fig. 5.1

As we neglect fermion masses, the contributions of all the diagrams are equal and we need not specify the fermion loop exchanged ($\mu + \bar{\nu}_\mu$ or $e + \bar{\nu}_e$), but multiply the contribution of each diagrams on Fig. 5.1 by two.

Taking into account time reversal invariance and the conservation of charge and muon number, we are thus left with a two-channel problem, the channels being represented by, e. g., $(\nu_\mu e)$ and $(\nu_e \mu)$.

Written in matrix form in the channel space the BS equation thus reads

$$\begin{pmatrix} G_{11} & G_{12} \\ G_{21} & G_{22} \end{pmatrix} = \begin{pmatrix} G_0 & 0 \\ 0 & G_0 \end{pmatrix} + \begin{pmatrix} 0 & G_0 K \\ G_0 K & 0 \end{pmatrix} \begin{pmatrix} G_{11} & G_{12} \\ G_{21} & G_{22} \end{pmatrix} \tag{5.3}$$

($G_{12} = G_{21}$ because of T-invariance and $G_{11} = G_{22}$ because of our model). This equation can be diagonalized by the orthogonal transformation

$$U = \frac{1}{\sqrt{2}} \begin{pmatrix} 1 & 1 \\ -1 & 1 \end{pmatrix}$$

where upon it takes the form

$$G^{\pm} = G_0 \pm G_0 \, K G^{\pm} \tag{5.4}$$

with

$$G^{\pm} = G_{11} \pm G_{12}. \tag{5.5}$$

The reader will immediately recognize that the selection rules and amplitudes of our model correspond to those of Feinberg and Pais [Phys. Rev. **131,** 2724 (1963); ibid. **133,** B477 (1964)], G_{11}, G_{12} corresponding to their "even" and "odd" diagrams.

The multichannel formulation (cf. MIRZA A. BAQI BÉG, Brookhaven preprint, 1963) without recourse to diagrams is essential in our case, as we shall find that the final solution is not expandable in powers of the coupling constant.

The spinorial structure of our equation is essentially the same as in Sec. 2; after multiplication by G_0^{-1} from the left, eq. (5.4) reads in momentum representation

$$\begin{aligned}(\tfrac{1}{2}\hat{E}^{(1)}+\hat{p}^{(1)})\,(\tfrac{1}{2}\hat{E}^{(2)}-\hat{p}^{(1)})\,\gamma_\mu{}^{(1)}\,(1+i\,\gamma_5{}^{(1)})\,\gamma_\nu{}^{(2)}\,(1+i\gamma_5{}^{(2)})\,G^{\pm}_{\mu\nu}\,(p,p')=\\ =\delta\,(p-p')\pm\frac{1}{(2\,\pi)^4\,i}\int dq\,\gamma_\varrho{}^{(1)}\,(1-i\,\gamma_5{}^{(1)})\,\gamma_\sigma{}^{(2)}\,(1-i\,\gamma_5{}^{(2)})\,K_{\varrho\,\sigma}\,(p,q)\,\cdot\\ \cdot\,\gamma_\alpha{}^{(1)}\,(1+i\,\gamma_5{}^{(1)})\,\gamma_\beta{}^{(2)}\,(1+i\,\gamma_5{}^{(2)})\,G^{\pm}_{\alpha\beta}\,(q,p'),\end{aligned}\tag{5.6}$$

where the superscripts 1 and 2 refer to the two propagating particles, E is the total four momentum, p, p', q are relative momenta in the final-, initial- and intermediate-states respectively.

We assume again:

$$G_{\alpha\beta}=g_{\alpha\beta}G.\tag{5.7}$$

Inserting (5.7) into (5.6) and taking the trace of the equation, we obtain

$$\left(\frac{1}{4}E^2-p^2\right)G^{\pm}\,(p,p')=\delta\,(p-p')\pm\frac{1}{(2\,\pi)^4\,i}\int dq\,K\,(p,q)G^{\pm}\,(q,p')\tag{5.8}$$

where $K\,(p,q)=K_\varrho{}^\varrho\,(p,q)$.

$K_\varrho{}^\varrho$ can be calculated in a straightforward way: it essentially corresponds to a self-energy diagram in electro-dynamics with masseless electrons. Omitting here the details of the calculation, we only notice that eq. (5.8), written in Euclidean metric, is diagonal in the $(n\,l\,m)$ representation. Expanding $G^{\pm}$ and K according to the functions

$$Z^m_{n\,l}\,\frac{J_n\,(p\,r)}{p\,r}$$

we arrive at the radial equation in co-ordinate space

$$\left(\frac{d^2}{dr^2}+\frac{3}{r}\,\frac{d}{dr}+k^2-\frac{n^2-1}{r^2}\mp V\,(r)\right)G^{\pm}\,(r,r')=\frac{\delta\,(r-r')}{(r\,r')^{3/2}}\tag{5.9}$$

with $k^2=\frac{1}{4}E^2$,

$$V\,(r)=\frac{2\,f^2}{\pi\,r^6}\int\limits_0^\infty x^4\,dx\,K_1\,(x)\equiv\lambda^2\,r^{-6}$$

where $\lambda^2=32\,\pi^{-1}\,f^2$. The spectral representation of K needs two subtractions (the fermion loop is quadratically divergent) so the subtracted waves are $n=1,2$. [The calculations leading to eq. (5.8) are described in detail in II.) We see that we arrived essentially at the same equation as for the broken ladder. Additional complications arise only from the fact that corresponding to the two channels, we have to do with an equation with both signs of the interaction. Bearing the analogy with the

Schrödinger equation in mind, we shall speak sometimes of an "attractive" and "repulsive potential". In order to facilitate the reading of this note, we repeat some steps of the calculation, already done in the previous sections.

6. Investigation of the Green function

First of all we set $G^{\pm}(r, r') = (r\,r')^{-3/2}\,\Gamma^{\pm}(r, r')$, thereby eliminating the first derivative from eq. (5.9). According to the procedure sketched in Sec. 1, we write the equation for the radial wave function

$$\left(\frac{d^2}{dr^2} + k^2 - \frac{n^2 - \frac{1}{4}}{r^2} \mp \frac{\lambda^2}{r^6}\right) u^{\pm}(r) = 0. \tag{6.1}$$

Equation (6.1) is exactly of the form of a non-relativistic Schrödinger equation with a singular potential, $l + 1/2$ being now replaced by n, supplemented with boundary conditions which are also the same as in non-relativistic quantum mechanics. In investigating eq. (6.1) we have to treat the "repulsive" and "attractive" potentials separately.

We begin by noting that for $k^2 = 0$ the equation can be integrated exactly.

Two independent solutions are

$$u_0^{+}(r) = \begin{cases} r^{1/2}\, I_{n/2}\left(\dfrac{\lambda}{2\,r^2}\right) \\ r^{1/2}\, K_{n/2}\left(\dfrac{\lambda}{2\,r^2}\right) \end{cases} \tag{6.2}$$

and

$$u_0^{-}(r) = \begin{cases} r^{1/2}\, H^{(1)}_{n/2}\left(\dfrac{\lambda}{2\,r^2}\right) \\ r^{1/2}\, H^{(2)}_{n/2}\left(\dfrac{\lambda}{2\,r^2}\right) \end{cases}. \tag{6.3}$$

For u_0^{+} it is obviously only $K_{n/2}$ which satisfies the boundary condition at the origin, while for u_0^{-} both solutions behave in the same way, they only differ by a phase. In order to proceed further, we set $k^2 > 0$, and calculate $u^{\pm}(r)$ in the WKB approximation.

(a) Repulsive case

There is always one classical turning point. Its position (b) can be easily found as a series in descending powers of k:

$$b = \left(\frac{\lambda}{k}\right)^{1/3} \left[1 + \frac{n^2}{6\,\lambda^2}\left(\frac{\lambda}{k}\right)^{4/3} + \dots\right]. \tag{6.4}$$

The singularity of the potential manifests itself in (6.4): b is essentially determined by the interaction and does not approach its "free value" — n/k — as $k \to \infty$. Setting

$$u^{+}(r) = A(r) \exp\left[i\, S(r)\right]$$

$$S(r) = \int_b^r d\xi\, (k^2 - n^2\,\xi^{-2} - \lambda^2\,\xi^{-6})^{1/2}, \tag{6.5}$$

$A(r)$ satisfies the equation

$$A'' + 2iA'S' + iAS'' = 0$$

and dropping A'', in the first approximation we obtain the standard expression

$$A_1(r) = (S'(r))^{-1/2}.$$

(Dashes mean differentiation with respect to r.)

Treating A'' as a perturbation, we obtain in the next approximation

$$A_2 = -\frac{i}{4}(S')^{-1/2} \log S'$$

and so on. This calculation shows explicitly that something unexpected does not happen in comparison with the zero energy solution; one can always choose a solution which vanishes at the origin; the solution is determined uniquely and in the leading order it behaves as

$$\exp(-\lambda\, 2^{-1/2} r^{-2})$$

near the origin. The treatment of the transition region is likewise standard.

(b) Attractive case

The effective potential: $U(r) = n^2 r^{-2} - \lambda^2 r^{-6}$ is not monotonous; it has a maximum at $r_0 = 3^{1/2} n^{-2} \lambda^2$. There are either two real turning points or none, depending on the magnitude of k^2. To find their positions, we approximate $U(r)$ by a parabola near its maximum:

$$U(r) \simeq \frac{3^{-1}.\, 2\, n^3}{\lambda}\left[3^{-1/2} - \frac{2n}{\lambda}(r - r_0)^2\right].$$

This gives for the turning points

$$b_{1,2} = r_0 \pm \frac{\lambda}{2n}\left(3^{-1/2} - \frac{3\lambda k^2}{2n^3}\right).$$

We see that there are real turning points, if

$$k^2 < \frac{2n^3}{3^{1/2}\lambda} \equiv k_0^2 n^3. \tag{6.6}$$

If k^2 is less than the limit given by eq. (6.6), the solution is unique; there are two real turning points; under the barrier ($b_2 < r < b_1$) the solution is given in terms of functions of the parabolic cylinder, while for $r < b_2$, $r > b_1$ the solution is given by linear combinations of the functions

$$u^-(r) \sim (S'(r))^{-1/2} \exp\left[\pm i \int_{b_{1,2}}^{r} p(\xi)\, d\xi\right],$$

$$p(\xi) = (k^2 - n^2 \xi^{-2} + \lambda^2 \xi^{-6})^{1/2}.$$

The behaviour for $r \to 0$ to leading order is

$$u^-(r) \sim \exp(\pm i \lambda 2^{-1} r^{-2})$$

in agreement with the zero energy solution (6.3). The situation is different for $k^2 > k_0^2$. (Although we do not believe that the ladder diagram we are considering here gives an adequate description of the scattering amplitude at such extremely high energies, nevertheless, it is instructive to investigate this case from a purely theoretical point of view.) If $k^2 > k_0{}^2$, there is no real turning point at all, the barrier is too low to prevent the system from "collapsing". In order to get closer to the usual procedure applied in field theory, we rewrite (6.6) as

$$n > 3^{1/6} \lambda^{1/3} (k^2/2)^{1/3}. \tag{6.6'}$$

For a fixed energy, the partial waves satisfying the inequality (6.6') are well defined. (For $k^2 < k_0{}^2$ all the physical partial waves are such.) As k^2 increases, the first, second, ... partial waves become divergent.

In fact, let us try to write down the expression for $u^-(r)$ for the "supercritical" domain. Introducing a cut-off ϱ at the lower limit in the expression of $S(r)$, we have to leading order

$$u^-(r) \sim \lim_{\varrho \to 0} \exp \left[\pm i \int_{\varrho}^{r} d\xi \, p(\xi) \right]$$

or for $\varrho^2 < r^2 \ll \lambda$,

$$u^-(r) \sim \lim_{\varrho \to 0} \exp \left[\pm i \lambda \int_{\varrho}^{r} \frac{d\xi}{\xi^3} \right].$$

Had we to do with a renormalizable theory, one would begin to separate the divergent part of $u^-(r)$ by one of the usual methods, e. g., by successive differentiation with respect to the external variable (N. N. BOGOLUBOV, D. V. SHIRKOV, Introduction to the theory of quantized fields; Interscience, N. Y. 1959; Ch. IV). It is immediately clear that in this way one arrives at an infinite number of arbitrary subtraction constants — which is clearly nonsense. We propose another way of getting rid of the infinities. One can namely make *one subtraction in the exponential*, by writing

$$S(r) = \int_0^r d\xi \, \{(k^2 - n^2 \xi^{-2} + \lambda^2 \xi^{-6})^{1/2} - \lambda \xi^{-3}\} - \frac{\lambda}{2 r^2} + S_0, \tag{6.7}$$

where S_0 is a real constant, *independent of n*. The subtraction constant S_0 affects but the phase of the wave functions $u^-(r)$, and it is common to all partial waves. It can be removed by a suitable redefinition of $u^-(r)$, having no effect on any physical consequence of the theory. In order to explore a bit further this result, let us expand the curly bracket in the expression (6.7) for small values of r (corresponding to large relative momenta in momentum representation). A simple calculation gives

$$S(r) = S_0 - \frac{\lambda}{2 r^2} + \frac{r^2}{2 \lambda} \left(\frac{k^2 r^2}{4} - n^2 \right) - \frac{r^6}{4 \lambda^3} \left(\frac{k^4 r^4}{5} - \frac{k^2 r^2 n^2}{2} + \frac{n^4}{3} \right) + \ldots \tag{6.8}$$

i. e., one obtains an expansion in descending powers of the coupling constant. Obviously, such a result cannot be produced by a straightforward application of perturbation theory, as there is a singularity at $\lambda^2 = 0$. In order to clarify the nature of the latter, we put $\lambda \xi^{-2} = y$; this gives

$$S(r) = \frac{1}{2} \int_{\lambda r^{-2}}^{\infty} dy \left\{ \left[\frac{k^2 \lambda}{y^3} - \frac{n^2}{y^2} + 1 \right]^{1/2} - 1 \right\}.$$

Hence we can see the singularity at $\lambda^2 = 0$:

$$S(r) \sim \frac{i\,n}{2} \log \lambda / r \qquad (\lambda \to 0) \tag{6.9}$$

or equivalently

$$u^-(r) = 0\,(\lambda^{\pm n/2}) \qquad (\lambda \to 0).$$

[The same is true for $u^+(r)$.] Let us remark that the same type of singularity in λ^2 is obtained from (6.2)—(6.3), so the nature of the singularity in the λ^2 plane is apparently independent of the energy.

Bearing these considerations in mind, we can unify the treatment of the repulsive and attractive cases by introducing an auxiliary variable σ into the differential equation (6.1):

$$\left(\frac{d^2}{dr^2} + k^2 - \frac{n^2 - \frac{1}{4}}{r^2} - \frac{\sigma \lambda^2}{r^6} \right) u(\sigma, r) = 0. \tag{6.1'}$$

The solutions $u(\sigma, r)$ are analytic in the complex σ plane, cut along the negative real axis. The repulsive interaction corresponds to the solution $u(1, r)$, while the attractive one to $u(-1 \pm i0, r)$; in the attractive case *real* solutions can be obtained by taking $u(-1 + i0, r) + u(-1 - i0, r)$ and $i^{-1} [u(-1 + i0, r) - u(-1 - i0, r)]$, respectively.

7. Subtraction Equations and Low Energy Limit

One of the most interesting features of the expression obtained for the ladder diagram considered in Secs. 3—4 was that the scattering amplitude was free of arbitrary subtraction constants. The reason of this interesting and, in a sense, astonishing result has been that due to the strong singularity of the interaction at $r = 0$, the boundary condition imposed on the wave function was independent of n and so we had $F_n(p) = \Phi_n = 0$ in the subtraction equation. Is this still true in the present, more complicated case?

For the "repulsive" channel (u^+) the equation which the wave function obeys is the same as that in Sec. 3, so all the results can be carried over without any change. Thus the second term on the right-hand side of eq. (4.3) vanishes identically and the expression of T^+ is free of arbitrary subtraction constants. The situation is more complicated in the "attractive" channel (u^-), because of the arbitrariness in the boundary condition. In order to decide which solution is to be chosen, we consider the zero energy limit of the scattering amplitude. Although the effective range expansion is divergent (the interaction does not fall off exponentially at large distances) its first few terms can be used as an asymptotic estimate for the amplitude as $k^2 \to 0$; in particular the scattering length is necessarily finite. At small energies only the lowest partial wave contributes ($n=1$), the scattering length can be found by expanding the zero energy solution (6.2) and (6.3) at $n=1$. Thus we find the scattering length of the "+" channel.

$$a^{+} = \lambda/2. \tag{7.1}$$

Let us observe now that for the *lowest partial wave* (and only for that!) the "four-dimensional" and "ordinary" partial amplitudes coincide. In fact, the coefficient of the lowest wave is $Z_{10}^{0}(\cos\Theta) = P_0(\cos\Theta) = 1$, where $\cos\Theta$ is the scattering angle in the c.m.s. Thus, unitarity requires that the scattering length of the $n=1$ eigenamplitude be finite and if complex, its imaginary part positive. This is still not sufficient to determine the boundary condition uniquely. We now *require* the uniqueness of the scattering amplitude (the absence of subtraction terms). This already fixes the solution to be chosen uniquely: in terms of the solutions of (6.1′) it is $u(-1+i\,0, r)$, giving a purely imaginary scattering length in the "−" channel

$$a^{-} = i\,\lambda/2. \tag{7.2}$$

One might wonder, why did we obtain complex scattering length. A glance at Fig. 7.1 makes the reason clear; as we consider massless particles, all the many particle cuts of the amplitude begin at $k^2=0$ together with the "elastic" one, so

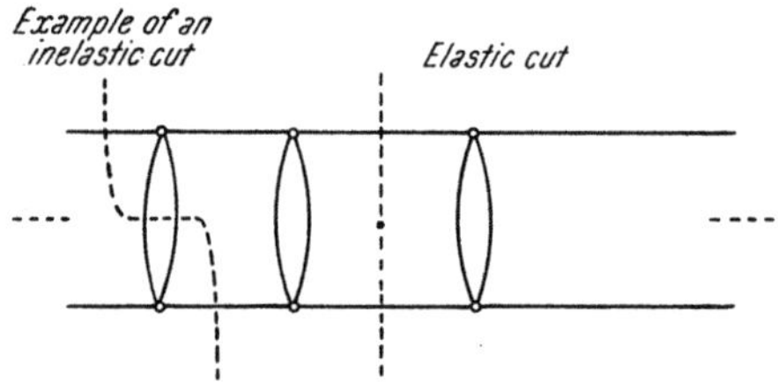

Fig. 7.1

that there is no region, where elastic unitarity would be satisfied. Nevertheless, it is interesting that the scattering length is proportional to the *first power* of the Fermi coupling constant; thus by comparing (7.1) with the first Born approximation to the scattering amplitude (which obviously can give a real scattering length only), one can see that our ladder reproduces it apart from a finite renormalization of the coupling constant. In order to be able to compare the present theory with that of Feinberg and Pais, we are going to sketch now the treatment of an uncrossed ladder with W-exchange by our method.

8. *W*-theory

We consider eq. (6.1) with the W-theory potential. An easy calculation shows that the potential arising from a single W-exchange is:

$V(r) = \frac{g\,m\,K_1(m\,r)}{r} + \frac{\delta(r)}{r^3}$. We first neglect the δ-function-part of the potential which will be taken into account later. In order to solve (6.1) in a closed form, we approximate the McDonald function $K_1(m\,r)$ as follows:

$$\frac{m\,g^2\,K_1(m\,r)}{r} = \begin{cases} g^2/r^2 & (r < m^{-1}) \\ 0 & (r > m^{-1}) \end{cases}$$

(g is the "semi-weak" coupling constant, m — the mass of W). This approximation takes into account exactly the singularity of the potential at the origin and simulates its rapid decrease at large distances. With this approximation we can readily write down the exact solution to eq. (6.1):

$$u^{\pm}(r) = r^{1/2}\,[J_n(kr) + \operatorname{tg}\bar{\delta}_n^{(\pm)}\,N_n(kr)] \quad (r > m^{-1})$$

and

$$u^{\pm}(r) = r^{1/2}\,J_{\bar{n}}(kr) \tag{8.1}$$

where

$$\bar{n} = (n^2 \pm 3g^2)^{1/2} \approx n \pm \frac{3\,g^2}{2\,n}.$$

Matching logarithmic derivatives of $u^{\pm}(r)$ at $r = 1/m$ we obtain

$$\operatorname{tg}\bar{\delta}_n^{(\pm)} = \frac{J_n(k/m)\,J_{\bar{n}}'(k/m) - J_{\bar{n}}(k/m)\,J_n'(k/m)}{J_{\bar{n}}'(k/m)\,N_n(k/m) - J_{\bar{n}}(k/m)\,N_n'(k/m)} \tag{8.2}$$

where $\bar{\delta}_n^{(\pm)}$ is the scattering phase shift for the potential without the δ function.

For $k/m \ll 1$, (8.2) reduces to:

$$\operatorname{tg}\bar{\delta}_n^{(\pm)} \sim \pm\left(\frac{k}{2m}\right)^{2n}\frac{\pi}{n^3}\left(\frac{3\,g^2}{4}\right). \tag{8.3}$$

In order to take into account the contact part of the interaction, we return to a momentum representation and follow the procedure outlined for the subtraction equations.

Let us consider eq. (5.8); we write the kernel K as a sum of two terms:

$$K = K_1 + K_2$$

where

$$K_1 = g^2\,\frac{3}{m^2 + (p-q)^2}$$

and

$$K_2 = g^2/m^2.$$

Call again G_1 the Green function of eq. (5.8) corresponding to the kernel K_1.

As before, we find:

$$T - T_1 = \frac{(1 + T_1 G_0)\,K_2(n)\,(1 + G_0\,T_1)}{1 - (1 + T_1 G_0)\,K_2(n)\,G_0}$$

where T_1 is the scattering amplitude calculated from G_1. We see that if T_1 can be expanded in powers of g^2 we will have, to lowest order:

$$T \approx T_1 + K_2(n).$$

$K_2(n)$ gives a non-vanishing contribution for $n=1$ only. Hence:

$$T \approx \begin{cases} T_1 + g^2/m^2 & (n=1) \\ T_1 & (n>1) \end{cases}.$$

Thus, taking into account (8.3), we obtain for the scattering length

$$a^{\pm} \approx \frac{g^2}{m^2}\,[1 - 3\pi/16] \approx 0.4\, g^2/m^2.$$

9. Application to purely leptonic processes at low energies

We see from the foregoing that either the Fermi- or the W-theory gives observable corrections to physical processes. As an example, we consider the lowest non-vanishing correction to μ-decay. Within the framework of our approximation we can immediately write down the corrected expressions for any leptonic process. The recipe is: replace the coupling constant squared in the lowest order expression by the modulus squared of the transition amplitude. In the scattering length approximation both for the Fermi- or W-theory the latter is given by

$$T_1^{\pm} \approx \frac{a^{\pm}}{1 - a^{\pm} k^2} \approx a^{\pm}\,(1 + a^{\pm} k^2).$$

The other partial waves give corrections of higher order. In the Fermi-theory we obtain (μ-decay being an "allowed" process) in the rest system of the muon:

$$\frac{\text{decay rate}}{\text{renormalized decay rate to lowest order}} = 1 - \frac{\bar{f}\mu^2}{2^{1/2}}\left(1 - \frac{2\,p_l}{\mu}\right)$$

where μ is the mass of the muon. [The electron mass and terms $0\,(k^2)$ have been neglected.] $\bar{f}$ is the renormalized Fermi coupling constant.

For the W-theory ladder we obtain:

$$\frac{\text{decay rate}}{\text{renormalized decay rate to lowest order}} = 1 + 0\left(\frac{G^4 k^4}{m^4}\right)$$

$$\left(k^2 = \mu^2\,(1 - 2\,p_l/\mu)\right)$$

where G^2 is the renormalized (observable) coupling strength related to the bare coupling constant by the relation

$$G^2 = g^2\,(1 - 3\,\pi/16) \approx 0.4\,g^2. \tag{9.1}$$

We note that had we neglected the $g_{\mu\nu}$ — term in W-kernel, we would have obtained — instead of (9.1) — the relation

$$G^2 = g^2\,(1 - \pi/16) \approx 0.8\,g^2.$$

This is very close to the value $G^2/g^2 = 3/4$ obtained by Feinberg and Pais.

Let us note the characteristic difference between the Fermi and the W ladders: in the former we obtain a correction of order k^2 while in the latter we do not.

Consider now neutrino-lepton scattering, e. g., the reaction $\nu_\mu + e^- \to \nu_e + \mu^-$. An easy calculation yields for R — the ratio of the cross-section to the renormalized lowest order result (including terms of order E^2):

$$R_F = \left(1 - \frac{\bar{f} E^2}{2^{5/2}}\right) \qquad \text{(Fermi theory)} \quad (9.2)$$

$$R_W = \left(1 + 0\left(\frac{G^4 E^4}{m^4}\right)\right). \qquad \text{(W-theory)} \quad (9.3)$$

For a "*forbidden*" process, like $\nu_\mu + e \to \nu_\mu + e$, the W-theory gives zero to order G^2, while the "Fermi ladder" gives

$$\bar{R}_F = \left(1 + \frac{\bar{f} E^2}{2^{5/2}}\right) \qquad (9.4)$$

where $\bar{R}_F$ is the ratio of the cross-section predicted by the "Fermi ladder" starting from an interaction with charged currents only, to that calculated in lowest order from the hypothetical interaction $\bar{f} \cdot (\nu_\mu \bar{\nu}_\mu) \times (e\, e)$. Unfortunately, even if this difficult experiment could be carried out, in which one would observe only the lowest order terms, one could not tell, whether the process took place due to an elementary interaction with neutral currents, or to those induced by higher order corrections. However, if one observes "zero" to order G^2, one would be led to believe on the basis of the present calculation that W-exchange predominates at low energies.

10. Scattering of leptons at high energies

We want to consider the scattering of leptons at high energies in the multiperipheral approximation. The diagram suitable for the description of such a process is the "broken ladder" Fig. 2.1 a. The procedure we follow is commonly known in the theory of strong interactions: one considers the ladder in the crossed channel, where $\sqrt{-E^2}$ plays the role of the invariant momentum transfer. If we restrict E^2 to small values, the low energy solution (3.16) for the partial wave amplitude can be used. First of all we have to determine the total scattering amplitude from the partial ones (3.16). Remembering the addition theorem for four-dimensional spherical harmonics (Appendix) and the well-known identity:

$$C^1_{n-1}(\cos \Theta) = \frac{\sin n\, \Theta}{\sin \Theta}$$

(C_ν^λ being a Gegenbauer function), we obtain the scattering amplitude in the form of a Fourier series:

$$T(z, k^2) = \left(\frac{\pi}{2}\right)^{1/2} (\sin \Theta)^{-1} \sum_{n=1}^{\infty} n \sin n\, \Theta\; T(n, k^2). \qquad (10.1)$$

Here, $z = \cos\Theta$, or expressed in Mandelstam's variables, $z = 1 + \frac{2s}{t}$, $t = 4k^2$. It is convenient to transform the series (10.1) into a contour integral:

$$T(z, k^2) = \left(\frac{\pi}{2}\right)^{1/2} (\sin\Theta)^{-1} \frac{i}{2} \int_C n\, dn\, \operatorname{ctg} \pi n \sin n\Theta\, T(n, k^2) \quad (10.1')$$

the contour C being shown on Fig. 10.1. We have already remarked that $T(n, k^2)$ is a moromorphic function of n in the whole complex n-plane, its poles lying for $k^2 < 0$ on the imaginary axis.

A simple investigation shows that if we put $n = R \exp i\varphi$ then for $R \to \infty$,

$$T(n, k^2) = 0\,(\exp(-R \log R \cdot \cos\varphi)).$$

Thus for $k^2 < 0$ the contour C can be deformed almost to the imaginary axis (C') (an infinitesimal sector being excluded).

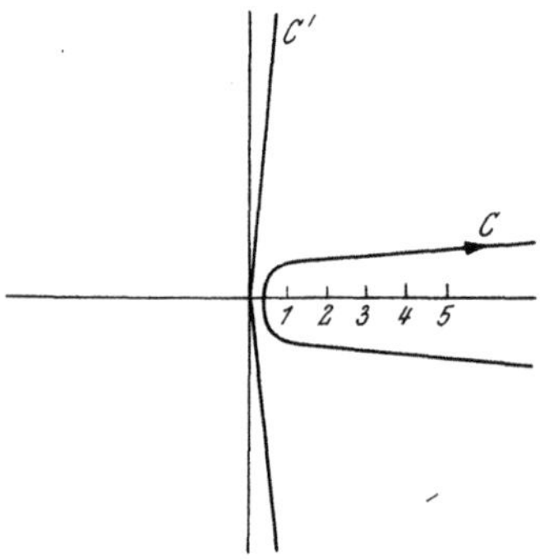

Fig. 10.1

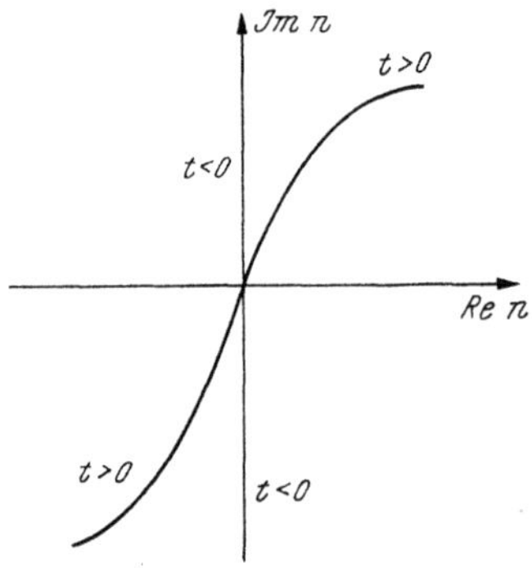

Fig. 10.2

Let us now obtain an approximate expression for $T(n, k^2)$ for small values of k^2 and nonintegral values of n. Making use of the familiar expansions of the cylindrical functions, after some rearrangements we arrive at the expression:

$$T(n, t) \sim \frac{e^{in\pi} \sin n\pi}{Y - 1}, \quad (10.2)$$

where

$$Y = \frac{\Gamma(1+n)}{\Gamma(1-n)} \frac{\Gamma\left(1+\frac{n}{2}\right)}{\Gamma\left(1-\frac{n}{2}\right)} \left(\frac{t\, e^{-i\pi}}{64}\right)^{-n}. \quad (10.3)$$

To find the poles of the amplitude, put $n = i\tau$; then

$$\frac{\Gamma(1+n)}{\Gamma(1-n)} \frac{\Gamma\left(1+\frac{n}{2}\right)}{\Gamma\left(1-\frac{n}{2}\right)} = \exp i\psi \qquad \psi = -3C\tau + 0\,(\tau^2),$$

where C is the Euler-Mascheroni constant.

$T(n, t)$ obviously has poles in the n-plane determined by the equation:

$$\psi - \tau \log \frac{t\, e^{-i\pi}}{64} = 2\pi N \tag{10.4}$$

$$N = 0, \pm 1, \pm 2, \ldots$$

or, approximately,

$$n = \frac{2\pi i N}{3C + \log \dfrac{t\, e^{-i\pi}}{64}}. \tag{10.5}$$

Thus we have an infinite number of poles; all of them have a common trajectory in the n-plane, as sketched on Fig. 10.2. For $t = 0$ the poles fill the whole imaginary axis, giving a continuous spectrum, as can be also checked directly.

These considerations show that the concept of Regge poles to determine the behaviour of the amplitude in the crossed channel is not a useful one in our case, because we have no "leading pole".

Nevertheless, the contour integral representation of the scattering amplitude (10.1′) can be used to determine the asymptotic behaviour for large positive values of s and small negative values of t. The integral in (10.1′) can be evaluated by means of the method of steepest descents.

The calculation is a bit tedious but quite elementary; we simply quote the resulting asymptotic expression of the scattering amplitude, as a function of s and t:

$$T(s, t) \sim \left(\frac{2}{\pi}\right)^{1/2} \frac{16\, t}{s\, (\log s)}\, M(x) \tag{10.6}$$

where

$$x = \frac{-\log s}{\log(-t)}$$

$$M(x) = x^{3/2} \exp(-x^{-1} \log x).$$

The quantities s and t are measured in units of f^{-1}. The formula (10.6) is valid for $s \gg 1$ and $|t| \ll 1$.

11. Conclusion

Coming to the end of my talk, I would like to follow the style of the novels where to guess the end of the last scene and to draw the conclusions is left to the reader. In fact, hoping that those who believe that this whole story from the beginning till the bitter end is a nonsense, have not attended the lecture at all, I would like to accept an extremely optimistic standpoint and say: in fact, what you have heard is but the beginning of a story. To develop and finish it is upon us.

What are the features of the present, after all very simpleminded, calculation which may be more general than the calculation itself?

I believe, they are the following:

(1) A nonrenormalizable theory, in spite of its apparent "wildness", may well exist. Due to the singularity in the coupling constant, however, one has to abandon a lot of the familiar concept which proved so useful e. g. in quantum electrodynamics (bare particles, perturbative expansion — therefore the original concept of Feynman diagrams etc.). Some other methods are badly needed, which do not use these concepts. (One would think perhaps about functional methods — but so far they have not proved very useful because of practical reasons.)

(2) If so, it may happen that the nonrenormalizable theories are even better than good old quantum electrodynamics, because in spite of their wild singularities (or rather because of them) they may not contain arbitrary subtraction constants at all.

(3) However, let me call your attention at this point to a difficulty, I already mentioned at the beginning of my talk. Unlike renormalizable theories, there is no direct equivalence between the Euclidean and Minkowskian formulations of the theory. In fact, our Green function has an essential singularity at $p_0 = \infty$, so one cannot deform freely the integration contours, as e. g. in the case of a scalar ladder, considered in the classical paper of WICK. We tacitly adopted a procedure of *starting* from the Euclidean theory, calculate scattering amplitudes, Green functions etc. and continue back the *final expressions* to their Minkowskian values. Is this procedure a germ of a general procedure, does it work only in the ladder approximation or is simply incorrect — I don't know. Nevertheless, let me say that there are some features of our expressions which are encouraging in this respect. First of all, SCHWINGER and SYMANZIK have shown (cf. e. g. K. SYMANZIK, MIT lecture, 1963) that the existence of a Euclidean field theory is a necessary condition of the existence of its Minkowskian version. Secondly, our scattering amplitude has a familiar analytic structure — apart from the singularity at infinity. Thirdly, about nine years ago, considering some nonrenormalizable model theories, ARNOWITT and DESER (Phys. Rev. **100**, 449, 1955) arrived at expressions which in their singularity structure are very similar to our ones.

Coming back to reality, let me finally mention some problems, which may be of some relevance for the physics of weak interactions.

(1) The induced neutral currents

Our calculations of Section 4 indicate that the problem of neutral currents is a more complicated one than believed so far. In fact, a closer examination of the structure of our equations shows that when we start from an elementary interaction containing charged currents only, higher order corrections of the Fermi theory necessarily induce an effective neutral current. (The only combination of currents reproducing itself after summation of the Fermi ladder is a purely neutral one.) The reason why we obtain neutral currents in the Fermi theory and not in the W-theory is that the perturbation expansion of the amplitude in the former case is badly divergent — even in the low energy approximation — therefore, unlike in the W-theory, it is not possible to separate "even" and "odd" diagrams. Nevertheless, due to the definite phases in the expression for the scattering amplitude, if one could measure Coulomb interference in a $\nu_\mu + e \to \nu_\mu + e$ scattering process, one could perhaps decide whether: (a) there are no neutral currents to lowest order; (b) there is an "elementary" neutral current; or (c) there is an induced neutral current as we obtained from our "straight ladder".

(2) Corrections to lepton scattering and decay

An essential prediction of the present theory is that there are no corrections to the leptonic processes calculated in lowest order. A glance at eqs. (9.2)—(9.4) shows that the correction terms contain at least one more power of the coupling constant. [Due to the approximation (1.7) the numerical coefficients of the correction terms are probably not accurate, since in assuming (1.7) we neglected terms of the same order as the correction.] In contradistinction to previous calculations in terms of diagrams — we obtain finite and unambiguous answer because the correction terms arise not from an expansion in the coupling constant but rather in an expansion with respect to the energy. (The theoretician dreams, of course, of having experimental results at hand, sufficiently accurate to test the presence of the higher order terms...)

(3) Concerning lepton scattering at high energies, one is perhaps astonished a bit to obtain a differential cross section, which *decreases* with energy. (Weak interactions *do not* "become strong" at high energies — as one believed?) If one looks a bit more carefully at the equations, one finds that the differential cross section actually *does* increase roughly up to the critical energy: ($E_{CM} \approx f^{-1/2} \approx 300$ GeV) and begins to decrease only at still higher energies. In view of the fact that experimentators do not seem to have 300 GeV or more in the CMS for the next years, this seems to be a rather academic problem at present and we have treated it mainly because of its theoretical interest.

Appendix

We list here some elementary properties of four-dimensional spherical functions used in the main text. In a canonical basis, the components of the four dimensional angular momentum $L_{\mu\nu}$ act as differential operators:

$$L_{\mu\nu} = i^{-1} (x_\mu \ \partial_\nu - x_\nu \ \partial_\mu).$$

We are looking for the simultaneous eigenfunctions of

$$\tfrac{1}{4} L_{\mu\nu} L_{\mu\nu} \text{ and } \underline{L}^2$$

where $\underline{L}^2$ is the three-dimensional angular momentum. On separating the four-dimensional Laplacean in polar coordinates:

$$\begin{aligned} x^0 &= r \cos \psi \\ x^3 &= r \sin \psi \cos \Theta \\ x^1 &= r \sin \psi \sin \Theta \cos \varphi \\ x^2 &= r \sin \psi \sin \Theta \sin \varphi \end{aligned}$$

the angular part of the operator reads:

$$L^2 \Phi \equiv \left\{ \frac{\partial^2}{\partial \psi^2} + 2 \operatorname{ctg} \psi \frac{\partial}{\partial \psi} + \frac{1}{\sin^2 \psi} \left[\frac{\partial^2}{\partial \Theta^2} + \operatorname{ctg} \Theta \frac{\partial}{\partial \Theta} + \right. \right.$$

$$\left. \left. + \frac{1}{\sin^2 \Theta} \frac{\partial^2}{\partial \varphi^2} \right] \right\} \Phi = -(n^2 - 1) \Phi \tag{A.1}$$

($n = 1, 2, 3, \ldots$). The Ansatz

$$\Phi = Y_l^m (\Theta, \varphi) \cdot y (\psi)$$

where $Y_l^m (\Theta, \varphi)$ are three-dimensional spherical functions, normalized to unity on the surface of the three-dimensional unit sphere, transforms (A.1) into:

$$\frac{d^2 y}{d\psi^2} + 2 \operatorname{ctg} \psi \cdot \frac{dy}{d\psi} - \frac{l(l+1)}{\sin^2 \psi} y + (n^2 - 1) y = 0. \tag{A.2}$$

Introducing the new variable: $x = \cos \psi$ and a new function by $y = (1 - x^2)^{-1/4} u(x)$ we get a Legendre equation:

$$(1 - x^2) u'' - 2 x u' + \left[n^2 - \tfrac{1}{4} - \frac{(l + \frac{1}{2})^2}{1 - x^2} \right] u = 0. \tag{A.3}$$

The regular solution of (A.3) is

$$u(x) = P_{n+\frac{1}{2}}^{-(l+\frac{1}{2})}(x)$$

or expressed with the help of the original variable:

$$y(\psi) = (\sin \psi)^{-\frac{1}{2}} \cdot P_{n+\frac{1}{2}}^{-(l+\frac{1}{2})}(\cos \psi). \tag{A.4}$$

Another, convenient form of (A.4) is in terms of Gegenbauer functions:

$$y(\psi) = (\sin \psi)^l \, C_{n-l-1}^{l+1} (\cos \psi). \tag{A.5}$$

The normalized functions are:

$$Z_{nl}^m (\psi, \Theta, \varphi) = 2^{l+\frac{1}{2}} \left(\frac{n \Gamma(n-l)}{\pi \Gamma(n+l+1)} \right)^{\frac{1}{2}} \cdot \Gamma(l+1) \cdot (\sin \psi)^l \cdot$$

$$\cdot C_{n-l-1}^{l+1} (\cos \psi) \cdot Y_l^m (\Theta, \varphi) \equiv p_{nl}(\psi) \cdot Y_l^m (\Theta, \varphi). \tag{A.6}$$

They satisfy:

$$\int_0^\pi \sin^2 \psi \, d\psi \int_0^\pi \sin \Theta \, d\Theta \int_0^{2\pi} d\varphi \, Z_{nl}^m Z_{n'l'}^{m'*} = \delta_{nn'} \delta_{ll'} \delta_{mm'}$$

i. e. the $\{Z_{nl}^m\}$ form an orthonormal system on the surface of the four-dimensional unit sphere.

The system of Z's is complete. The addition theorem follows immediately from those of the Gegenbauer functions and of the three-dimensional spherical functions:

$$Z_{n0}^0 (\gamma, \cdot, \cdot) = \sum_{l=0}^{n-1} \sum_{m=-l}^{l} Z_{nl}^m (\psi_1, \Theta_1, \varphi_1) \, Z_{nl}^m (\psi_2, \Theta_2, \varphi_2) \tag{A.7}$$

$$\cos \gamma = \cos \psi_1 \cdot \cos \psi_2 - \sin \psi_1 \cdot \sin \psi_2 \cdot \cos \omega$$

$$\cos \omega = \cos \Theta_1 \cdot \cos \Theta_2 - \sin \Theta_1 \cdot \sin \Theta_2 \cdot \cos (\varphi_1 - \varphi_2).$$

(The dots in the argument of Z_{n0}^0 mean that the corresponding angles are arbitrary, as the function is independent of them.)

Finally, we notice the obvious, but important property:

$$Z_{nl}^m = 0 \text{ for } l > n - 1.$$

Racah-Coefficients of SU_3 for $8 \otimes 8 \otimes 8$ and $8 \otimes 10 \otimes 8$

By

Margarete Krammer

Institut für Theoretische Physik, Universität Graz

With 1 Figure

Racah-coefficients (6-j symbols) play an important role in the theory and physical applications of the rotation group. For the coupling of three angular momenta to a certain total angular momentum the Racah-coefficients imply a unitary transformation between two systems of basis vectors [1].

Similarly for some current symmetry group in present day elementary particle physics one can define recoupling coefficients. If the direct product $\mu_1 \otimes \mu_2 \otimes \mu_3$ is reduced in two different ways: $(\mu_1 \otimes \mu_2) \otimes \mu_3$ and $\mu_1 \otimes (\mu_2 \otimes \mu_3)$, then the Racah-coefficients are defined as follows:

$$\psi\begin{pmatrix}\mu_1\,\mu_{23,\beta}\,\mu_{\gamma'}\\ \nu\end{pmatrix} = \sum_{\mu_{12},\alpha,\gamma} \psi\begin{pmatrix}\mu_{12,\alpha}\,\mu_3\,\mu_\gamma\\ \nu\end{pmatrix} \langle(\mu_1\,\mu_2)\,\mu_{12,\alpha}\,\mu_3\,\mu_\gamma \,|\, \mu_1\,(\mu_2\,\mu_3)\,\mu_{23,\beta}\,\mu_{\gamma'}\rangle$$

$$\langle(\mu_1\,\mu_2)\,\mu_{12,\alpha}\,\mu_3\,\mu_\gamma \,|\, \mu_1\,(\mu_2\,\mu_3)\,\mu_{23,\beta}\,\mu_{\gamma'}\rangle = \tag{1}$$

$$= \sum_{\substack{\nu_1\nu_2\nu_3\\ \nu_{12}\nu_{23}}} \begin{pmatrix}\mu_1\,\mu_2\,\mu_{12,\alpha}\\ \nu_1\,\nu_2\,\nu_{12}\end{pmatrix} \begin{pmatrix}\mu_{12}\,\mu_3\,\mu_\gamma\\ \nu_{12}\,\nu_3\,\nu\end{pmatrix} \begin{pmatrix}\mu_2\,\mu_3\,\mu_{23,\beta}\\ \nu_2\,\nu_3\,\nu_{23}\end{pmatrix} \begin{pmatrix}\mu_1\,\mu_{23}\,\mu_{\gamma'}\\ \nu_1\,\nu_{23}\,\nu\end{pmatrix}.$$

(Notation and phase conventions as introduced by de Swart [2].)

When we split the Clebsch-Gordan-coefficients of the SU_3 in their isoscalar factors and the C.-G.-coefficients of the rotation group

$$\begin{pmatrix}\mu_1\,\mu_2\,\mu_{12,\alpha}\\ \nu_1\,\nu_2\,\nu_{12}\end{pmatrix} = \left(\begin{matrix}\mu_1 & \mu_2\\ I_1\,Y_1 & I_2\,Y_2\end{matrix}\,\middle|\,\begin{matrix}\mu_{12,\alpha}\\ I_{12}\,Y_{12}\end{matrix}\right) \cdot C^{I_1\;I_2\;I_{12}}_{I_{1z}\,I_{2z}\,I_{12z}} \tag{2}$$

(1) becomes

$$\langle(\mu_1\,\mu_2)\,\mu_{12,\alpha}\,\mu_3\,\mu_\gamma \,|\, \mu_1\,(\mu_2\,\mu_3)\,\mu_{23,\beta}\,\mu_{\gamma'}\rangle =$$

$$= \sum_{\substack{I_1\,I_2\,I_3\,I_{12}\,I_{23}\\ Y_1\,Y_2\,Y_3\,Y_{12}\,Y_{23}}} \left(\begin{matrix}\mu_1 & \mu_2\\ I_1\,Y_1 & I_2\,Y_2\end{matrix}\,\middle|\,\begin{matrix}\mu_{12,\alpha}\\ I_{12}\,Y_{12}\end{matrix}\right) \left(\begin{matrix}\mu_{12} & \mu_3\\ I_{12}\,Y_{12} & I_3\,Y_3\end{matrix}\,\middle|\,\begin{matrix}\mu_\gamma\\ I\,Y\end{matrix}\right) \left(\begin{matrix}\mu_2 & \mu_3\\ I_2\,Y_2 & I_3\,Y_3\end{matrix}\,\middle|\,\begin{matrix}\mu_{23,\beta}\\ I_{23}\,Y_{23}\end{matrix}\right) \times$$

$$\times \left(\begin{matrix}\mu_1 & \mu_{23}\\ I_1\,Y_1 & I_{23}\,Y_{23}\end{matrix}\,\middle|\,\begin{matrix}\mu_{\gamma'}\\ I\,Y\end{matrix}\right) \langle(I_1\,I_2)\,I_{12}\,I_3\,I \,|\, I_1\,(I_2\,I_3)\,I_{23}\,I\rangle. \tag{3}$$

$\langle(I_1\,I_2)\,I_{12}\,I_3\,I \,|\, I_1\,(I_2\,I_3)\,I_{23}\,I\rangle$ are the Racah-coefficients of the rotation group which can be found in tables [3].

The Racah-coefficients (3) of SU_3 for $8 \otimes 8 \otimes 8$ and also $8 \otimes 8 \otimes 10$ and $8 \otimes 10 \otimes 8$ have been calculated and compiled in tables I, II and III in terms of orthogonal matrices. The columns are denoted by $\mu_{12,\alpha}$ and the rows by $\mu_{23,\beta}$. The indices are to be understood as follows:

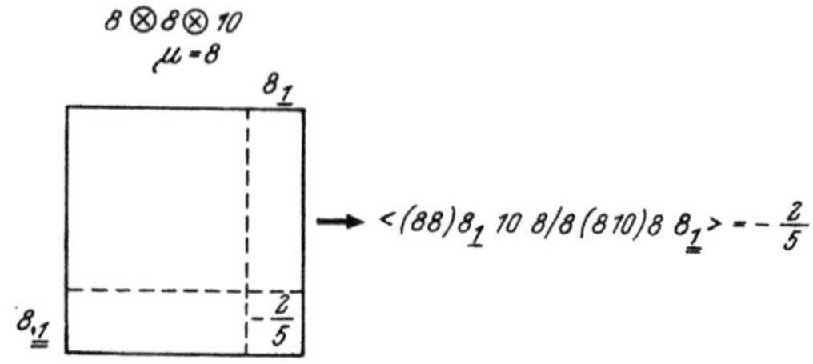

Fig. 1. The comma in $8_{,1}$ means that the index 1 does not refer to $\mu_{23,\alpha} \in (\mu_2 \otimes \mu_3)$ but to $\mu_{\gamma'} \in (\mu_1 \otimes \mu_{23})$

The coefficients in tables II and III are related to each other by the following sum rule, similar to the rotation group:

$$\langle (8\,10)\,\mu'_{12}\,8\,\mu_\delta | 8\,(10\,8)\,\mu_{23}\,\mu_{\delta'} \rangle = \\ = \xi_1\,(8\,10\,\mu_{23}) \cdot \xi_1\,(8\,\mu'_{12}\,\mu_\delta) \sum_{\mu_{12},\alpha} \xi_1\,(8\,8\,\mu_{12,\alpha}) \times \\ \times \langle (8\,8)\,\mu_{12,\alpha}\,10\,\mu_\gamma | 8\,(8\,10)\,\mu_{23}\mu_{\delta'} \rangle \langle (8\,8)\,\mu_{12,\alpha}\,10\,\mu_\gamma | 8\,(8\,10)\,\mu'_{12}\mu_\delta \rangle . \tag{4}$$

In analogy to the symmetry properties of the 6-j symbols there exist relations between the elements of the various orthogonal matrices. The elements of the crossing matrix [4] are also related to the coefficients defined above. All relations are easily derived with help of (14.1), (14.8) and (14.12) of ref. [2].

The author wishes to thank Professor URBAN for continued guidance and encouragement and also for giving her the opportunity to attend the "III. Internationale Universitätswoche für Kernphysik" at Schladming. She is also grateful to Professors JOOS and MAYER for helpful discussions.

References

1. See e. g.: A. R. EDMONDS, Angular Momentum in Quantum Machanics, Chap. 6, Princeton University Press (1957).
2. DE SWART, J. J., Rev. Mod. Phys. **35**, 916 (1963).
3. e. g.: M. ROTENBERG et al., The 3-j and 6-j Symbols. The Technology Press, MIT, Cambridge, Mass. (1959).
4. DE SWART, J. J., Nuovo Cimento **31**, 420 (1964).

Table I. $8 \otimes 8 \otimes 8$

$\mu = 1$

$\mu_{23} \backslash \mu_{12}$	8_2	8_1
8_2	1	0
8_1	0	1

$\mu = 8$

$\mu_{23} \backslash \mu_{12}$	27	10*	10	$8_{2,2}$	$8_{2,1}$	$8_{1,2}$	$8_{1,1}$	1
27	$\frac{7}{40}$	$\frac{\sqrt{30}}{40}$	$\frac{-\sqrt{30}}{40}$	$\frac{\sqrt{6}}{4}$	0	0	$\frac{-3\sqrt{6}}{20}$	$\frac{3\sqrt{3}}{8}$
10	$\frac{\sqrt{30}}{40}$	$\frac{1}{4}$	$\frac{-1}{4}$	0	$\frac{1}{2}$	$\frac{-1}{2}$	$\frac{-\sqrt{5}}{5}$	$\frac{-\sqrt{10}}{8}$
10*	$\frac{-\sqrt{30}}{40}$	$\frac{-1}{4}$	$\frac{1}{4}$	0	$\frac{1}{2}$	$\frac{-1}{2}$	$\frac{\sqrt{5}}{5}$	$\frac{\sqrt{10}}{8}$
$8_{2,2}$	$\frac{\sqrt{6}}{4}$	0	0	$\frac{1}{2}$	0	0	$\frac{1}{2}$	$\frac{-\sqrt{2}}{4}$
$8_{2,1}$	0	$\frac{1}{2}$	$\frac{1}{2}$	0	$\frac{1}{2}$	$\frac{1}{2}$	0	0
$8_{1,2}$	0	$\frac{-1}{2}$	$\frac{-1}{2}$	0	$\frac{1}{2}$	$\frac{1}{2}$	0	0
$8_{1,1}$	$\frac{-3\sqrt{6}}{20}$	$\frac{-\sqrt{5}}{5}$	$\frac{\sqrt{5}}{5}$	$\frac{1}{2}$	0	0	$\frac{-3}{10}$	$\frac{-\sqrt{2}}{4}$
1	$\frac{3\sqrt{3}}{8}$	$\frac{-\sqrt{10}}{8}$	$\frac{\sqrt{10}}{8}$	$\frac{-\sqrt{2}}{4}$	0	0	$\frac{-\sqrt{2}}{4}$	$\frac{1}{8}$

$\mu = 10$

μ_{23} \ μ_{12}	27	10	8_2	8_1
27	$\frac{1}{10}$	$\frac{\sqrt{15}}{10}$	$\frac{\sqrt{30}}{10}$	$\frac{3\sqrt{6}}{10}$
10	$\frac{\sqrt{15}}{10}$	$\frac{-1}{2}$	$\frac{\sqrt{2}}{2}$	$\frac{-\sqrt{10}}{10}$
8_2	$\frac{\sqrt{30}}{10}$	$\frac{\sqrt{2}}{2}$	0	$\frac{-\sqrt{5}}{5}$
8_1	$\frac{-3\sqrt{6}}{10}$	$\frac{\sqrt{10}}{10}$	$\frac{\sqrt{5}}{5}$	$\frac{-2}{5}$

$\mu = 10^*$

μ_{23} \ μ_{12}	27	10*	8_2	8_1
27	$\frac{1}{10}$	$\frac{\sqrt{15}}{10}$	$\frac{\sqrt{30}}{10}$	$\frac{-3\sqrt{6}}{10}$
10*	$\frac{\sqrt{15}}{10}$	$\frac{-1}{2}$	$\frac{\sqrt{2}}{2}$	$\frac{\sqrt{10}}{10}$
8_2	$\frac{\sqrt{30}}{10}$	$\frac{\sqrt{2}}{2}$	0	$\frac{\sqrt{5}}{5}$
8_1	$\frac{3\sqrt{6}}{10}$	$\frac{-\sqrt{10}}{10}$	$\frac{-\sqrt{5}}{5}$	$\frac{-2}{5}$

$\mu = 27$

μ_{23} \ μ_{12}	$27,_2$	$27,_1$	10*	10	8_2	8_1
$27,_2$	$\frac{1}{2}$	0	$\frac{\sqrt{6}}{12}$	$\frac{-\sqrt{6}}{12}$	$\frac{\sqrt{6}}{3}$	0
$27,_1$	0	$\frac{3}{10}$	$\frac{\sqrt{70}}{20}$	$\frac{\sqrt{70}}{20}$	0	$\frac{\sqrt{14}}{5}$
10	$\frac{\sqrt{6}}{12}$	$\frac{\sqrt{70}}{20}$	$\frac{2}{3}$	$\frac{-1}{6}$	$\frac{-1}{3}$	$\frac{-\sqrt{5}}{5}$
10*	$\frac{-\sqrt{6}}{12}$	$\frac{\sqrt{70}}{20}$	$\frac{-1}{6}$	$\frac{2}{3}$	$\frac{1}{3}$	$\frac{-\sqrt{5}}{5}$
8_2	$\frac{\sqrt{6}}{3}$	0	$\frac{-1}{3}$	$\frac{1}{3}$	$\frac{-1}{3}$	0
8_1	0	$\frac{\sqrt{14}}{5}$	$\frac{-\sqrt{5}}{5}$	$\frac{-\sqrt{5}}{5}$	0	$\frac{1}{5}$

(8_1 is symmetric
8_2 is antisymmetric)

$\mu = 35$

μ_{23} \ μ_{12}	27	10
27	$\frac{1}{2}$	$\frac{\sqrt{3}}{2}$
10	$\frac{\sqrt{3}}{2}$	$\frac{-1}{2}$

$\mu = 35^*$

μ_{23} \ μ_{12}	27	10*
27	$\frac{1}{2}$	$\frac{\sqrt{3}}{2}$
10*	$\frac{\sqrt{3}}{2}$	$\frac{-1}{2}$

$\mu = 64$

	27
27	1

Table II. $8 \otimes 8 \otimes 10$

$\mu = 8$

μ_{23} \ μ_{12}	27	10*	8_2	8_1
27	$\frac{1}{10}$	$\frac{-\sqrt{15}}{10}$	$\frac{\sqrt{30}}{10}$	$\frac{3\sqrt{6}}{10}$
10	$\frac{\sqrt{15}}{10}$	$\frac{1}{2}$	$\frac{\sqrt{2}}{2}$	$\frac{-\sqrt{10}}{10}$
8, 2	$\frac{\sqrt{30}}{10}$	$\frac{-\sqrt{2}}{2}$	0	$\frac{-\sqrt{5}}{5}$
8, 1	$\frac{-3\sqrt{6}}{10}$	$\frac{-\sqrt{10}}{10}$	$\frac{\sqrt{5}}{5}$	$\frac{-2}{5}$

$\mu = 10$

μ_{23} \ μ_{12}	27	8_2	8_1	1
35	$\frac{\sqrt{15}}{20}$	$\frac{\sqrt{7}}{4}$	$\frac{\sqrt{35}}{20}$	$\frac{\sqrt{7}}{4}$
27	$\frac{-\sqrt{7}}{20}$	$\frac{-\sqrt{15}}{20}$	$\frac{-9\sqrt{3}}{20}$	$\frac{3\sqrt{15}}{20}$
10	$\frac{\sqrt{210}}{20}$	$\frac{\sqrt{2}}{4}$	$\frac{-3\sqrt{10}}{20}$	$\frac{-\sqrt{2}}{4}$
8	$\frac{-\sqrt{42}}{10}$	$\frac{\sqrt{10}}{5}$	$\frac{-\sqrt{2}}{5}$	$\frac{-\sqrt{10}}{10}$

$\mu = 10^*$

μ_{23} \ μ_{12}	27	10
27	$\frac{\sqrt{10}}{5}$	$\frac{-\sqrt{15}}{5}$
8	$\frac{\sqrt{15}}{5}$	$\frac{\sqrt{10}}{5}$

$\mu = 64$

μ_{23} \ μ_{12}	27	10*
35	$\frac{\sqrt{6}}{4}$	$\frac{\sqrt{10}}{4}$
27	$\frac{\sqrt{10}}{4}$	$\frac{-\sqrt{6}}{4}$

$\mu = 1$

	10*
8	-1

$\mu = 28$

	10
35	1

$\mu = 35^*$

	27
27	1

$\mu = 81$

	27
35	1

$\mu = 27$

μ_{23} \ μ_{12}	27	10*	10	8_2	8_1
35	$\frac{\sqrt{3}}{6}$	$\frac{\sqrt{5}}{6}$	$\frac{-\sqrt{14}}{12}$	$\frac{\sqrt{35}}{12}$	$\frac{-\sqrt{7}}{4}$
27, 2	$\frac{\sqrt{70}}{20}$	$\frac{-\sqrt{42}}{12}$	$\frac{-\sqrt{15}}{12}$	$\frac{5\sqrt{6}}{24}$	$\frac{3\sqrt{30}}{40}$
27, 1	$\frac{-3\sqrt{6}}{20}$	$\frac{-\sqrt{10}}{20}$	$\frac{\sqrt{7}}{4}$	$\frac{3\sqrt{70}}{40}$	$\frac{-\sqrt{14}}{40}$
10	$\frac{\sqrt{210}}{30}$	$\frac{\sqrt{14}}{6}$	$\frac{\sqrt{5}}{6}$	$\frac{\sqrt{2}}{12}$	$\frac{3\sqrt{10}}{20}$
8	$\frac{2\sqrt{21}}{15}$	$\frac{-\sqrt{35}}{15}$	$\frac{\sqrt{2}}{3}$	$\frac{-2\sqrt{5}}{15}$	$\frac{-2}{5}$

$\mu = 35$

μ_{23} \ μ_{12}	27	10	8_2	8_1
35, 2	$\frac{\sqrt{10}}{10}$	$\frac{\sqrt{6}}{12}$	$\frac{5\sqrt{3}}{12}$	$\frac{3\sqrt{15}}{20}$
35, 1	$\frac{\sqrt{5}}{5}$	$\frac{-\sqrt{3}}{3}$	$\frac{\sqrt{6}}{6}$	$\frac{-\sqrt{30}}{10}$
27	$\frac{-\sqrt{10}}{10}$	$\frac{\sqrt{6}}{4}$	$\frac{\sqrt{3}}{4}$	$\frac{-3\sqrt{15}}{20}$
10	$\frac{\sqrt{15}}{5}$	$\frac{1}{2}$	$\frac{-\sqrt{2}}{4}$	$\frac{-\sqrt{10}}{20}$

Table III. $8 \otimes 10 \otimes 8$

$\mu = 8$

μ_{23} \ μ_{12}	27	10	8, 2	8, 1
27	$\frac{-1}{10}$	$\frac{\sqrt{15}}{10}$	$\frac{-\sqrt{30}}{10}$	$\frac{3\sqrt{6}}{10}$
10	$\frac{\sqrt{15}}{10}$	$\frac{1}{2}$	$\frac{\sqrt{2}}{2}$	$\frac{\sqrt{10}}{10}$
8, 2	$\frac{-\sqrt{30}}{10}$	$\frac{\sqrt{2}}{2}$	0	$\frac{-\sqrt{5}}{5}$
8, 1	$\frac{-3\sqrt{6}}{10}$	$\frac{-\sqrt{10}}{10}$	$\frac{\sqrt{5}}{5}$	$\frac{2}{5}$

$\mu = 10$

μ_{23} \ μ_{12}	35	27	10	8
35	$\frac{1}{8}$	$\frac{\sqrt{105}}{40}$	$\frac{\sqrt{14}}{8}$	$\frac{\sqrt{70}}{10}$
27	$\frac{-\sqrt{105}}{40}$	$\frac{-37}{40}$	$\frac{\sqrt{30}}{40}$	$\frac{\sqrt{6}}{10}$
10	$\frac{\sqrt{14}}{8}$	$\frac{-\sqrt{30}}{40}$	$\frac{3}{4}$	$\frac{-\sqrt{5}}{5}$
8	$\frac{-\sqrt{70}}{10}$	$\frac{\sqrt{6}}{10}$	$\frac{\sqrt{5}}{5}$	$\frac{-1}{5}$

$\mu = 10^*$

μ_{23} \ μ_{12}	27	8
27	$\frac{1}{5}$	$\frac{2\sqrt{6}}{5}$
8	$\frac{2\sqrt{6}}{5}$	$\frac{-1}{5}$

$\mu = 1$

	8
8	-1

$\mu = 27$

μ_{23} \ μ_{12}	35	27, 2	27, 1	10	8
35	$\frac{1}{24}$	$\frac{\sqrt{210}}{48}$	$\frac{-\sqrt{2}}{16}$	$\frac{\sqrt{70}}{24}$	$\frac{\sqrt{7}}{3}$
27, 2	$\frac{\sqrt{210}}{48}$	$\frac{-5}{16}$	$\frac{3\sqrt{105}}{80}$	$\frac{11\sqrt{3}}{24}$	$\frac{-\sqrt{30}}{30}$
27, 1	$\frac{\sqrt{2}}{16}$	$\frac{-3\sqrt{105}}{80}$	$\frac{57}{80}$	$\frac{-3\sqrt{35}}{40}$	$\frac{\sqrt{14}}{10}$
10	$\frac{\sqrt{70}}{24}$	$\frac{11\sqrt{3}}{24}$	$\frac{3\sqrt{35}}{40}$	$\frac{-1}{12}$	$\frac{-\sqrt{10}}{15}$
8	$\frac{\sqrt{7}}{3}$	$\frac{-\sqrt{30}}{30}$	$\frac{-\sqrt{14}}{10}$	$\frac{-\sqrt{10}}{15}$	$\frac{1}{15}$

$\mu = 35$

μ_{23} \ μ_{12}	35, 2	35, 1	27	10
35, 2	$\frac{1}{8}$	$\frac{\sqrt{2}}{4}$	$\frac{7}{8}$	$\frac{\sqrt{6}}{8}$
35, 1	$\frac{\sqrt{2}}{4}$	0*	$\frac{-\sqrt{2}}{4}$	$\frac{\sqrt{3}}{2}$
27	$\frac{-7}{8}$	$\frac{\sqrt{2}}{4}$	$\frac{-1}{8}$	$\frac{\sqrt{6}}{8}$
10	$\frac{\sqrt{6}}{8}$	$\frac{\sqrt{3}}{2}$	$\frac{-\sqrt{6}}{8}$	$\frac{-1}{4}$

$\mu = 28$

	35
35	1

$\mu = 35^*$

	27
27	1

$\mu = 64$

μ_{23} \ μ_{12}	35	27
35	$\frac{1}{4}$	$\frac{\sqrt{15}}{4}$
27	$\frac{\sqrt{15}}{4}$	$\frac{-1}{4}$

$\mu = 81$

	35
35	1

* We assume $\langle (8\ 10)\ 35\ 8\ 35_1 \mid 8\,(10\ 8)\ 35\ 35_1 \rangle = 0$ which yields a constraint for the reduction of $8 \otimes 35 = \ldots \oplus 35_1 \oplus 35_2 \ldots$ and removes the ambiguity.

Weak Interactions without Change of Strangeness (Experimental)

By

H. Faissner, CERN, Genève*

With 42 Figures

In this series of lectures I want to talk mainly about the high energy neutrino experiments performed at CERN in which I participated during the last years. Some of the material I am going to present has been published in the Proceedings of the Conferences at Brookhaven [1], Hamburg [2] and Siena [3—5]. The techniques of producing and detecting neutrinos, as well as the theory of neutrino reactions, are extensively described in a recent yellow CERN-Report [6]. During the last three months the analysis of the data has been considerably improved and refined, so that there is growing confidence in the results. The new data is being prepared for publication. At the end I would like to say a few words about what has happened at lower energies, i. e. in μ-capture and related problems.

In order to give you a fair impression of how the data is obtained, let me go into some experimental detail. The points I will discuss are as follows:

1. Production of neutrino beams.

2. Detection of neutrinos and their positive identification.

3. Qualitative questions, namely whether there are one, two or more different kinds of neutrinos, and what selection rules they obey.

4. Elastic processes. These are, if you want, inverse μ-capture and β-decay. I will add a few remarks on inelastic reactions.

5. The question of the *existence of the intermediate boson W.*

6. Some recent *results at lower momentum transfers.* This would deal mainly with μ-capture, where a number of very brilliant experimental investigations have lately been carried out.

In my talk I shall concentrate on experimental methods and facts, and leave the theory to the theoreticians.

---

* Now at Technische Hochschule Aachen, Aachen, Germany.

1. Production of Neutrinos

1.1 Production modes

All neutrino beams use as a source the weak decay of mesons:

$$\pi^+ \to \mu^+ + \nu_{\mu\,\pi}, \quad \pi^- \to \mu^- + \bar{\nu}_{\mu\,\pi}. \tag{1}$$

In every high energy machine capable of producing an intense pion-beam of several GeV energy we have also high energy kaons produced. The kaons decay, amongst other channels, into muons. I shall make an index μ and K to the associated neutrino.

$$K^+ \to \mu^+ + \nu_{\mu\,K}, \quad K^- \to {}^-\mu^- + \bar{\nu}_{\mu\,K}. \tag{2}$$

This serves only as a definition of the neutrinos emitted in these decays. It is, of course, clear that ν and $\bar{\nu}$ are respectively particle and antiparticle. For a long time it was tacitly assumed that the neutrino known from β-decay:

$$n \to p + e^- + \bar{\nu}_e \tag{3}$$

is identical with the neutrino due to meson decays:

$$\nu_e \equiv \nu_{\mu\,\pi} \equiv \nu_{\mu\,K}. \tag{4}$$

The Brookhaven experiment [7] proved the first equality in (4) wrong. The second equality, however, holds and I will provide the evidence for this later.

The purpose of writing down the production modes (1) and (2) was to show that we have in principle four different kinds of neutrinos (counting neutrinos and antineutrinos separately) due to $\pi^\pm$ and $K^\pm$-decays. In the CERN-beam the π and K neutrinos have been considerably enhanced whereas the two types of antineutrinos were suppressed. This is accomplished by focussing the positive and defocussing the negative parent particles. But about this later. When you don't have this focussing devices at your disposal then the experimental set-up is as you can see in Fig. 1.

This was the way the first neutrino experiment was done at the Brookhaven National Laboratory near New York. We have somewhere the beam of the protons in the proton-synchrotron. You put in the way of the internally circulating protons a target, and this target produces the secondary particles. Unfortunately one can only observe under a finite angle. After the target there is a magnet unit of the synchrotron and therefore one has to observe at an emission angle Θ_{em} of about 8 degrees. We know that meson-production is strongly peaked in the forward direction. Therefore under these conditions the intensity of π's and K's is down by about a factor of 10, as compared to the case where one could observe in the forward direction. The resulting neutrino flux is easily estimated from the known decay-time of the meson and the distance it has travelled. If you look at the pion decay length involved, you find that it is typically of the order of 100 m. (The relativistic time dilatation makes them decaying later than if they were at rest.) Consequently only a small fraction can be expected to decay.

The distances are somewhat fixed by the requirement that the absorber in front of the detector absorbes the μ's which are also created. Now I give you a simple prescription, how to absorb μ's. If you have μ's of an energy k GeV you put k meters of iron in their way and they will stop. (It just so happens that 1 meter of iron is equivalent to ≈ 1 GeV energy loss.)

Since a few muons can be produced close to the energy of the primary protons, one puts as many meters of iron as the accelerator energy reads in GeV. If one is safe against μ's one is a fortiori safe against everything else. You can easily calculate that the attenuation of any strongly interacting particle is much stronger than the attenuation of the muons.

Next you find that you get optimal intensity if you make the decay distance from the target to the beginning of the shielding (d) and the absorber thickness (a) equal. Therefore, at 25 GeV machine energy the detectors have to be at a distance of 50 m from the target. If nothing is done about the pions, except taking them as they come out from an

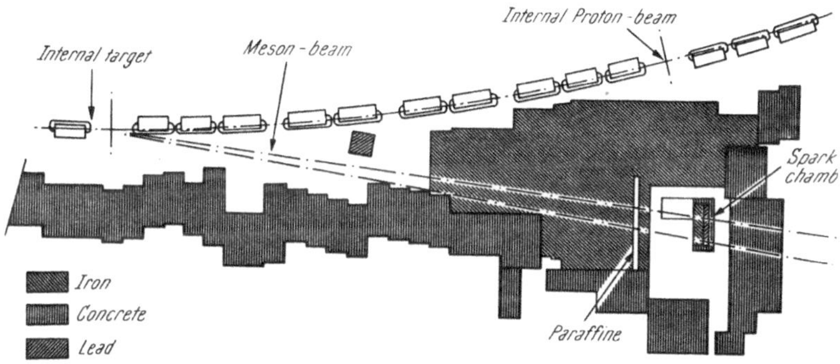

Fig. 1. Layout of a neutrino beam derived from a target inside a proton synchrotron (first Columbia-Brookhaven experiment [7]).

internal target the order of magnitude of the resulting neutrino-flux may be estimated as follows: at 25 GeV primary energy and an observation angle of 8 degrees we have due to the interaction of a primary proton with the nuclei of the target about one meson per steradian and interacting proton. The total number of protons accelerated in the machine is typically some 10^{11} per pulse, about half of which interact. Then you need only to consider the fraction of mesons decaying and the solid angle under which you see the detector. It has a typical size of 1 m^2. Therefore you ask what is the flux through 1 m^2, and you will find something of the order of 10^7 neutrinos/m^2 and pulse. The pulses come usually at intervals of 1 second, or one every 2 or 3 seconds, depending on the machine energy. The cross-section for the neutrino process under consideration is of the order of 10^{-38} cm^2. Doing the computation we arrive at the conclusion: in 1 m^3 of iron we have 2×10^{-5} neutrino interactions per machine pulse. Since the day is worth about 30,000 pulses we have, under the conditions explained, about 1 neutrino event per day and cubic meter of iron.

There was some talk about the fact that at CERN there was less intensity with an internal target than at Brookhaven. This can be explained by the combination of a few minor technical differences: At the Brookhaven AGS the distance between the magnet units was roughly 3 m, as compared to 1 m for the CERN machine. CERN therefore had to put up with a more restrictive aperture limitation, which at the end turned out to be murderous. It is not more than a factor 2 or 3 what you lose, but in this game even a factor 2 is important. Further you have to stop the μ's. Given the density of the absorber the distance is fixed. Now the Americans were in the fortunate position that they had a battleship to scratch. They could make their absorber out of iron. Switzerland has no navy, and no battleships to scratch, so we had to use concrete instead of iron and the distance was increased by a factor of 1.5. Since the decay and the absorber distance should be the same, the whole apparatus had to go away. It is easy to see that one really loses with the solid angle, which means with the square of the distance from the target. So iron instead of concrete brought again a good factor of 2. If you now multiply things together, even allowing for the fact the CERN-PS runs at a higher intensity than the Brookhaven-AGS, you wind up with a ν-flux produced with an internal target from the CERN machine which is down by a factor 3 compared to the Brookhaven flux. Because of this CERN decided to stop the experiment.

The ν-flux comes mainly from the ν's from π-decay, which are μ-neutrinos. The π-electron decay, which would produce electron type neutrinos, is down by almost exactly a factor of 10^{-4}, as compared to the $\pi-\mu$ decay and can be neglected. The electron decay mode cannot be neglected in the case of the K. Let me give you some information about the composition of the beam. It is typically about

90% $\nu_{\mu\pi}$
9% $\nu_{\mu K}$
1% ν_{eK} from $K^+_{e3} \to e^+ + \nu_e + \pi^0$ (and $K^0{}_2$-decays).

The numbers apply to the CERN beam. In the old Brookhaven beam the relative contribution of ν_K's was somewhat higher.

I should emphasize that this ν-flux is, if one takes it seriously, a very difficult thing to calculate. The approximations we have made in this lecture are: Everything flies with the velocity of light, and the decay angle is neglected. For the π this angle is of the order of the difference of the μ and π mass, which is 30 MeV, divided by the primary energy of the π (about 3 GeV), hence we find

$$\Theta_{\pi\nu} = \frac{30}{3000} = 10 \text{ mrad} = 0.6 \text{ degrees.}$$

For most practical applications you can forget about this small angle, and safely assume that the ν flies in the parent-direction. (In the geometry we had in "old"-times this was a serious effect. There was an aperture-limitation for the π's, the π-beam was small in size, and the neutrinos ran out.) With the same argument you find, if you have a K decaying, a correspondingly larger angle. The K-μ-mass difference (it is 400 MeV) divided by 3000 MeV, is by no means very small!; it comes to 130 mrad $\approx 8°$. It would therefore be a very inefficient way of making a ν_K-beam, if one focussed the K-mesons into a pencil beam. The neutrinos would just run out. What one needs is a large-area beam, where neutrinos running out are compensated by others coming in.

KUMMER: How do you get the number of π's?

FAISSNER: Both, experimentally and by calculation. In the older layout, with an internal target, it was relatively easy, because you had only to make a few appropriate holes in the shielding around the machine and to observe under a given angle (with a magnet and particle-analysis in the usual way) how many π's and K's are emitted per circulating proton. Everything else, which comes in, is pure geometry, namely solid angles and distances. So the calculations are in principle quite easy and reliable.

With the focussing devices, which are in use or under construction now at both Laboratories, at CERN and Brookhaven, the situation is more difficult. There we have very complicated trajectories of the particles. So even if we had a very good and precise measurement of the production-spectrum, there were still considerable difficulties in the calculation of the neutrino-flux. I come back to this point later and I emphasize already at this point that lack of knowledge about the ν-spectrum at present is one of the main obstacles to understand our results quantitatively. There is a program under consideration to measure the meson-flux from the focussing horn directly, but this will take some time. It poses an unprecedented problem to counter people, since all particles are emitted within 2.2 μsec!

1.2 High intensity ν-beam, Part 1: Ejection of the proton beam

Let me describe now the two steps which are necessary, if you want to make a high intensity ν-beam.

One step I have mentioned already, namely the necessity of observing the produced mesons under a small angle with respect to the incoming protons, let us say under zero degrees. This gives almost an order of magnitude in intensity, depending on what you believe about the production spectrum. Incidentally this very sharp decrease was not anticipated when one believed in a statistical model for particle production. It is in part due to the fact that π- and K-production goes to a large fraction by some sort of peripheral processes, and therefore really goes in the forward direction. It turned out that this was quite helpful for some aspects of the neutrino experiment. But in order to observe under 0° there is no other way but to get the beam out.

Therefore the ejection of the internal proton beam is the first step in order to get a high intensity ν-beam. This had been planned at CERN already a long time ago [8]. What do we do in order to eject the internal proton beam out of the synchrotron? In principle it is very easy (Fig. 2). In the synchrotron vacuum-chamber of 15×5 cm you have a proton beam which is not circulating as an uninterrupted current but which by the phase-focussing in the synchrotron is split up into so-called bunches. You have over the circumference of 628 m twenty of these bunches. In space they are something like 3 m long, which corresponds to a length in time of 10 nsec. After these 10 nsec of protons you have 100 nsec of nothing; then comes the next proton bunch and so forth. In one of the straight sections between two synchrotron-magnets, there is a special magnet unit, the kicker-magnet. Its field rises within 100 nsec from zero to the full strength of ≈ 3 kGauss. Therefore the preceding bunch was undisturbed but the next one gets already the full "kick".

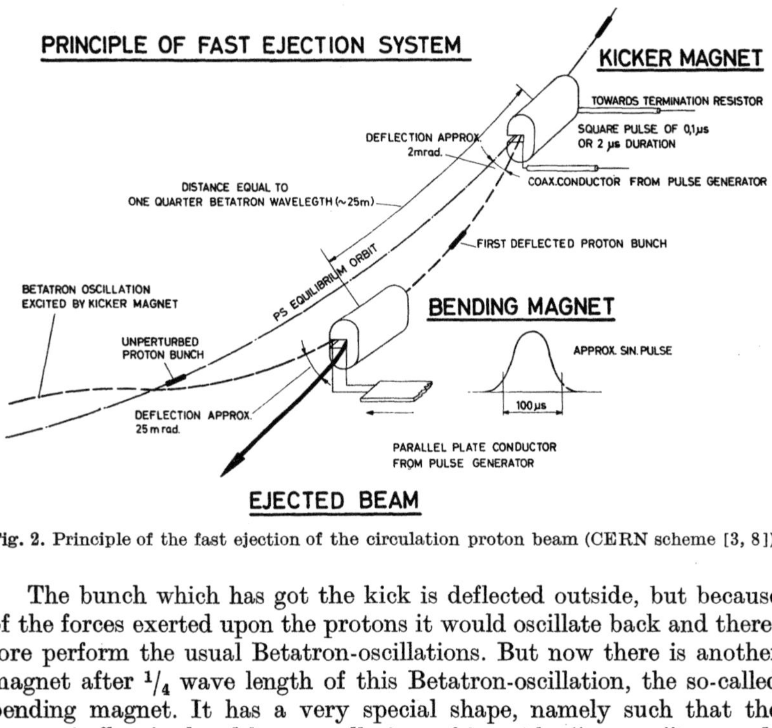

Fig. 2. Principle of the fast ejection of the circulation proton beam (CERN scheme [3, 8]).

The bunch which has got the kick is deflected outside, but because of the forces exerted upon the protons it would oscillate back and therefore perform the usual Betatron-oscillations. But now there is another magnet after $^1/_4$ wave length of this Betatron-oscillation, the so-called bending magnet. It has a very special shape, namely such that the magnetic flux is closed by a small piece of iron (the "septum") towards the inside, so that protons which had been circulating before did not see any field. The first deflected bunch just comes into this channel and is guided out of the synchrotron field. Now one can do two things: either one just kicks out one bunch and leaves the other nineteen in to circulate further, or one ejects the full beam and has all twenty bunches marching out, one after the other. By definition this system of ejection is supposedly 100% efficient provided it does work at all. Since last May 12th it has been demonstrated that the system does work. (Incidentally it turned out that the rise of the kicker-pulse, which is the most difficult problem, is only fast enough, if one applies a very high voltage.

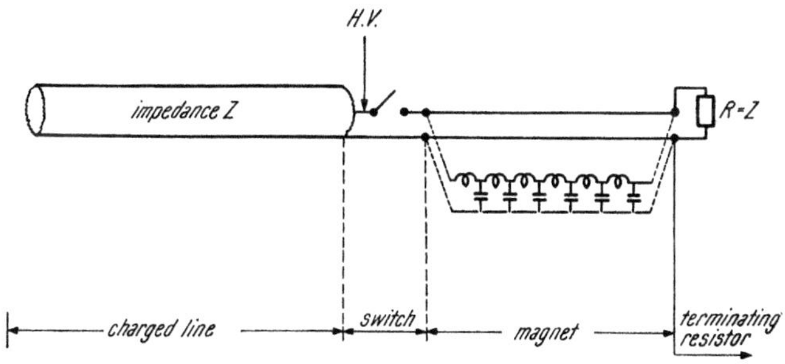

Fig. 3. Block diagram of the CERN kicker magnet.

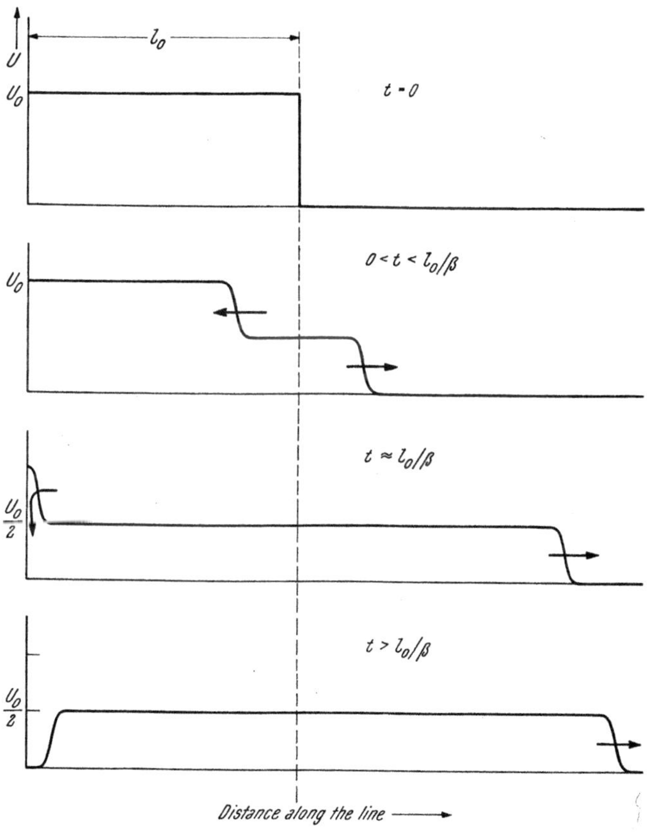

Fig. 4. Generation of an approximately rectangular current pulse by discharging a distributed line.

This is somewhat dangerous for the safety of the whole system as it may cause breakdown. Therefore the philosophy was to work with a smaller voltage, to lose the first bunch, and to eject 19 bunches to a hundred percent in a stable way.)

Fig. 3 indicates how the kicker-magnet is manufactured: it is shaped as part of a 10 Ω wave guide. A charged piece of cable, also of 10 Ω impedance and about 1 μsec long, is discharged via a fast switch (a spark-gap) into the magnet. The shape of the travelling wave is shown in Fig. 4. At the end, we have a rectangular-shaped wave of a length of somewhat more than 2 μsec, which is exactly what we want. This is the time one must have the magnet on voltage in order to guide the whole circulating proton beam out (the revolution-time is 2.2μsec). The next figure (Fig. 5) shows in the same scale the structure of the ejected proton beam. The width of the peaks is very small in comparison with the distance between them (≈1/10).

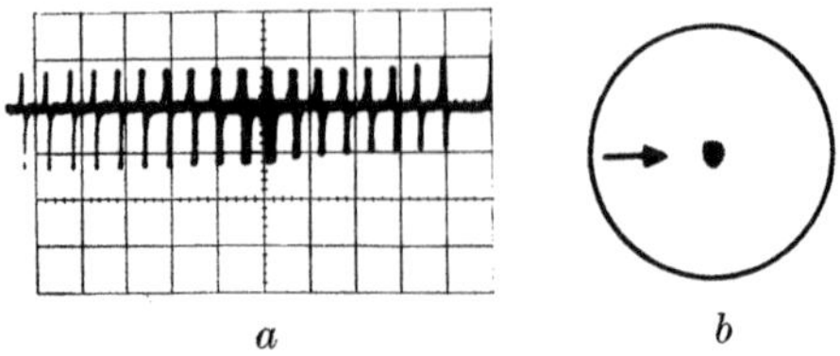

Fig. 5. Structure of the ejected proton beam, a) in time: 18 proton bunches of ≈ 10 nsec length spaced 105 nsec apart, b) in space: radiograph of the ejected beam at the position of the target (diameter ≤ 1 mm.)

Fig. 6. View of the kicker magnet, shortly before mounting inside the vacuum chamber which is visible in the lower part of the photo.

Fig. 6 gives a view of the actual kicker-magnet shortly before being mounted into its vaccum tank. If one had that kicker-magnet in its operating position all the time one could not accelerate the protons, because shortly after injection into the PS, they fill the whole vacuum-chamber. Therefore one can bring in this magnet only after one has accelerated the protons to almost the full energy. Then, because of the radial focussing, the protons are contracted to a diameter of less then 1 cm. For that you have to wait for 3/4 of a second, and then you have to ram in this magnet with a substantial force in order to move it in

Fig. 7. The bending magnet in its vacuum chamber.

place in $^1/_{10}$ sec. This is done by a system of hydrodynamic rams (the weights of the two magnets are 300 and 500 kg resp., the accelerations are of the order of 8 g).

The bending-magnet may be seen in Fig. 7. It is 1.50 m long, and has a field strength of 15 kGauss. Its pulsing is much less critical than the one of the kicker-magnet. Because of the septum we can switch it in long before the kicker-magnet is fired, without disturbing the circulating protons. Therefore, one simply discharges, through ignitrons, a condensor battery into the bending magnet, and adjusts things such as to have essentially one sinus-shaped half-wave of 150 μsec length. The kicker-magnet is triggered when this sinus is about at maximum. After

Fig. 8. Part of the beam transport between ejection and magnetic horn. The horn is inside the concrete blockhouse in the background; some magnet units of the proton synchrotron are visible in the right-hand part of the picture.

the protons have been bent out of the guiding field of the synchrotron, they are brought, via a street of bending magnets and quadrupole lenses (Fig. 8), onto an external target.

What do we require for this external target: the target should be thick with respect to the primary protons, but thin with respect to the emitted mesons (because otherwise these get reabsorbed in the target and are lost). Since the incoming protons are already very well focussed by the magnetic lenses (diameter of the beam ≈ 1 mm, negligible angular-divergence) you can make a target which is thick for the protons and thin for the emitted mesons: you make it of very small diameter (4 mm), fairly long (≈ 20 cm), and reasonably dense (Cu gives the best overall ν-flux). In order to have the target fully efficient you must of course have the proton-beam falling right upon its center, so you must adjust the thing to within 1 mm. This is done in the following way: There is a scintillation screen which has a hole in the middle (2 mm diameter) and a television-camera watching the screen. You can see where the ejected beam hits the screen. Now you adjust your magnets in such a way that the scintillation-spot disappears through the hole. In the same moment a radiation signal will appear in some scintillation counters watching the radiation emerging from the target.

1.2 High intensity ν-beam, Part 2: The neutrino-horn

This is a very ingenious device which was constructed, brought into operation and supervised for two million pulses by SIMON VAN DER MEER [9] and his associates [23].

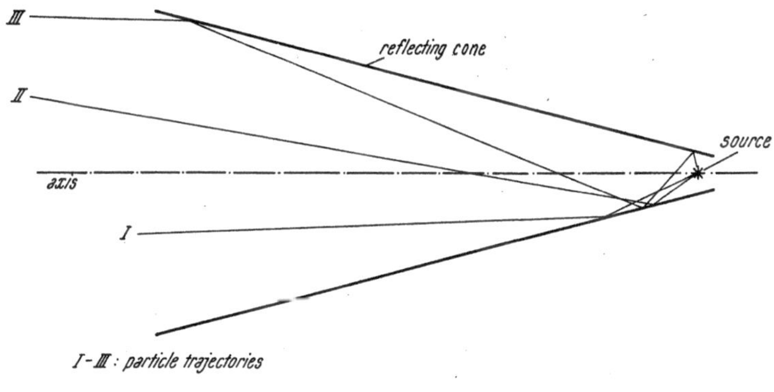

Fig. 9. Principle of a horn working with successive reflections.

It brought another order of magnitude in the ν-flux by the following idea: We have already all principles of how to produce an energetic high-intensity ν-beam. We just have to produce as many mesons as we possibly can, and direct them towards the detector. This is done by a sort of mirror, which reflects particles emitted at large angles back to the primary proton direction (Fig. 9). Mirrors of this type do exist in form of magnetic fields. VAN DER MEER's idea was coming about by an acoustical analogon, the megaphon. What he invented is some sort of electro-magnetic trumpet.

The principle how it works can be seen in Fig. 10. The horn is a double-walled funnel, through which a very strong current flows. For symmetry reasons the magnetic field lines are circles around the horn

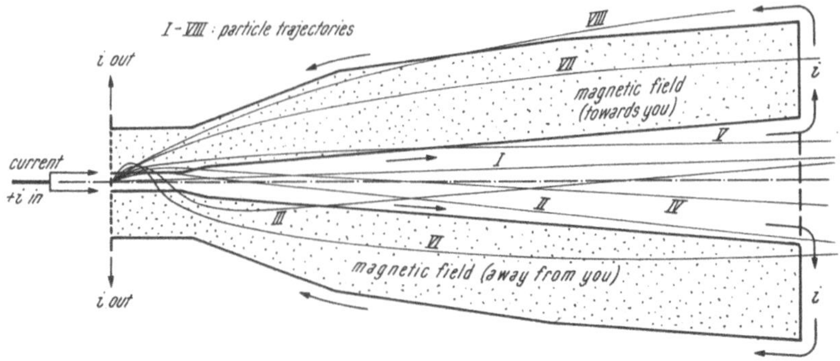

Fig. 10. Principle of a magnetic horn.

Fig. 11. Photograph of the actual horn. Electrical connections through the discs at the mouthpiece of the horn.

axis. In practice (Fig. 11) the outer conductor has a shape different from the inner, conical one. This has the following reason: the mesons in this case need not cross the conductor-material as often as in the horn of Fig. 9. Accordingly the intensity won't be decreased as much by absorption as in the case above.

We may summarize the action of the horn as follows: It transforms a real particle source of small spatial dimension and large angular spread into a virtual one of large dimension and small angular spread. Clearly, this works well for one particle momentum only. The horn used in the experiment was designed for a momentum of about 6 GeV/c. The improvement in the rate of elastic neutrino reactions is a factor of 7.

Furthermore one gets a very clean ν-beam as opposed to an $\bar{\nu}$-beam, because positive particles give neutrinos and negative ones give anti-

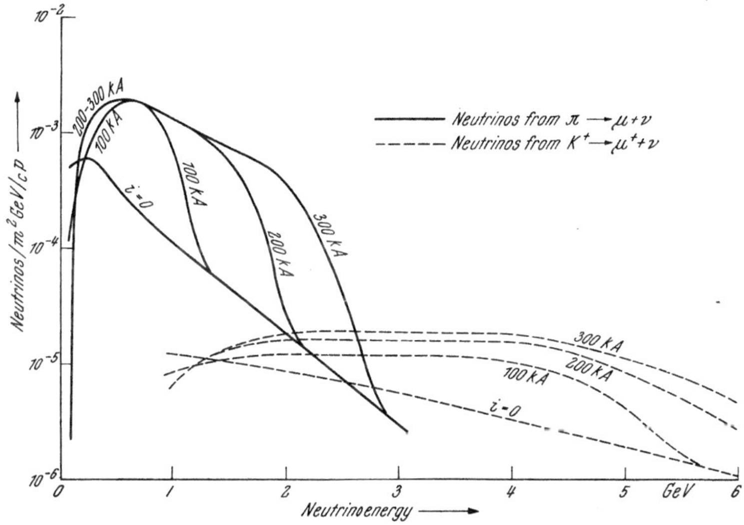

Fig. 12. Neutrino spectra from the horn, computed for different values of the current.

neutrinos. Therefore by defocussing the negative parents of antineutrinos we have a beam, which according to the calculations contains 97% neutrinos and only 3% antineutrinos. VAN DER MEER calculated the spectrum by making a numerical integration by a Monte-Carlo-method over the particle paths, putting in empirical values of this magnetic field.

Fig. 12 gives one of the first ν-spectra calculated by VAN DER MEER, which demonstrates what the horn really does. Plotted is the ν-flux at the detector in some units (actually neutrinos per m² and incident proton) versus the ν-momentum. According to the fact that we have two different sorts of parents, namely π's and K's we have two entirely different contributions to the ν-spectrum. You can see that the curves (note the logarithmic scale) fall down very steeply at the high-energy part of the spectrum. This is a pity. What we really want to do is high energy ν-physics, and what we are forced to do is intermediate-energy ν-physics around 1 GeV or so. You see that about $^1/_2$ of the ν's have an energy less than 1 GeV. You also see the increasing number of high energy ν's with increasing current in the horn. This is the third desirable property of the horn: It makes the neutrino spectrum harder.

The softness of the ν-spectrum has a trivial explanation. You are up in a certain sense against the second law of thermodynamics: You start with 25 GeV energy concentrated on one proton, which is a very unprobable state. Therefore, as soon as it gets the possibility of interacting with a target, it tends to split its energy into small pieces. Therefore the π's have a spectrum concentrated around 3 GeV, which makes already almost a factor of 10 in loss. And then, unfortunately, in the decay of a π into μ and ν, the heavy particle, namely the μ, is bound, in the average, to get the lion-share of the energy, just because of momentum conservation. Therefore the maximum energy you get out for the ν's from π-decay is only 40% of the π-energy: the ν-spectrum is shifted to ≤ 1 GeV. The K's give a much higher energy to the neutrino, because the μ is as light as the ν compared to the K. These high energy ν's (about 10% of the total number) from K-decay are important for high energy processes e. g. for the production of intermediate bosons.

1.3 The CERN neutrino lay-out

Fig. 13 is a drawing of the CERN neutrino lay-out as of June 13th, 1963, the day the experiment began. About 6000 tons of iron and more than 3000 tons of concrete have been used for the shielding. The iron is a (small) part of the Swiss Strategic Reserve, which was kindly lent to us by the Swiss Authorities. The 25 m of iron on the direct line-of-sight to the target are sufficient to prevent any muon from reaching the detectors directly. Shielding against stray muons scattering *around* the central iron core was not as easy. It is quite easy, indeed, to calculate how much one has to flare out the shielding in order to depress muons, which multiple scatter still around, to a negligible level. It is not easy, however, to get the required iron shielding into place — notably when it comes to dig into the floor! Eventually we dug as deep, as the civil engineers would allow us to do without causing the two pillars of the PS-building (black in Fig. 13) to collaps. Since this was only 80 cm, we extended this iron re-inforcement of the floor far into the decay-tunnel and also into the detector area. Another difficulty arose from the existing PS-shielding bridge in between the two above-mentioned pillars, which is made of light concrete, which cannot be altered, and which does not fit very well into the shielding lay-out. We helped ourselves by a careful iron-lining of the decay tunnel, a funnel-shaped shielding around the horn, and by having the iron shield going up to a height of almost 6 m in front of the detectors.

This iron wall can be seen in the background of Fig. 14. It turned out later that even this was not enough. Only after I had filled the whole empty space between the iron ingots and the tubes visible in

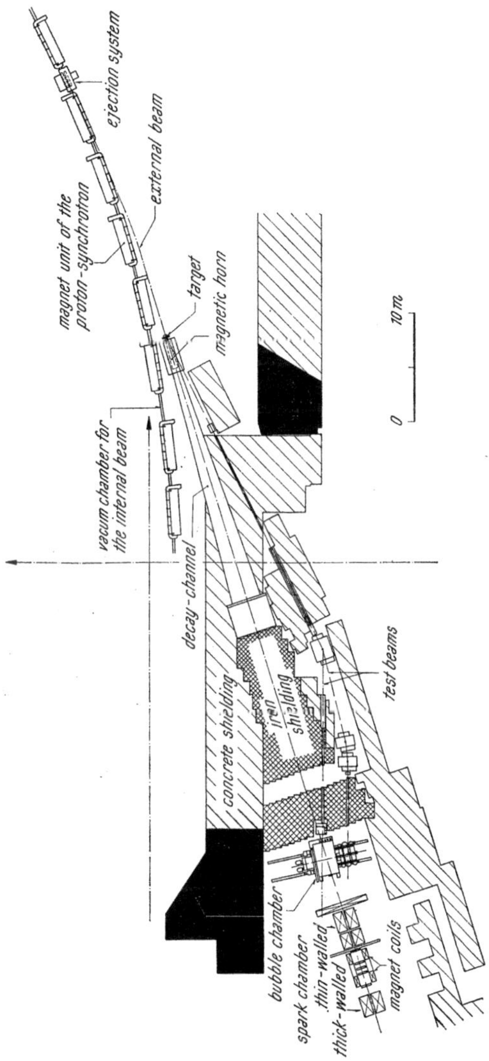

Fig. 13. Layout of the CERN neutrino experiment as of 13 June, 1963 (when the runs started).

Fig. 14 with heavy concrete, the shielding was safe. Already before this happened, another weak region had been spotted by the spark chamber people, who simply looked at the direction of muons entering the chamber, extrapolated back and found a 50 cm wide vertical slit between iron and concrete shielding, which one simply had forgotten to fill up.

Another facility may also be seen in Fig. 14: The worker in the picture is fiddling around with a pipe pointing into the bubble chamber through which a test beam can be guided. During neutrino search the pipe is filled with mercury. A similar (though simpler) arrangement allows another beam to enter the spark chamber area. Two bending magnets necessary for this and a beam stopper may be recognized on the left-hand-side of Fig. 14. But this brings us already to the next point we have to treat: the detectors.

Fig. 14. Part of the iron shielding. To the left, some bending magnets and collimators, used for test beams.

2. Detection and identification of neutrino-events

2.1 The bubble chamber

The heavy liquid bubble chamber used in the experiment was one of the largest now existing in the world. It is 1.15 m in diameter, 0.5 m deep and filled with 500 litres freon (CF_3Br) of 1.5 gm/cm^3 density. The weight of the liquid is therefore 750 kg. For detailed analysis a fiducial volume of about 220 litres is available.

The chamber is in a strong magnetic field of 27 kGauss. The measuring accuracy is as follows: The momentum of a muon or pion of 250 MeV/c can be determined from the curvature of the track with a relative error of 8%. The main limitation is multiple scattering. The maximum detectable momentum is >15 GeV/c. A distinction between pions and protons on the basis of ionization (i. e. bubble density) is possible up to a momentum of 700 MeV/c.

The liquid has a short radiation lenght X_0 of 11 cm. This is important for the conversion of gamma rays. Also electrons will produce showers. The shower energy can be inferred from the shower properties, in particular from the total track lenght of electrons (of both signs).

The accuracy of this method is about 30% at ≈ 1 GeV primary shower energy [10]. The accuracy is mainly limited by fluctuations in the shower losses. In a liquid of longer radiation lenght the energy of an electron could of course be much better determined by magnetic curvature. But since the conversion lenght of high energy gammas is $^9/_7\ X_0$, one has to make a choice: In the neutrino experiment one chose to have a high conversion probability for gammas. This offers the possibility of detecting and measuring neutral pions. The price one has to pay, is a reduced accuracy in momentum measurement for muons, mesons and protons, and a virtually complete loss of momentum information for electrons.

The large size of the chamber, together with the high density of the liquid, makes also an interaction analysis possible. The interaction lenght of relativistic strongly interacting particles in freon is 58 cm. Therefore, the average probability that a strongly interacting particle will show an interaction is about 60%. This is not large enough to decide between muon and pion in a single event. But it does permit a statistical decision about the fraction of muons and pions in a sample of events. In some cases a μ^- will stop in the chamber. Then it will display its typical electron-decay in about 50% of the cases.

Obviously the CERN heavy liquid bubble chamber is a very good instrument for doing neutrino physics, as was recognized years ago [11]. It's only severe draw-back is the size which limits both, statistics and retaining power of the generated particles. There will be much improvement in the next years: The CERN chamber will be doubled in size, and chambers with a weight of 10 tons are under discussion.

2.2 The spark chamber

In contrast to a universal instrument, as the bubble chamber is, a spark chamber is a specialized object. The neutrino spark chamber, for instance, has only one virtue: it is large. It can rapidly collect statistics about the more popular neutrino reactions, and one can hope to see also rare ones. Moreover, it's large depth makes it very suited for the identification of muons.

As shown in Fig. 15 the spark chamber set-up consists of three separate sections: The first one is called the "production region". There we have about 20 tons of spark chamber units composed of thin brass and aluminium plates. Each modular spark chamber unit [12] has three 5 mm thick plates of 1 m height and 1.60 m width; the gaps are 1 cm wide (Fig. 16). Between these units are many plastic scintillation counters to ensure high detection efficiency ($\approx 60\%$). The chamber is triggered when two adjacent plastic counters [13] within the chamber or the liquid counter L_1 [14] at the end give a signal. Usually we demand

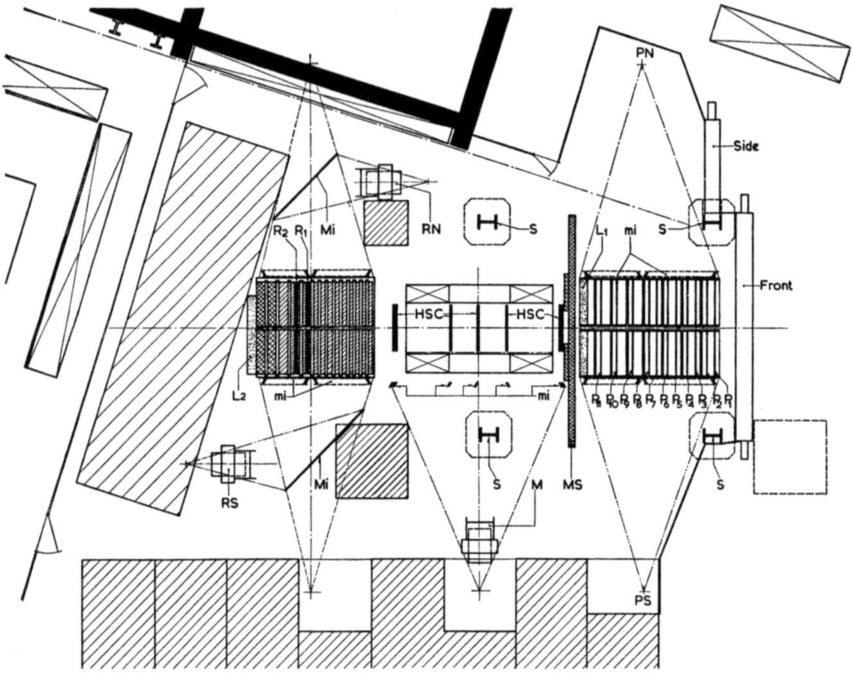

Fig. 15. Plan view of the spark chamber set-up.

$P_1 - P_{11}$, R_1, R_2	Plastic Counters	RN, RS, M	Stereo Cameras (range chamber north, south and magnet)
L_1, L_2	Liquid YES Counters	MS	— Magnetic Shielding Wall
Front, Side	Anti Coincidence Counters	HSC	— Hodoscope Spark Chambers
+	Virtual Camera Positions	mi	— Small Mirror Pairs
PN, PS	Camera Pairs (production region north, and south respectively)	Mi	— Large Mirrors
		S	— Roof Support
		(hatched)	— Pb Walls
		(dark)	— Fe Walls

anticoincidence with the liquid counter in front, in order to exclude charged particles coming from the shielding.

Next to the production region follow two Helmholtz-type coils supplying a magnetic field of about 4 kGauss strength. In this magnetic field the paths of roughly 10% of the particles emerging from the production region are sampled. With help of 7 thin-plate aluminium spark chambers the momentum of a passing particle can be measured with reasonable accuracy (15% at 1 GeV/c). In any case, up to a maximum detectable momentum of 10 GeV/c, the sign of the charge can unambiguously be determined.

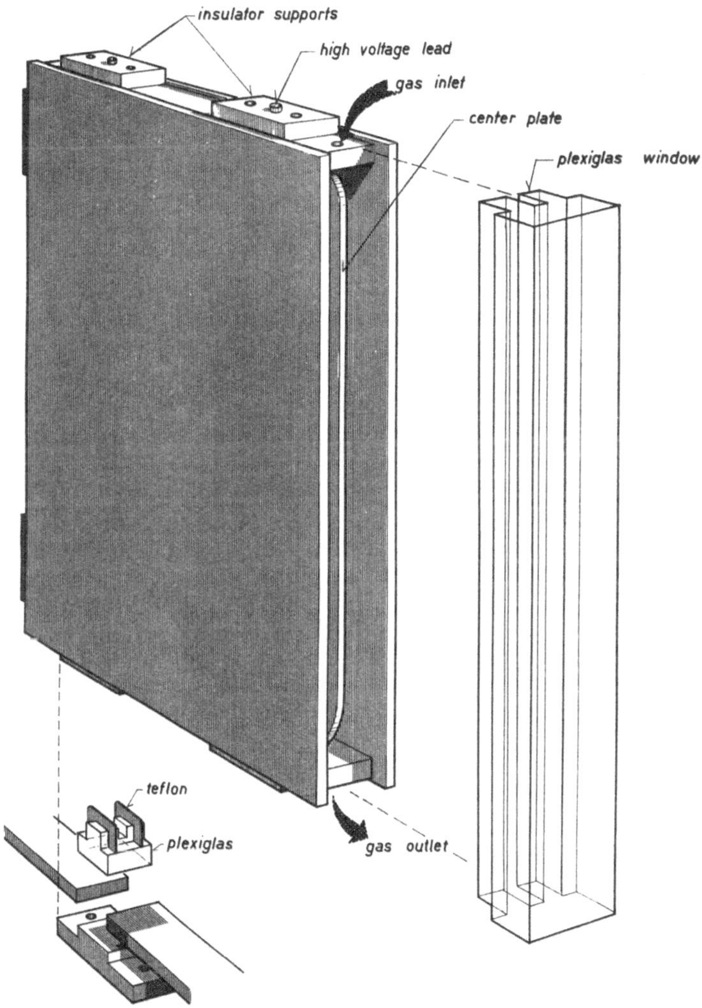

Fig. 16. Construction of the modular three-plate spark chamber units.

The third and last part, the range chamber, is a thick walled spark chamber assembly, simply a number of lead and iron walls of increasing thickness from 5 to 20 cm, with 3-plate spark chamber modules between two subsequent walls. The whole thing is 3 m long and weighs about 50 tons. In order to detect also events which originate in this chamber, we installed two plastic counters (R_1 and R_2 in Fig. 15) in and a liquid counter (L_2 in Fig. 15) behind the chamber. Muons and muon pairs are then detected with about 20% efficiency. The purpose of the range chamber is to make pions (and other "stronglies") interact and muons either penetrate or stop.

There are also anticoincidence counters on top of the chambers. They were indispensible when the experiment was run using the longish burst from the internal target. Even now, when we make use of the short duration (2.2 μ sec) of the pulse, they can help to discriminate against cosmic rays and stray muons from the machine. But they are not essential.

The spark chambers are stereophotographed, which is not quite as easy as it sounds. Rather sophisticated mirror arrangements have been necessary for having the cameras at accessible places. A pair of small mirror strips in front of each spark chamber module permits one to look into the gaps.

2.3 The most frequently observed events

Let us first recall what we expect to find. It is clear that a ν from π-decay must be capable of producing muons. This is because π-decay is in a sense equivalent to μ-capture. Therefore the inverse process must also be present in nature.

Because of lepton conservation the ν_μ defined as coming from π^+-decay should produce a μ^- which is a lepton (and not an antilepton). The reaction we are expecting to see is therefore:

$$\nu_\mu + \mathrm{n} \rightarrow \mu^- + p. \tag{5}$$

The process is referred to as "the elastic reaction". Kinematically this is really the case and the change of charge is inherent to weak processes. Inelastic processes are those, where in place of the proton there is some complex of strongly interacting particles. This one expects to happen quite often: The proton may always shake off a number of mesons, provided it has sufficient energy.

The identification of ν_μ-events makes use of the fact that a negatively charged muon should be created in each event. The muon has no other interactions with matter but weak and electromagnetic. Effects of weak interactions cannot be observed over short path lengths. Electromagnetic forces cause multiple scattering and slowing down of the particle, effects which can easily be seen.

Now let me describe some typical events which were found very frequently.

If you look at the bubble chamber photos, you have for instance the following event (Fig. 17), which consists of one long range particle which has negative charge; (the magnetic field bends negative particles counter-clock-wise). In addition you see something which is by its short range, heavy ionisation, and opposite curvature in the magnetic field, identified as a proton.

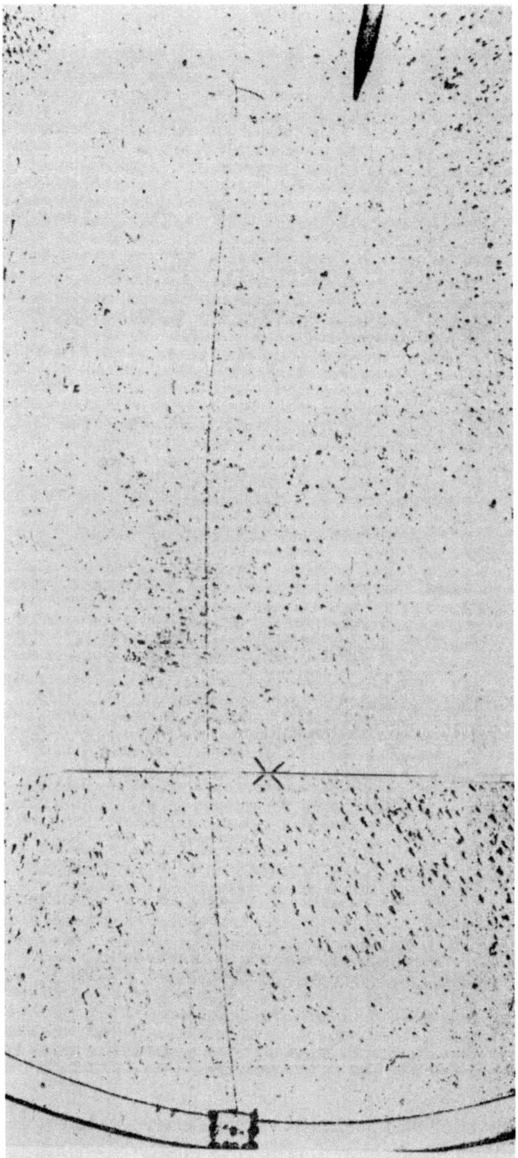

Fig. 17. One of the first candidates for elastic muon production in the bubble chamber: a negatively charged relativistic particle, emitted close to the forward direction, and three accompanying proton recoils.

If you look very closely you can see two additional protons. They are the result of secondary nuclear reactions of the original recoil proton and you should not worry about that. Of course a thing like that makes a kinematical reconstruction of the event a little difficult. In the far future of ν-physics one will certainly use hydrogen or deuterium chambers, provided one gains another factor of 10 in the ν-intensity. In comparison

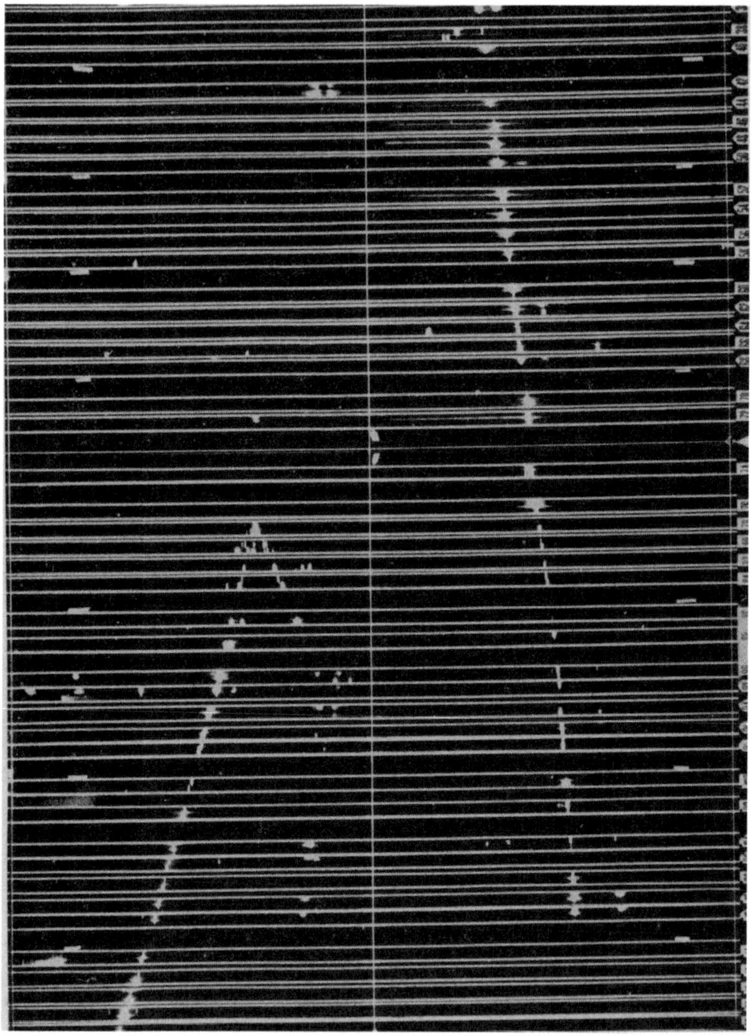

Fig. 18. Two candidates for elastic muon production in the spark chamber, photographed accidentally in the same picture. (Neutrinos come from the right in these pictures).

to the "old" days of ν-experiments we already gained a factor of 200. Ejection of the p's brought a factor of 7, the ν-horn another factor of 7 and higher machine intensity (7×10^{11} p's per pulse) again a factor of 4.

The same sort of events are seen on spark-chamber-photographs; one long-ranged particle with a tracklength of approximately 1 m (about 3 geometrical interaction lengths) or more, doing nothing but multiple scattering, and a short track, compatible with a proton. Sometimes there are two events in one picture (Fig. 18) due to a sort of chance-coincidence of two ν-events, (which incidentally shows the high intensity of the ν-beam). One reaction (the earlier one) triggered the chamber and the second ν-event was photographed by chance. The fact that the two events are not simultaneous, is easily recognized by a nice effect: The brightness of the earlier sparks diminishes very rapidly

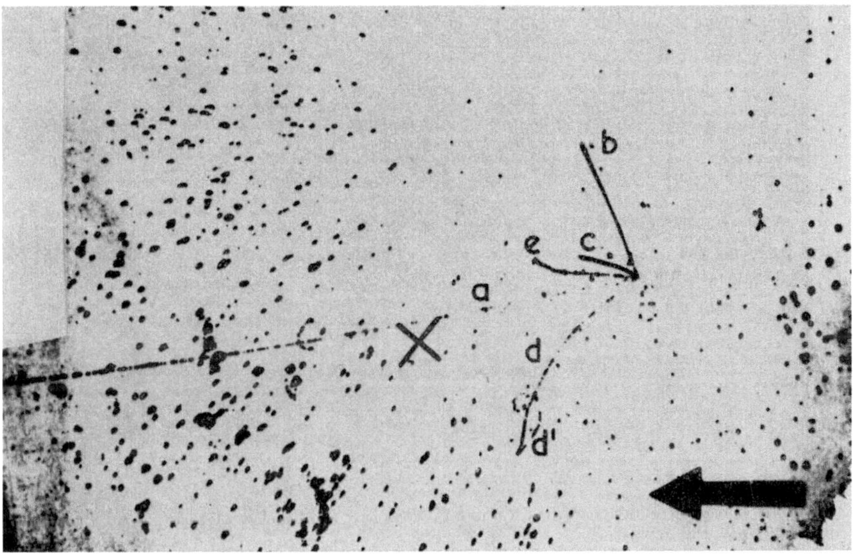

Fig. 19. Inelastic neutrino reaction in the bubble chamber. The tracks are identified as follows: a $= \mu^-$, b and c = protons, d – d' $= \pi^+ \to \mu^+ \to e^+$ decay chain, e $= \pi^-$.

if another event is initiated (robbing effect). By the brightness of the sparks one can therefore tell which particle came first.

There is some advantage of spark-chamber tracks with respect to the bubble-chamber tracks: The paths are much longer, and hence the spark chamber is the better tool to study nuclear interactions.

Another important type of events has the following form, seen very clearly in the bubble chamber (Fig. 19): a long ranged μ^- (note that the field direction is reversed with respect to Fig. 17), a few fairly long ranged p's, a negative particle which has minimum ionisation and stops within the chamber and is, by all probability, a π^-. There is also a beautiful case of a positive relativistic particle, a π^+, which looses energy as seen very clearly from the increasing bubble density, then stops and decays into a particle of very short range making a curly track, which undoubtedly is an electron. Therefore you have a full decay chain: pion, muon, and electron in one picture. Its high resolution makes the bubble chamber a unique tool for studying events of this

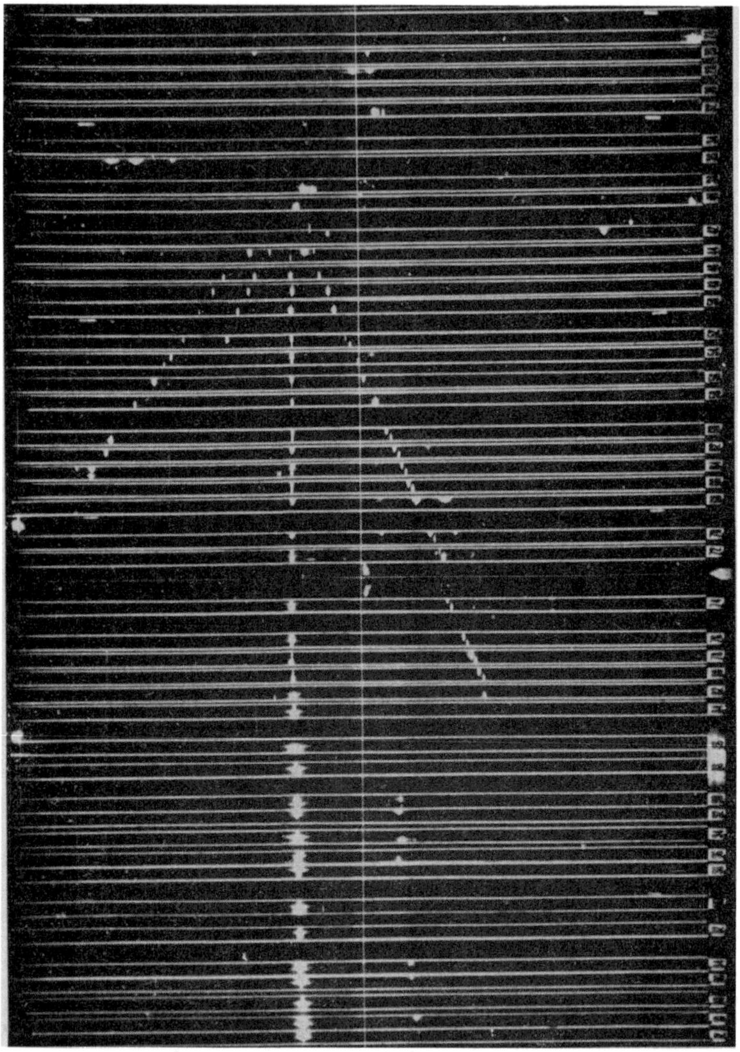

Fig. 20. Inelastic neutrino reaction in the spark chamber. Judging from the large amount of matter transversed the forward particle is likely to be the muon. The other particles cannot be identified with confidence.

complexity with very little ambiguity indeed. A similar event in the spark chamber would look like that of Fig. 20:

There seems to be a long ranged μ, two π's and a short ranged p; but nobody can be sure that the short path did not for instance belong to another π. Therefore an exact identification of particles in complicated events is not possible here.

The idea of having two kinds of detectors is: The spark chamber gives rapid information and good signatures for muons, whereas the bubble chamber, though limited by statistics, has a much greater power for detailed analysis.

2.4 Idenficitation of ν-events in the bubble chamber

One measures the visible energy E_{vis} which is obtained by summing up the energy of all the particles in the event you see. This visible energy should be quite close to the primary ν-energy. However it might be somewhat less because p's or π's re-interacting in a nucleus can free a few

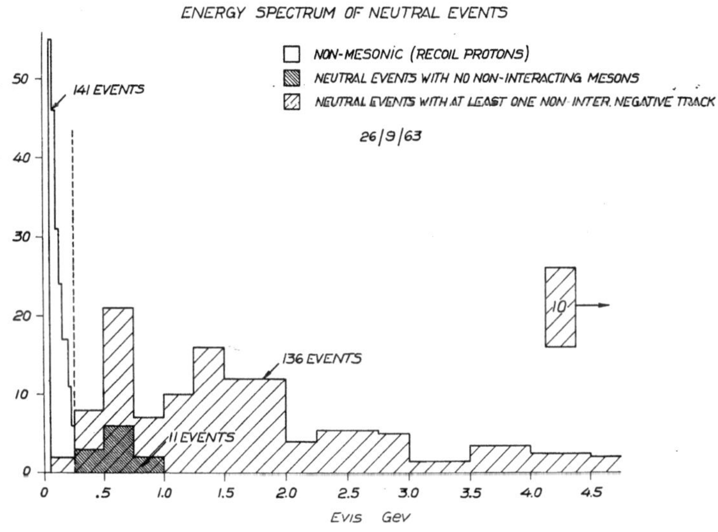

Fig. 21. Distribution of visible energies of stars initiated by neutral primaries in the bubble chamber. The low energy peak is due to neutrons; high energy events are almost exclusively neutrino-induced.

neutrons which are only in part reabsorbed in the chamber. Very often it is hard to tell whether a n-star observed close to an apex has something to do with that apex or is just some sort of background. Also gamma rays have a finite probability to escape. Consequently E_{vis} is only an approximate measure for the incoming ν-energy. Fig. 21 shows a plot of the number of events versus the visible energy. At the high energy end we have 136 events, positively identified as being ν-induced, and a peak at low energies comprising recoil protons or small stars, which altogether make 141 events. (The numbers refer to the time of the Siena-conference. In the meantime quite a few events have been added but the conclusions drawn remain unchanged).

Events in the low energy group have no candidate for a μ^- and are due to the n-background. That the high energy group is almost

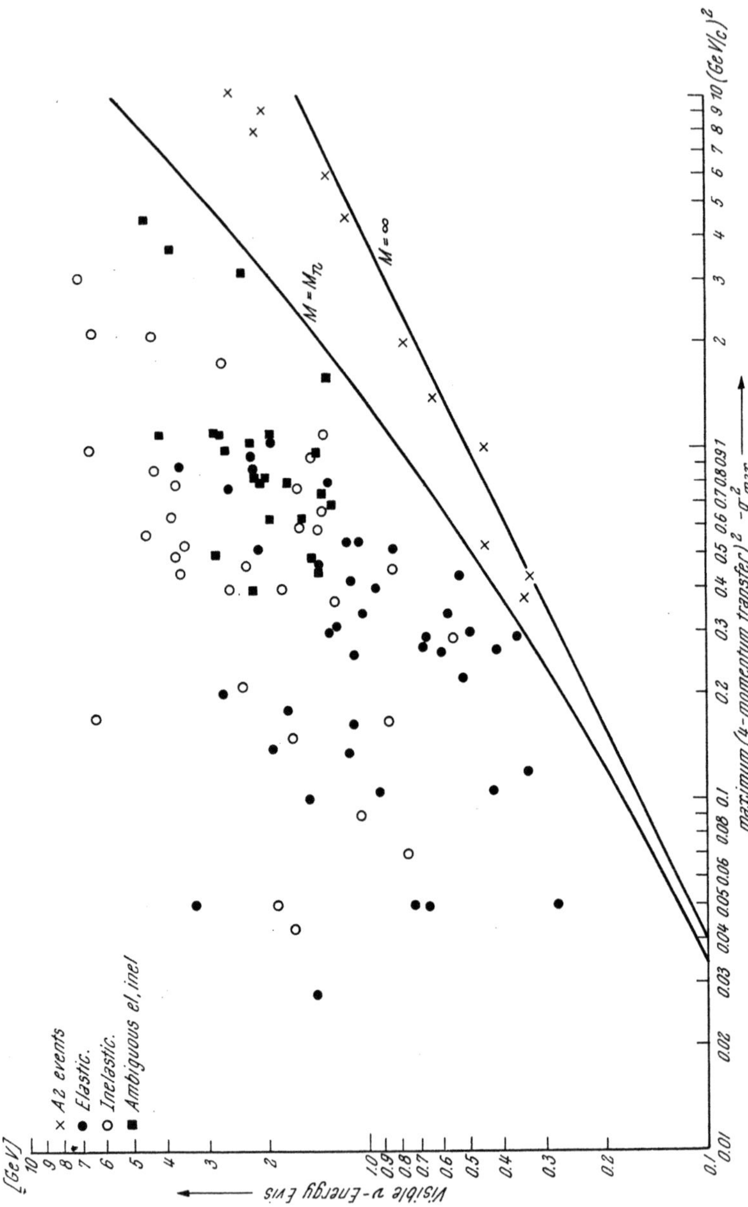

Fig. 22. Correlation between q^2 and visible energy of neutrino events observed in the bubble chamber. The upper curve gives the maximum possible q^2 for a mass-less particle of energy E, colliding with a nucleon of Fermi-momentum 250 MeV/c. (The lower curve holds for an infinitely heavy target.)

entirely neutrino induced was proved by looking at the charges and the interactions. The long range negative particles, which may be μ^- or π^- altogether gave a total pathlength $l_{tot} = 55$ m $= 80\Lambda_0$. ($\Lambda_0 =$ nuclear mean free path for relativistic π's). So, if there had been π's only, they would have been entitled to 80 interactions. What indeed was seen were 17 interactions over that total track-length. Out of these 17 there were 11 with another negative particle which could have been a μ^-, since it did not interact. Thus, there are 6 events left without a candidate and therefore this 6 are really background, just the high energy tail of the neutron background. Now, if you have seen 6 of them interacting then you can calculate how many you have not seen and, therefore, how many π^-'s are mistaken for μ^-'s. The answer is: four. Therefore the n-contamination is of the order of 3%, which is negligible for most practical questions.

The bubble chamber group tried also to find out whether the initial particle had indeed zero mass. For this they use some kinematics. They looked for the 4-momentum-transfer to the initial nucleon. This q is defined as the (space-like) difference between the 4-momentum of the final nucleon (plus mesons, if any) P' and the 4-momentum of the initial nucleon P. Of course, by energy-momentum conservation this is equal to the difference of the lepton 4-momenta:

$$q = P' - P = -(p_\mu - p_\nu). \tag{6}$$

It is related to the incident ν-energy E_ν and the angle of emission Θ of the μ^- in the following way:

$$-q^2 = \frac{(2\,E_\nu \sin \Theta/2)^2}{\left(1 + \frac{2\,E_\nu}{M_p} \sin^2 \Theta/2\right)} \qquad \begin{array}{l}(M_p = \text{Nucleon mass,} \\ \quad m_\mu \text{ neglected}).\end{array} \tag{7}$$

q^2 can be computed from the knowledge of momentum and angle of the μ, completely independent of nuclear re-interaction, and can therefore be done for all events provided the lepton is recognized.

Now you say: $\sin^2 \Theta/2$ can be at maximum 1 and therefore one can give at a fixed q a lower limit for E_{vis}. Permitting $M_p \to \infty$ one gets

$$E_\nu{}^2 = E_{\text{vis}}^2 \geq -\frac{q^2}{4}.$$

(Remark: such an equation would not hold if a neutron were the incident particle.) So you correlate two independent kinematical qualities, E_{vis} and q^2 (Fig. 22). Indeed all 136 "good" events are above the line obtained with a finite nucleon mass!

The 11 events lying in the forbidden region are those where one is uncertain about the lepton in the sense that it could have been a π^+ entering the chamber and making a star. From the plot just presented all these events can be rejected.

2.5 Identification of neutrino-events in the spark chamber

Clearly, the detailed analysis of every event is not possible in the spark chamber. There is no magnetic field in the chambers, there is no possibility to measure ionization, and even the classification of what a track or a shower is, is to some extent arbitrary. Therefore the analysis of spark chamber events is primarily statistical in nature, exploiting the behaviour of a *sample* of events, rather than of single events.

Let us first discuss, how we reject the low energy neutron induced stars and recoil protons showing up in the bubble chamber. The answer is simple: we just don't see them. Remember that the chamber is triggered by a coincidence between two adjacent plastic counters. There are 28 to 34 g/cm² of material between them, corresponding to 170 to 190 MeV kinetic energy T_p for a proton if it travels in the forward direction. This is an absolute minimum for their detection; only for a projected range of double the counter spacing (i. e. for $T_p > 270$ MeV) the detection efficiency reaches unity. Therefore the bulk of the low energy peak observed in the bubble chamber is simply suppressed. (This shows, incidentally, that a counter triggered spark chamber is not a good

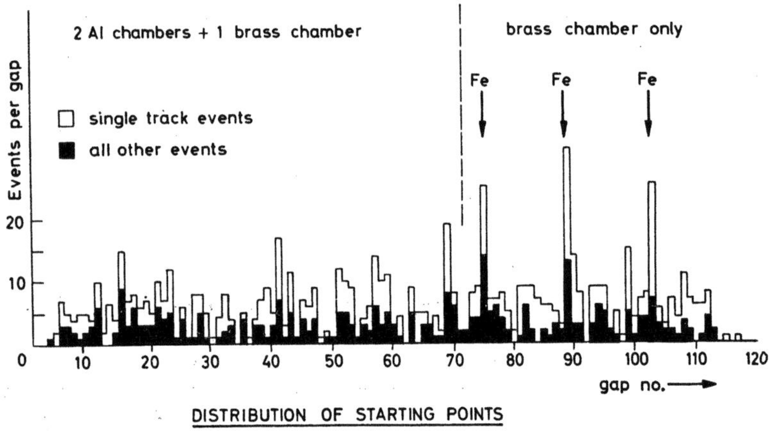

Fig. 23. Distribution of starting points of spark chamber events along the neutrino direction. At the places marked "Fe" were 3 cm thick iron plates.

instrument for detecting processes like anomalous ν-p-scattering of the type $\nu + p \to \nu + p$!) Of course one has to pay a price for this nice property of the instrument: one has a triggering bias also for muons. But since a muon has, at a given momentum, a much larger range than a proton, it is much less severe. Moreover, it can be calculated with any required degree of accuracy.

The next thing is, to look for the interactions on the long track, which is there in most of the events. Already a superficial look at the pictures reveals that they simply don't show any. To be more specific: We have collected almost 4000 events in the thin-walled chamber. The majority of them does have a "long" track of the type shown in the pictures*. In general the length is limited by the chamber-dimension: about 85% of these tracks leave the chamber. Anyway, if we assume as

* See Figs. 18 and 20.

an average track length about half a meter, we have collected about 2 km of muon track length, and the number of interactions seen is exceedingly small: It corresponds to about 200 geometrical interaction lengths per interaction. The same is true for the ≈ 3000 events observed in the thick-walled "range"-chamber. Let me note in passing that a thick-walled chamber is very useful for interaction studies. It is true: one does see only part of the interactions visible in a bubble chamber — what one sees in a thick-walled spark chamber are essentially scatters by large angles. But this is approximately compensated by the enormous thicknesses one forces the particle to transverse.

The next type of distributions studied concerns the distribution of the events in space. If the events are really neutrino induced, they must be initiated uniformly along the neutrino direction in the sense that the same amount of matter produces the same amount of events. Since the thin-walled chamber is 6 geometrical interaction lengths thick (the "range"-chamber even 12) events produced by neutrons coming from the front would have shown an enormous absorption effect. The measured distribution does not show this effect. It is quite uniform, and the non-uniformities indeed observed have to do with inefficient gaps, unequal counter spacing, and the fact that in front of 3 of the gaps there are 3 cm thick iron plates, which indeed induced three significant peaks in the distribution (Fig. 23).

There are other distributions showing uniformity: the distribution of starting points projected onto the (y, z)-plane perpendicular to the ν-direction (Fig. 24), and also the distribution of azimuthal angles Φ. In particular the latter is a good proof for not having a substantial distribution from events induced by neutrons coming from the sides. As far as the uniformity in the (y, z)-plane is concerned one should expect some concentration of neutrino flux in the center of the chamber. Up to now we did not see this effect — (we just did not look very hard). Van der Meer predicts it to be most pronounced for the high energy neutrinos, and when we shall have some time, we shall try to measure in this way the intensity profile of our neutrino beam.

Having convinced ourselves about the reasonable behaviour of our events in space, we turned to the fourth component, i. e. we looked at their distribution in time. We first made sure that they were uniformly distributed in time in the sense that each of the 20 (or rather 19) bursts gave us the same number of events. For this we wrote on every spark chamber event the number of the bunch in which it occured. (This can be done very simply, for instance by having a counter watching the horn, registering the number of radiation bunches seen, on a scaler near the chamber, and blocking the line from the moment on, the triggering counters had given a "ν"-signal.)

The only non-uniform distribution obtained is shown in Fig. 25. It gives (cum grano salis) the distribution of the time-of-flight of the initiating particles. More precisely, it is the distribution of the time-differences between "ν-like"-counter signals and the passage of a neutrino-

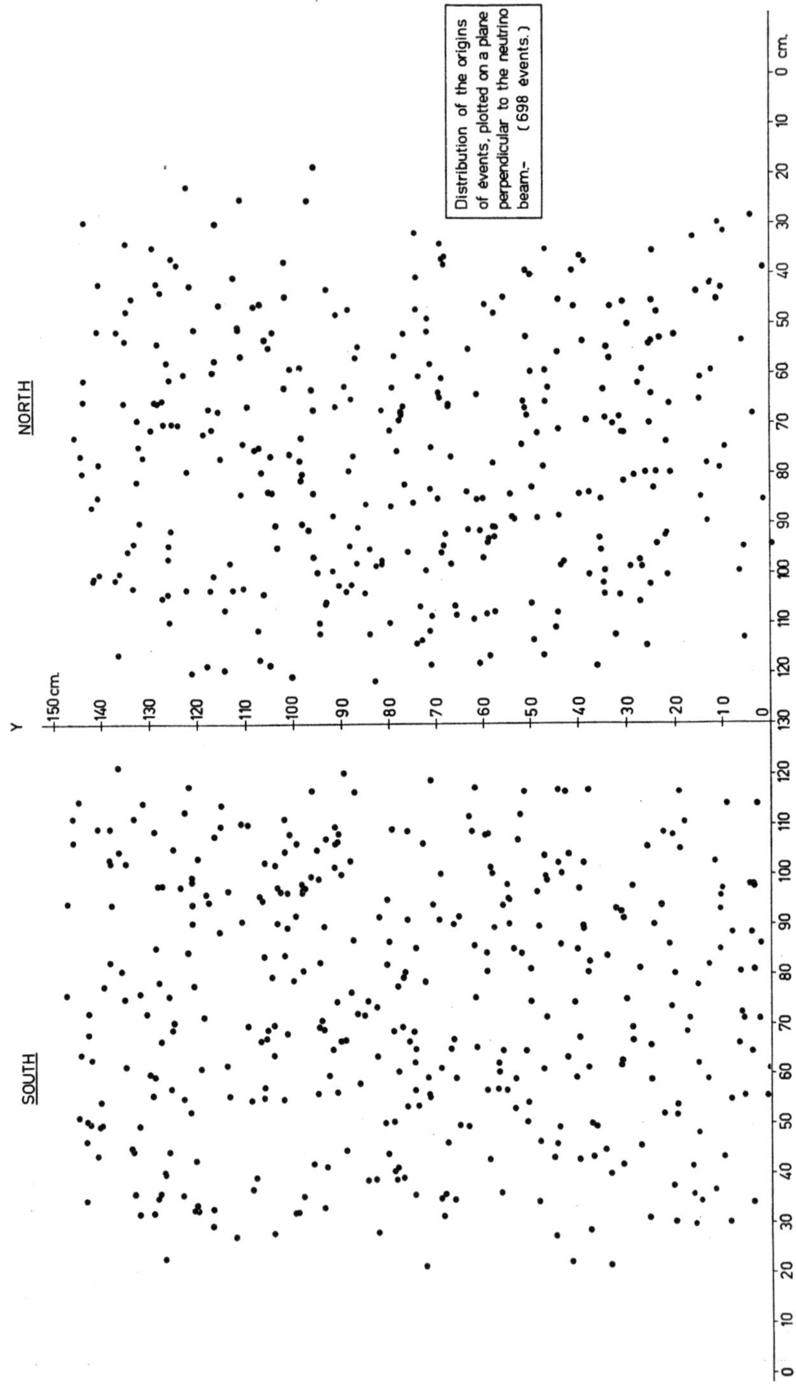

Fig. 24. Distribution of starting points of spark chamber events, projected onto a plane perpendicular to the neutrino direction.

bunch through the chamber (with an artificial delay of 46 nsec inserted into the signal line). The time-information about the ν-bunch one gets in practice from the phase of the accelerator radio frequency [15]. The distribution of counter-signals (upper part of the figure) has a peak at the position corresponding to particles travelling with the speed of light. (We know this position from calibration measurements with muons which managed to get around the shielding.) The tail on the left hand side corresponds to later arrival times and is attributed to non-relativistic neutrons; the constant background comes from very slow neutrons,

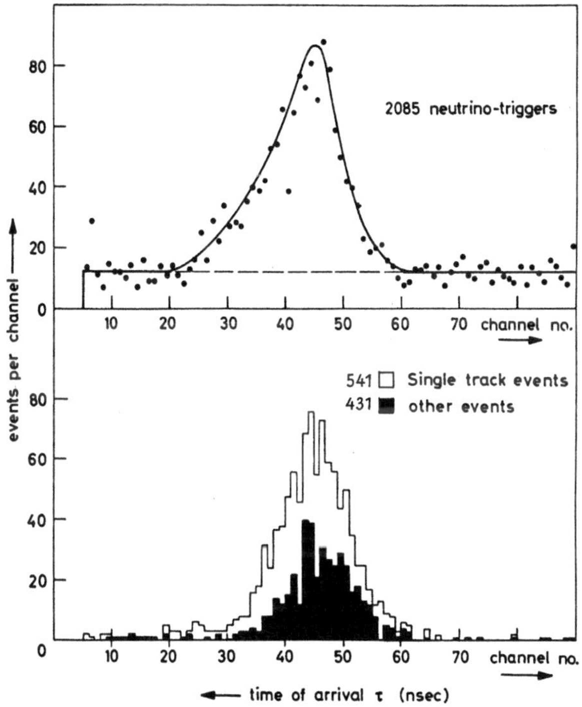

Fig. 25. Distribution of arrival times, relative to the phase of the neutrino bunches. Upper part: counter signals, lower part: events induced inside the spark chamber.

which are seen by the triggering liquid counters (for which we did not require a coincidence!). Cosmic rays contribute only about 1% of the triggers, since we are sensitive only for 2.5 μsec every 3 seconds. It is comforting to see that the constant background disappears, when we look at the distribution of *events* initiated in the chamber (lower part of the figure). Also the tail is reduced. If we require now a "ν-event" to have within ± 12 nsec the correct timing, the remaining background from neutrons comes out to be $\leq 4\%$ — in perfect agreement with the contamination inferred by the bubble chamber group.

A final remark might be in order: What I tried to describe in this chapter is neither exciting nor does it give already physically significant results. However, it is the basis the whole neutrino physics rests upon. In all modesty: To have proved that their samples contain only a few percent neutron admixture is something the CERN neutrino groups can be proud of.

3. Properties of neutrinos. Conservation laws

3.1 The one- or two-neutrino question

The first and very convincing answer was found at Brookhaven by SCHWARTZ, LEDERMANN, STEINBERGER et al. [7]. The experiment was done under the following conditions: internal target, relatively longish bust and a terrific amount of background from all sides, including cosmic rays. They observed 29 clear single muons, after having subtracted 5 which were presumably due to cosmic rays. Based on this 29 muons you would have predicted 22 electrons, assuming universality for the coupling, that means the same cross-sections for muons and electrons, under the hypothesis

$$\text{A:}\ \nu_{\mu\pi} = \nu_{\mu K} = \nu_e.$$

But instead of the expected 22 they observed 6 and these were presumably nuclear showers. In a spark chamber a shower is not necessarily of electromagnetic origin; small neutron induced stars with a few prongs and maybe a γ nearby are also capable of producing something which one would call a shower. But even if one admitted the 6 showers as possible cases of electron production: 22 is different from 6, and there are in fact two neutrinos. The second alternative is

$$\begin{aligned} \text{B}: \nu_e &\neq \nu_{\text{Meson}},\ \text{but} \\ \nu_{\mu\pi} &= \nu_{eK} \\ \nu_{e\pi} &= \nu_{\mu K} = \nu_e. \end{aligned}$$

That is the so-called neutrino-flip hypothesis [16]. The idea is to consider the coupling of a pair of leptons in a way analogous to strangeness. You say a μ or ν_μ is sort of a strange particle, and a K^+ is a strange compound namely $(p\overline{\Lambda})$ under strong interactions and also a strange compound, namely $(\mu^+\bar{\nu}_e)$ under weak interactions. However, the Brookhaven experiment gave an indication that also the neutrino-flip is not very probable, because they would have expected 3—5 large electron showers ($^1/_2$ m long), which one could not easily have missed.

The last possibility, the truth, is the following:

$$\begin{aligned} \text{C}: \nu_e &\neq \nu_\mu \\ \nu_{\mu\pi} &= \nu_{\mu K} = \nu_\mu \\ \nu_{e\pi} &= \nu_{eK} = \nu_e. \end{aligned}$$

It says, that the coupling of charged and neutral leptons in K- and π-meson is identical. The data for the ratio (elastic processes only) obtained in the CERN spark chamber are (s. Fig. 26).

expected		observed
A	50%	
B	8%	$(1.6 \pm .4)\%$
C	1% (uncertainty $\approx 50\%$).	

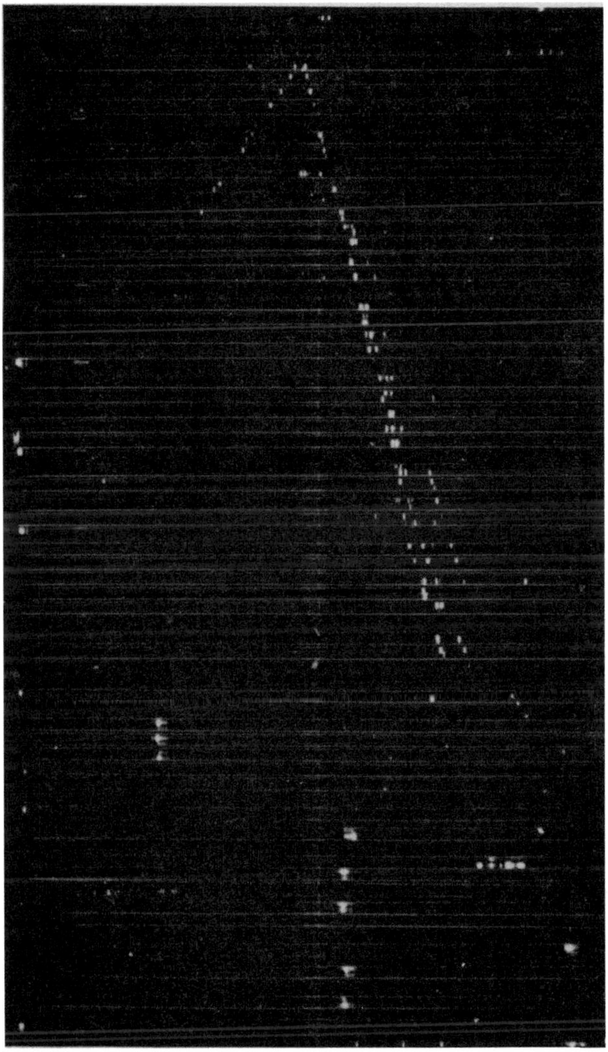

Fig. 26. Candidate for elastic electron production observed in the spark chamber.

The expected numbers were calculated with the K-meson contribution to the CERN neutrino beam and for the detection probabilities estimated for the CERN spark chamber. The expected rates differ somewhat from the Brookhaven ones. The observed rate clearly excludes A and B.

The same conclusion may be derived from the bubble chamber result: 3 events with an electron plus one track (π ?!), no clear elastic e versus 68 elastic μ's. Under hypothesis B you would have seen 4 or 5 high energy showers, which have not been seen. So B is excluded with good confidence. There are also events with high energy muons which, by kinematical reasons, could only be produced by neutrinos from K-decay. This observation is sort of a direct proof that $\nu_{\mu K} = \nu_\mu$.

The conclusion therefore is: we have two two-component neutrinos ν_μ, ν_e; you assign them a muonic-number M in the way

$$(\mu^-, \nu_\mu) \text{ have } M = 1 \quad (-1 \text{ for antileptons})$$

whereas all other particles, including e, ν_e, have $M = 0$. This new quantum number M is rigorously conserved.

Note added in proof: In the meantime also the bubble chamber group has seen one very clear case of elastic electron production. Having unambiguously eastablished that K_e-neutrinos produce electrons is very significant, since some theoreticians recently attempted to forbid them from doing so!

3.2 "Lepton conservation" in the sense $\nu_\mu \to \mu^-$ but $\nu_\mu \not\to \mu^+$

We use here lepton conservation in the sense that a definite helicity implies a definite charge. With the conventional assignment the "left-handers" ν_μ and μ^- are leptons, the "right-handers" ν_μ and μ^+ are antileptons, and a reaction $\nu_\mu \to \mu^+$ is forbidden. Of course this statement is also true for electrons and ν_e. This type of "lepton conservation" does not exceed conservation of helicity already imposed by the V-A theory [17].

To test this prediction is easy and has been done in both, bubble chamber (B. C.) and spark chamber (S. C.). The results are:

B. C.: 8 candidates for μ^+ out of 136 elastic events (a few of them could be π^+).

S. C.: $(92 \pm 4)\%$ are indeed negative particles (measured in the magnet between the two chambers), based also on a sample of about 100 events. Contained in these 8% there are also π^+ and p.

So the two instruments agree quite well. Because of the antineutrino contamination of the beam we expect the contribution from positive muons to be about 3%. Therefore this is a test for lepton conservation (in the sense stated above) good to $\lesssim 5\%$.

The reversed experiment was done by reversing the horn polarity, which gives an enriched $\bar{\nu}_\mu$-beam. The spark chamber measured $(65 \pm \pm 15)\%$ positive particles. The explanation of this 65% compared to the 92% from above is purely instrumental and given by the production rates for π^+ and π^- (production $\pi^+/\pi^- \approx 2$) and because of the fact that the cross-section $\nu/\bar{\nu} \approx 2$. Therefore the relative chance to detect a $\bar{\nu}$ is smaller than for ν, and the two numbers are not in disagreement.

3.3 Lepton conservation in the sense: $\nu_\mu \not\rightarrow$ Mesons

The spark chamber is helpless in this case. The bubble chamber has identified cases with pions (π^+, π^-, $+$nucleons). The bubble chamber has neutral stars of this type, which are in fact the high energy tail from this peak at low energies in the distribution of visible energy (Fig. 21), attributed to neutrons. The low energy stuff was cut away by demanding a visible energy $E_{\text{vis}} > 1$ GeV for this test, and there were 2 candidates of this type out of 101 ν-events with $E_{\text{vis}} > 1$ GeV. Therefore lepton conservation in this sense is good to $(2 \pm 2)\%$.

3.4 Anomalous ν-scattering

The great contribution comes from the bubble chamber. They searched for reactions of the type $\nu_\mu + p \rightarrow \nu_\mu + p$, where the proton can be replaced by a $(N\pi)$ complex. They looked for protons, which had a kinetic energy larger than 250 MeV. None were observed, whereas the number of normal elastic neutrino-events, where the ν changes charge and becomes a muon, is, with the same cutoff, 31. The conclusion is, that this neutral current is $\lesssim (3 \pm 3)\%$ of the normal charge changing weak reaction.

You can make a theory to exclude these high momentum transfers and explain the scattering process by assuming a coupling of ν_μ and p due to an intermediate field X (this X has a very low mass now). Therefore you cut off momentum transfers at say 100 MeV/c. A supplementary experiment was done with this possibility in mind [18]. 1.7 tons liquid scintillators were exposed and it was looked at everything which could be a recoil proton with $E_p \geq 15$ MeV. This is a very low limit corresponding to 170 MeV/c in momentum transfer. Because of neutron background we arrived at the conclusion that $\sigma_{\nu p} \lesssim 2 \times 10^{-37}$ cm^2, which does not look a good limit in weak interactions, but which is only a factor of 3 worse than the bubble chamber result, if phrased in terms of $\nu - X$ coupling strength.

Since we have positive evidence that there are no anomalous interactions one really questions the nature of the muon-neutrino. One cornerstone would be a measurement of its mass which, however, has not yet been achieved to some degree of accuracy. At present the limit is 3 MeV. Unfortunately all experiments which could possibly measure the rest mass ν_μ are exceedingly difficult, because you must look under conditions, where the ν_μ, if it had a mass, would stand still in the lab.-system and all the other particles would take the maximum amount of momentum.

3.5 The meaning of lepton conservation

In the textbooks you read that there is in nature a preference of lefthanded neutrinos and righthanded antineutrinos, and no preference of the contrary. Believing in symmetry of nature one can make the hypothesis: we have a second two-component neutrino v_μ, which is indeed a righthanded ν with lepton number $L=+1$ and helicity $H=+1$, whereas the normal electron-neutrino has $H=-1$, $L=+1$. With this assignment you get a 4-component neutrino theory (if you really want to!). Question: is this in conflict with the known experimental data? The answer is no. One makes the following assignments: the helicities as known and redefine the lepton number (which is now a new quantum number on its own):

$$L=+1:\ \nu_e,\ v_\mu,\ e^-,\ \mu^+$$
$$L=-1:\ \text{antiparticles.}$$

With this you find the relation between the old and new muon-neutrino: $v_\mu=\bar{\nu}_\mu$. You can easily verify that nothing changes as far as predictions are concerned [17].

Note added in proof: As the author found out only now, NISHIJIMA [19] had proposed precisely this scheme as early as in 1957! He deduced even, that ν's from π-decay cannot produce electrons; (actually he spoke of ν's from μ-capture, but this makes no difference). Finally he explained why the long-searched and never-found modes $\mu\to e+\gamma$, $\mu+p\to e+p$ etc. are absent. So he had the full truth in his hands (even in a much more elegant formulation than it is customary nowadays) — and yet his paper passed by almost entirely unnoticed.

4. Elastic production of muons and electrons. Inelastic processes

4.1 Angular distribution of muons from $\nu_\mu+\mathrm{n}\to\mu^-+p$ (spark chamber)

We want to look for reactions of the type

$$\nu_\mu+\mathrm{n}\to\mu^-+p. \tag{5}$$

Under the assumption of a specific model, i. e. with an intermediate boson, this reaction is represented by the diagram

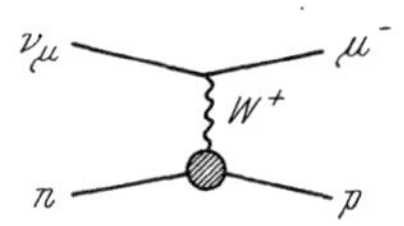

In describing the structure of the $(\mathrm{n}W^+p)$ vertex, which comes about by strong interaction, we employ, of course, the conserved vector current theory. This means we put the two vector form factors equal to the electromagnetic ones:

$$F_V=F_{WM}=F_{EM}. \tag{8}$$

For the axial vector part we use a parametrization of the following type [20]:

$$F_A=\frac{1}{\left(1+\frac{q^2}{M_A{}^2}\right)^2} \tag{9}$$

M_A an adjustable parameter.

3. We consider the boson mass M_W another free parameter.

4. Nuclear effects (Pauli principle etc.) are taken into account, and so is the finite detection probability due to the momentum cut-off introduced by the counter spacing.

The result is given in Fig. 27. This is the measured diffraction pattern of the extended structure of the nucleon, compared with two theoretical predictions. Another correction should be mentioned: The cut-off in *angle* introduced by the finite counter spacing. It is serious only for angles larger than 30°, and has been considered in the experimental histogram. As can be seen, the curve with $M_W = \infty$ and $M_A = 5$, GeV does not fit at all. It is for the first time that a definite structure effect in the axial vector part of the weak interaction has been seen. Quantitatively: An axial vector form factor equal to the vector one, i. e. $M_A = M_V = 0.84$ GeV, seems not to be a bad choice.

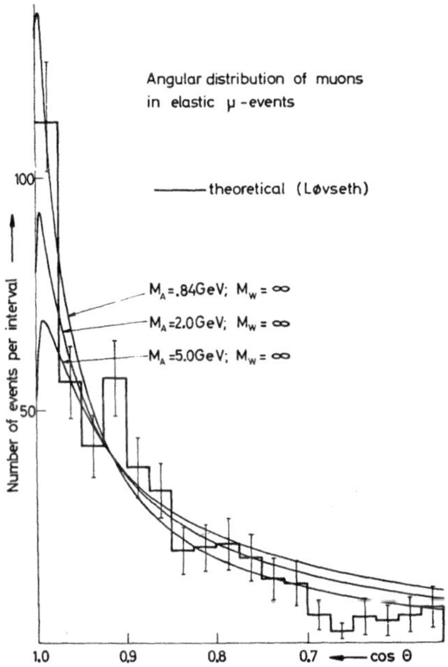

Fig. 27. Angular distribution of elastically produced muons, deduced from non-mesonic events initiated in the thin-plate aluminium chambers (final data from 1963 and 1964 together [22]).

In these curves one has put the boson mass $= \infty$. But what does the boson mass of 1.3 GeV do? Qualitatively it is clear. We superimpose a second type of structure inherent to weak interaction to what we have already from the strong one. Therefore the nucleon does look more diffuse, than it were if it had only the structure due to strong interactions. As a result the angular distribution becomes more peaked as before. If one insists in having axial vector and vector form factors still equal (for which there is no reason whatsoever), then the agreement with the experiment is destroyed. Instead, one gets agreement again, if one choses M_A equal to a few GeV. Even $M_A = \infty$ could not be rejected any more. Evidently, one has to clear up the boson question, before an unambiguous analysis of the elastic reaction can be made. But even putting this question aside, there is another big unknown in the business, namely the neutrino spectrum, which enters the theoretical predictions in a serious way.

The bubble chamber group is in a better position. They can give a distribution which does depend only little on the neutrino spectrum, namely the distribution of the squares of the transferred four-momentum q. As has been explained in the last section, q^2 may be derived from the measurement of the muon's 4-momentum alone, which is not affected by nuclear re-interactions. One has to assume though, that the target neutron was at rest. This assumption may be dropped if one computes q^2 according to eq. (7) from the muon angle and the total visible energy of the event E_{vis}, (which one assumes to represent the primary neutrino energy E_ν). Both methods lead to very similar distributions. The one shown in Fig. 28 was obtained with the second one.

The q^2-distribution has an immediate physical significance: it is just the Fourier transformed of the spatial distribution of the weak charge, we were talking about before. The data shows the depression of higher momentum transfers characteristic for a finite spatial extension. Under the assumption of an infinitely heavy boson it favours a cut-off in the axial vector form factor around 1 GeV, provided all ambiguous events are included. Specifically, $M_A<1.35$ GeV with 68% confidence. Already $M_A=1.8$ GeV would be a bad fit. If all ambiguous events are considered inelastic, $M_A<1$ GeV with 69% confidence. In this case the most probable value of M_A turns out to be 0.85 GeV, practically identical with the cut-off in the vector part. This is the value deduced earlier [4]. The curve corresponding to the extreme assumption of a point-like axial vector part: $M_A=M_W=\infty$, is excluded with high confidence.

A few qualifications should be added to these statements: The appearance of "ambiguous events" in the bubble chamber data does show that the identification of elastic events is not quite as easy as I tried to make you believe. True enough — it suffices to show that no additional pion has been produced. But just that is difficult. For instance, the ambiguous events indicated in Fig. 28 are those, where in addition to the μ^--candidate a positive particle was emitted with a momentum higher than 700 MeV/c. At momenta as high as that, protons and pions can no larger be distinguished by ionisation. Therefore all these particles *could* be protons — and the events correspondingly elastic. But they could be also mesons from inelastic processes. Here even the bubble chamber people have to resort to statistical methods: they tried δ-ray counting — and found that most of the tracks seem to be really protons. Strictly speaking, even this does not *prove* that the events are elastic; they could be inelastic with the pion re-absorbed in the parent nucleus! Fortunately, the situation is not quite as desperate as it sounds. In the bubble chamber one has several means to check if the kinematics does correspond to an elastic reaction or not. The simplest one was already implicitely mentioned: One computes the primary neutrino energy from the muon momentum, and compares with the observed visible energy. For most of the non-pionic events the agreement was good. About 10% were rejected, because the deviation exceeded two standard deviations. From the observed pion spectrum (see Fig. 31) and the known absorption cross sections, one concludes that the contamination with inelastic events should not considerably exceed 10%.

There is no time to go into further details of the art, how to select elastic events. They have been mentioned in the Siena Conference

Reports [4, 5], and they will be more fully treated in forthcoming papers of both, the bubble chamber [21] and the spark chamber group [22]. About the latter only as much as that: it is obviously the more difficult to select a good sample, the less information one has. As long as most of the muons leave the spark chamber we don't know their momentum, and we can only try to differentiate between associating pion or proton by multiple scattering *vs.* range. The situation will be improved in the 1964 experiment, where a large fraction of the muons initiated in the thin plate chamber will be stopped and measured in the magnetic cham-

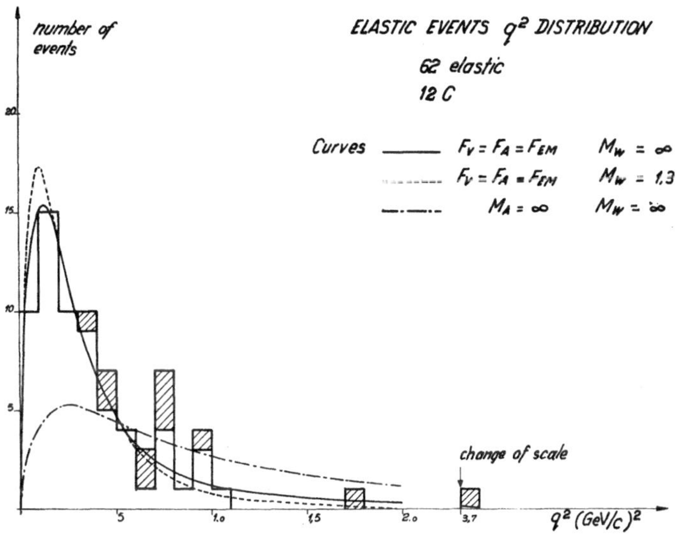

Fig. 28. Distribution of q^2 of non-mesonic bubble chamber events. The shaded areas refer to events where the recoil proton could not be positively identified by ionization.

ber behind. Nevertheless, even in the old experiment the rate of elastic μ-events agrees so well with the better founded bubble chamber rate that the inelastic contamination, in my opinion, cannot exceed $\approx 20\%$.

4.2 Rate of elastic muon production. The question of the neutrino spectrum

The last remark brings us to a very touchy question, which I managed to avoid so far, namely: Do we actually observe the number of elastic neutrino events we expect according to the various assumptions about the form factors and the supposedly known spectrum? — At present the answer is flatly: No! — We see too many. This discrepancy was first recognized by the spark chamber group, who reported at the Siena Conference [5] an elastic rate about twice as high as expected on the basis of the "Lee-Yang" cross section [20] (i. e. $F_A = F_V = F_{EM}$) —

which otherwise fitted angular- and q^2-distribution so well. It was for this reason that the experimental distributions (Figs. 27 and 28) were not compared to an absolute theoretical prediction; we rather preferred to normalize experiment and theory to the same area, that is the same total number of events.

This result was soon confirmed by the bubble chamber group. But they went one decisive step further: Relying on the good fit obtained to the observed q^2-distribution, by having $F_A = F_V$, they used the Lee-Yang cross section to *deduce* a neutrino spectrum from the observed rate. Since this has been criticized, let me qualify what they really did: They used a certain parametrization of their measured q^2-distribution, integrated at a given E_ν over q^2, took this number to divide the observed rate at this energy E_ν and called the remaining number an experimentally determined ν-flux at the energy E_ν. Of course, the procedure has two weak points: At high q^2 you really don't known the form factors — but we know that high q^2-events are rare (please look at Fig. 28); for the total rate they are quite unimportant. At low q^2 instead we run into trouble with nuclear physics; we really don't know how the free-nucleon cross section is modified in detail by nuclear effects. But a simple Fermigas model computation indicates that this is, integrated over all q^2, only a 10—20% effect. As long we are worrying about factors of two, the simple Fermigas model, I think, is good enough. Anyway what the bubble chamber people got out was, for neutrino energies >1 GeV, a practically constant enhancement of about a factor 1.5 to 2 with respect to the then-latest van der Meer spectrum [3,9]. To the extent that this enhancement is energy-independent, the comparison between theory and experiment, done in Figs. 27 and 28 was in order, since a constant factor drops out by the normalization. However there was in addition a spike around 0.5 GeV, indicating either an unexpected excess of low energy neutrinos — or some unexpected behaviour of cross sections. Considering our present state of knowledge of the spectrum, I think most of us are inclined to blame all deviations on the spectrum rather than on physics. But when we really know our beam, the observation of the bubble chamber group may turn out to be very significant.

But how do we explain the factor of 2 all over the spectrum in the first place? I really don't know. Admittedly, the particle trajectories in the magnetic horn are complicated. But we all believe in Maxwells equations, and in the ability of VAN DER MEER et al. [3,9] to make good use of them. The weak point seems to lie in the meson production and/or re-absorption and re-generation data. We can only hope that we get soon better data — and that we may be able some day to measure the meson spectra from the horn directly (but this is going to become tough!).

Note added in proof: It is with great relief that the present writer adds this note. Very fortunately the new meson production data of DEKKER et al. in their final form [23], lead VAN DER MEER to a spectrum which now agrees perfectly well with the experimental bubble chamber spectrum over all energies — save the spike at 0.5 GeV, which still survived (Fig. 29). One will have to take this peak serious, although the author still believes that it might find an explanation through ternary, quaternary and so on pions produced by primary, secondary etc. particles not only in the target but also all over the shielding of the decay tunnel.

An obvious improvement in the experimental determination of the ν-spectrum has been suggested by M. Block [24]: one should restrict oneself to values of q^2 where one is safe from the disturbing effects

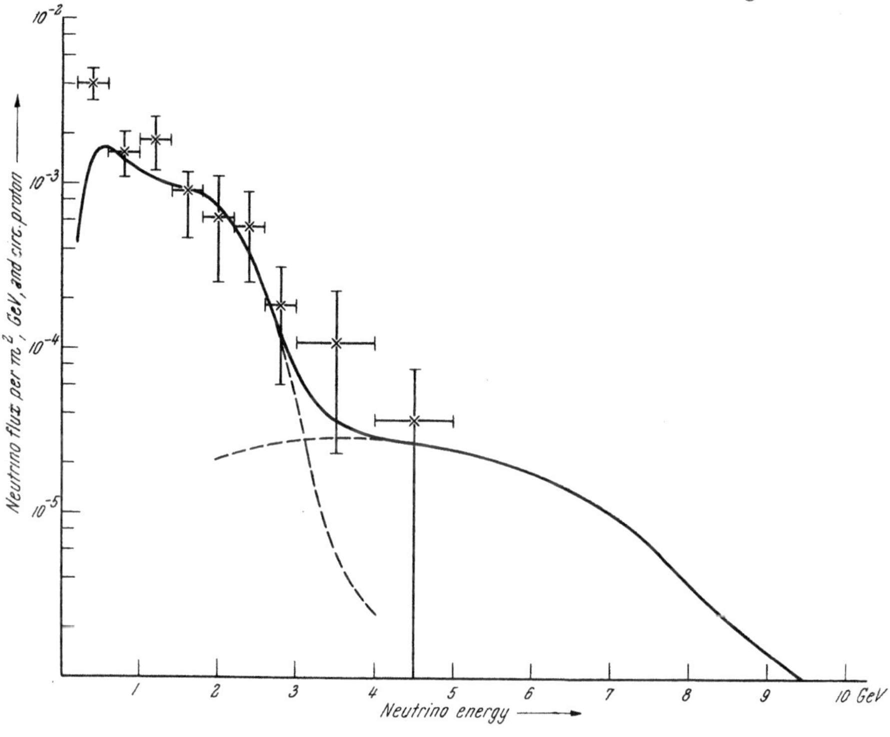

Fig. 29. Neutrino spectrum deduced from the energy distribution of elastic bubble chamber events, assuming a theoretical cross-section [20] with $M_A = 1$ GeV (final data [21]). The curves refer to van der Meer's calculations [9].

described above, say to $0.1 < q^2 < 0.2$ (GeV/c)^2. Also spark chamber events, recognized as elastic in the thin-walled chamber (plus charge measurement in the magnet!) with the μ being measured in the magnetic chamber (by range), should be useful in this respect.

4.3 Elastic electron production

The qualitative significance of having seen elastic electron production:

$$\nu_e + n \rightarrow e^- + p \qquad (10)$$

was already pointed out in the preceding chapter. Considering the rate of about 1% with respect to elastic muon production, it does mean that the neutrino from K_e-decays is identical to the electron neutrino, familiar from nuclear beta decay. (Of course, our data does not rule out part of the K_e-neutrinos being of a different, exotic variety — but who would believe in something like that!)

But now we want to go further: We want to see, if electron-neutrinos scatter (under charge exchange) off nucleons the same way, muon-neutrinos do. This is a test for muon-electron-universality in weak interactions considerably more detailed than the one currently available, namely the electron to muon ratio in pion decay. It tests the equivalence of a $(\nu_e\, e)$-pair to a $(\nu_\mu\, \mu)$-pair over a wide range of momentum transfers. (This equivalence would be trivial, if each of these pairs would constitute the same intermediate boson W!)

This task requires first of all the selection of a sample of elastic electron events. For reasons of statistics this can only be done, at present, with the spark chamber. Analogous to the moun case we would require one clear electron-type shower and at most one associated track which is compatible with a proton. The danger of getting some inelastic events mixed in is about as great here as it was there. One has one additional trouble with the electrons: an associated track may be buried inside the shower and thus escape observation. However, we do see the events with the associated tracks *not* buried inside the shower and can estimate how frequently this would happen. We feel this effect is small and under control. On the other hand, we can measure the energy of a shower with reasonable accuracy. We can therefore apply kinematical tests to a higher fraction of events than in the muon case. The sample of "elastic" electron events should therefore be at least as clean as the corresponding elastic muon sample. At present we have a preliminary sample of about 40 events.

A further non-trivial step in the analysis is the evaluation of the detection biases involved. This is not very easy, and has been done so far only in a very approximate way.

Much work has gone, instead, into the measurement of shower energies. Since our chambers are many radiation lengths deep, we measure shower energies by total absorption. As a measure of the shower energy E_0 one may simply take the total number of sparks seen, N. This quantity has no simple physical meaning. It is somewhat related to the total path length, which goes linearly with the shower energy. A linear relation between E_0 and N cannot be expected. There are a number of instrumental effects which make us seeing less sparks than we would be entitled to according to physics. An obvious one is the so-called "robbing effect", namely a reduction of detection efficiency for one particle by the presence of other sparks. Surprisingly enough, this effect is small. As we have proved 2 years ago [25], one can detect a particle in the presence of up to 12 other sparks with an efficiency almost equal to the single-particle efficiency of 90—100% provided the sparks are some mm apart. In the meantime we have measured detection efficiencies of about 50% in the presence of about 50 sparks (!) in the same gap. (By the way, I don't understand why the sparks disturb eachother so little — but let the gas discharge experts worry about it!)

But there are effects which cause losses: If two electrons cross a gap with a distance $\lesssim$ the spark width, cleary they cannot give two separate sparks. Since the angles involved in bremsstrahlung and pair production are small, (namely of the order m_e/E), this is a large effect. A gamma-produced electron pair always starts off with 1 spark per gap. We

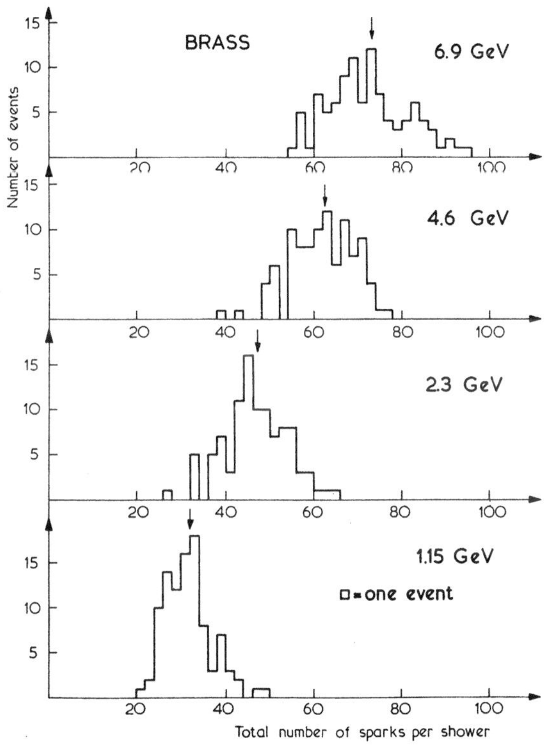

Fig. 30. Distribution of total spark numbers in showers initiated by electrons of known energies. The energy can be inferred from spark counting with an average accuracy of 25 – 30 %.

have to wait until multiple scattering has brought the two electrons sufficiently apart from eachother. From this observation, namely that the lateral spread of a shower is essentially due to multiple scattering, we appreciate immediately that the loss, we are discussing, is the more severe the higher the energy is. This is also true for another trivial but important effect, namely the covering of sparks by other ones in front of them. As a result, the calibration curve N *vs.* E_0 bends severely over towards higher energies, as evidenced by one of our many calibration measurements given in Fig. 30.

The accuracy of the method is about 25% around 1 GeV. It gets worse towards lower energies, because one has fewer sparks, and also towards higher energies because of the non-linearity of the calibration curve. Of course, all these statements depend on the way one samples the shower: At a given shower energy we measure more accurately in aluminium than in brass — provided the shower was really absorbed. Unfortunately a shower initiated in pure aluminium has a high chance to run out. In this case we help ourselves by following the shower curve as far as we can, and extrapolating with the known shape of the shower tail. Integrating the shower curve rather than taking the total spark number N, is also advisable, if the set-up is very inhomogeneous. One application of studying the shower curves I just mention: Very different shower phenomena, for instance one electron versus two overlapping gammas from a π^0, may be distinguished from eachother with some probability.

The last quantity we need to pin the event kinematically down, is the angle of emission of the electron. In aluminium this is easy to measure, in brass it is not, since the shower widens up so fast. A computer programm was written [26], which allows us to determine the shower axis by a fairly sophisticated best fit. At present we use a ruler with a very flexible portable computer on line: eyes plus human brain.

The preliminary results from these studies are as follows:

1. The rate of elastic electron to muon production of presently $(1.6 \pm \pm 0.4)\%$ does demonstrate that the integrated cross sections for the two processes cannot differ very much, (say not more than by a factor of 2), since the expected number is about 1%. To the extent, that we have the respective detection efficiencies properly taken into account, this should be a good number, since the admixture of inelastic events should be about the same in both cases.

2. The observed spectrum of electron energies follows the ν_e-spectrum expected from K_{e3}-decay quite closely. This shows that the cross section is, as expected, practically flat at these energies (some GeV).

3. The angular distribution shows a similar peaking as observed for muons (Fig. 27), demonstrating the finite extension of the nucleon.

4. Energy and angle measurement together will allow us perhaps to deduce a q^2-distribution and compare it to the moun-one (Fig. 28). This would be a very good test for $(\nu_e\, e) = (\nu_\mu\, \mu)$, almost independent of the spectrum, and very detailed. We have not yet quite succeeded, mainly because of the systematic uncertainties involved.

It seems, however, quite safe to me to predict, that electron-neutrinos see the nucleon the same shape as muon-neutrinos do — (except, perhaps, for the influence of the induced pseudoscalar term; but this is another story).

4.4 *Inelastic reactions (without change of strangeness)*

Here all the relevant information comes from the bubble chamber group. They found that almost precisely 50% of all neutrino interactions are inelastic, i. e. involve the production of one ore more meson. (The spark chamber had originally a somewhat lower figure, about 40% [1, 5], but recent improvements in the distinction between elastic and inelastic events brought it in close agreement with the bubble chamber value.) The relatively most important channel is the production of one pion.

About one-pion-production one might have one's prejudices. One cannot help thinking that the bulk of the one-pion-events should correspond to excitation of the target nucleon to the 3-3-resonance N^*. As

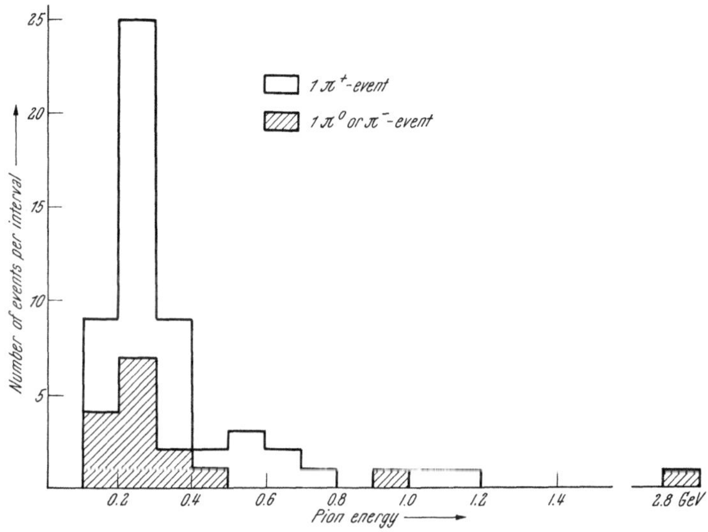

Fig. 31. Energy spectrum of pions in inelastic one-pion events (bubble chamber data).

a matter of fact, there are a number of facts supporting this hypothesis: The spectrum of single positive pions (and also of single π^0's) has a broad peak around 300 MeV (Fig. 31), just where one would expect to find the decay π's from a slowly moving N^*. It decreases rapidly towards higher energies. If one evaluates the effective mass of the system: proton plus pion (Fig. 32), one gets a peak close to the N^* mass (1238 MeV). Finally, the isobar-production process would give a π^+/π^0 ratio of 5 (peripheral pion production instead 1:1). The value reported to the Siena Conference was 4 ± 1.5 — a very good confirmation of the N^*-formation process indeed! [Incidentally, since they have now a pure aluminium chamber also the spark chamber group started studying single π^0 production, see Fig. 33 — and the bubble chamber picture (Fig. 34) for comparison!]

In the mean-time a more refined study made the situation much less clear. As Hegel [27] has told us, every "thesis" provokes an "anti-thesis". The anti-thesis in this case is: there are no N^*'s at all. The statement is based on the fact, that the one-pion-events do not show the kinematical behaviour (angular correlations between protons and pions etc.) expected for N^*'s.

To my mind the Hegelian synthesis will be somewhat like this: At low pion energies the bulk of the events does indeed correspond to N^*-formation. The failure of the more subtle kinematical tests is to be blamed on re-interaction of the π or/and the p inside the nucleus. But there are other processes as well; notably the high energy tail of the π-spectrum may come from a peripheral process. — Note that this is

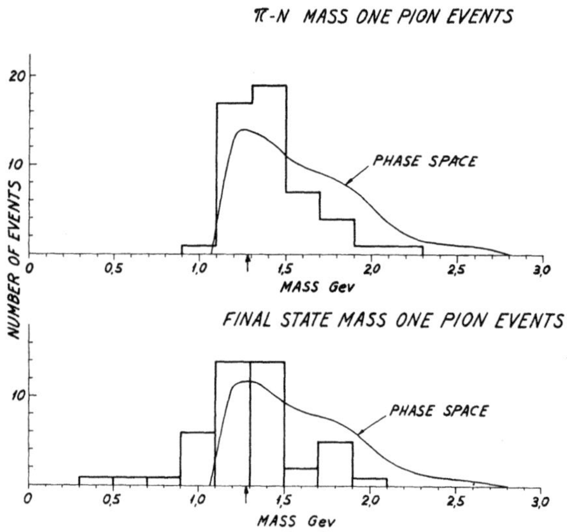

Fig. 32. Invariant mass of the pion-proton system in one-pion events. The curve refers to phase-space.

a completely un-authorized statement — (but where would we be, if we always waited for the proper authorization!).

About the many-pion events I shall be even briefer than about the one-pion events. There is no simple model at present to explain them. In particular, it has not been anticipated, how fast their cross section rises with energy. To the Siena Conference an inelastic cross section, rising quadratically with neutrino energy, has been suggested [4]. The statement should be understood with some reserve: the fast rise of the cross section is well established only between 0.5 and 7 GeV; (and a *linear* rise would do almost as well). It is not implied that the quadratic rise goes on and on.

There are some strange events, though, at energies above 10 GeV: One at 12, another one at 14 GeV. There are a number of charged and

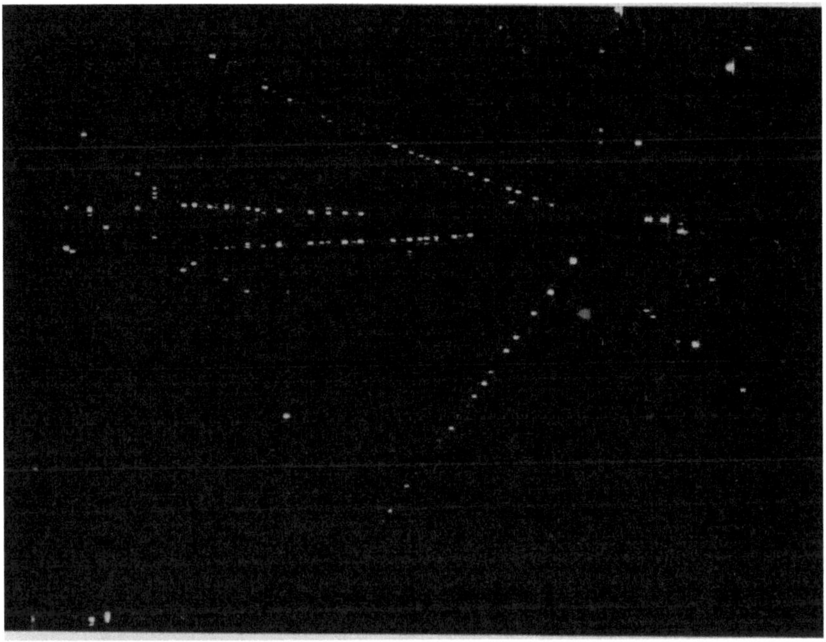

Fig. 33. Inelastic ν_μ event with single π^0 production in the spark chamber.

Fig. 34. Inelastic ν_μ event with single π^0 production in the bubble chamber.

several neutral pions — more than half a dozen altogether — and of course a possible μ^- (even e is not excluded). One sees these events also in the spark chamber; there a measurement of the shower energy brings one again to about 12 GeV. At energies >10 GeV the neutrino flux is down almost by three orders of magnitude as compared to the region between 4 and 9 GeV (see Fig. 12). The mere observation of this very high energy events is therefore a significant fact: it shows that their cross section is enormous. This argument has been invoked to testify that "the inelastic cross section" still rises at energies >10 GeV. To me this seems not necessarily a good point. It is not excluded that these events have nothing to do with normal inelastic pion-producing processes. As a matter of fact, even at the lower energies some of the many pion events could be due to production and pionic decays of intermediate bosons.

4.5 Strange particle production

Strictly speaking, this item goes beyond what I have promised to present. But I think it is important, therefore I am going to mention it at least. Strange particle production has been observed in the bubble chamber, namely 1 case of $\Lambda^0 + K$, one single K^0, and one single Λ. In all of these events the total visible energy was >4 GeV. It is not excluded that *all* of these events were due to associated production via the *strong* interaction; the high visible energy makes this indeed very probable. There is therefore no positive indication of an *elastic* weak reaction with change of strangeness, like:

$$\nu_\mu + \mathrm{n} \rightarrow \mu^- + \Sigma^+ \tag{11}$$

and this is a very significant result. This reaction is characterized by $\Delta S = -\Delta Q$, and it's absence pleases people like me very much.

But what does this absence mean in numbers? Well, let us be very naive, and let us multiply observed total number of strangeness conserving elastic reactions by the suppression factor known from leptonic decays of hyperons; (this is what CABIBBO [28] calls $\sin^2 \Theta$). We would have expected about half a dozen of reactions of type (11), whereas none was seen. To me this is already a good indication that nature does not like $\Delta S = -\Delta Q$ transitions. I know that the naive argument needs refinement; one has to put in form factors and the like. The final answer will come with our anti-neutrino-run. There the $\Delta S = +\Delta Q$ reactions

$$\bar{\nu}_\mu + p \rightarrow \mu^+ + \Lambda^0 \tag{12}$$

$$\bar{\nu}_\mu + \mathrm{n} \rightarrow \mu^+ + \Sigma^- \tag{13}$$

have to occur, and then the significance of the absence of reaction (11) can be assessed in numbers. Needless to say that the study of strangeness-changing neutrino interaction is of extreme interest also beyond the question of the $\Delta S = \Delta Q$-rule. This has been particularly stressed by M. BLOCK [29].

5. The search for the intermediate boson

5.1 The production of W's by high energy neutrinos

It has been speculated how nice it would be if the weak interaction, (say between ($\mu\,\nu_\mu$) and ($e\,\nu_e$), i. e. μ-decay) would be transmitted, in complete analogy with electrodynamics, by an intermediate boson W [30].

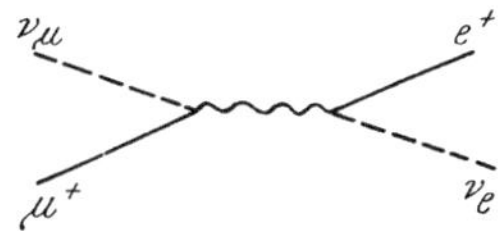

As has been emphasized by several authors [31], the cleanest way to produce them would be by high energy neutrinos, in a third order process analogous to bremsstrahlung

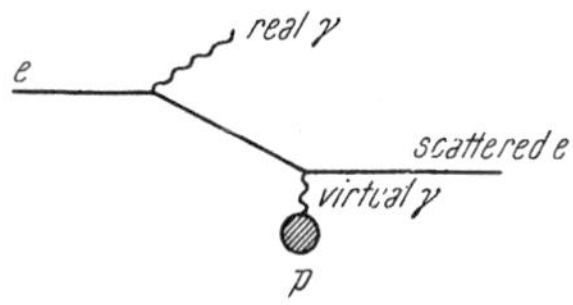

by weak bremsstrahlung

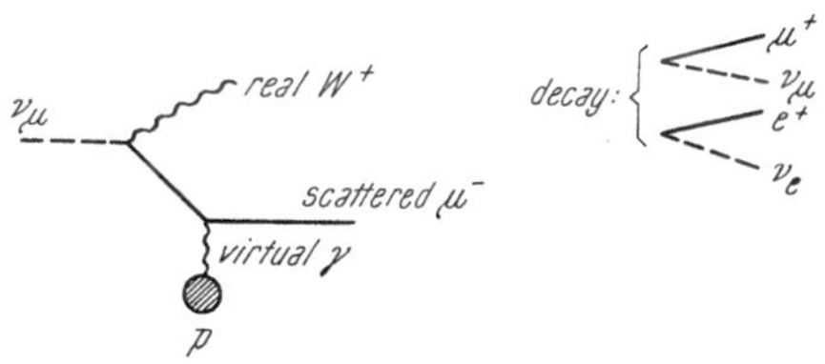

A signature of boson production by ν_μ and its subsequent decay would be a μ^- from the production stage, together with either a μ^+ or an e^+ from the decay. Note that the cross section is proportional to the weak four-fermion coupling constant G rather than G^2 (as the usual weak processes). It is therefore large, typically 10^{-37} cm^2 — provided one is far enough above the production threshold.

5.2 The search for muon pairs

a) The search for muon pairs $\mu^+\mu^-$ encounters a terrific background from reactions which can also give two tracks — even with the correct sign of charge. We list them in order of their importance:

$$\mu^-\pi^+, \mu^-p, \mu^-K^+.$$

Note also that in the bubble chamber a fast π^+ is easily confused with a μ^+. The discrimination has to be based on nuclear interactions and for this one has to use amounts of track length, which present day bubble chambers cannot easily supply. Consequently, what I am going to describe now is spark chamber work.

b) Fortunately, the spectrum of π^+ from inelastic reactions is soft, as has been pointed out, presumably because of the influence of the 3-3-resonance (see Fig. 31). We obtain therefore an already appreciably purified sample, if we require: *2 tracks*, and track length $l > 150$ g/cm² brass equivalent (gBE)* for the longer track, and $l > 80$ gBE for the shorter one. Any further track must have $l < 50$ gBE and there are no showers. An example of an accepted event is given in Fig. 35. In order to be able to analyse interactions, it was also required that both tracks have a projected angle with respect to the neutrino direction of $< 45°$. Amongst the 3750 events, found in the thin-walled spark-chambers, 200 satisfied these criteria.

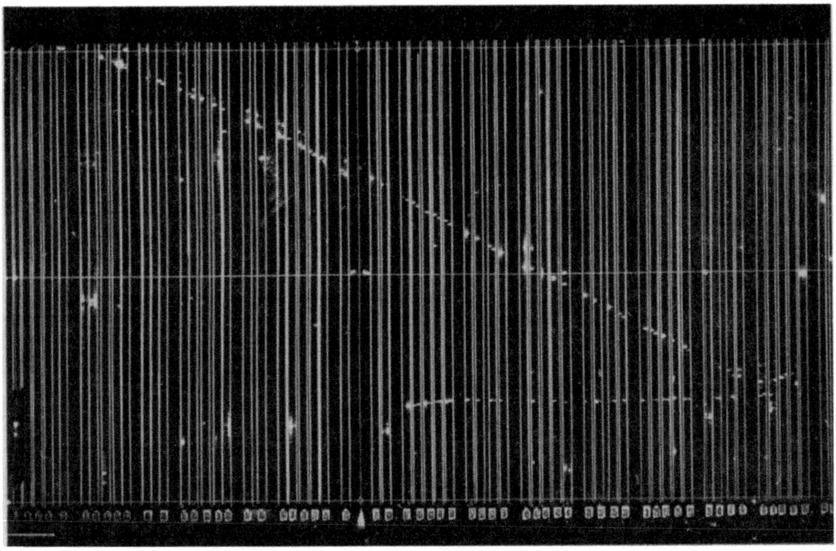

Fig. 35. Two-track event of the type used in the search for muon pairs produced in the spark chamber.

c) In order to decide, if this sample does contain muon pairs or not, we apply a method due to SOKRATES [32] (ca. 400 a.C.n.): *We assume* that there are *no muon pairs* in the sample, deduce the consequences and confront them with reality.

d) Having made explicitely this assumption, namely that each of the 2-track events contains at least 1 strongly interacting particle, we

* Since we want to treat nuclear interactions we scale track length (g/cm²) with $A^{2/3}$. Note that the geometrical interaction length Λ_0 is 80 g/cm² for Al, 105 g/cm² for brass, and 150 g/cm² for Pb. Important: All measured ranges are understood (in this paragraph 5.2) as projected on the neutrino direction. Average true ranges were obtained later in the analysis by dividing by $\overline{\cos\Theta}$.

expect to see interactions. In order to make this an operational statement, we must define what an interaction is. We do it as follows. A visible interaction is:

1. A scattering by $\geq 10^\circ$ in any of the two projected views we photograph. For the definition of a track before and after the scattering we require at least 4 sparks in $3^1/_2$ subsequent 3-plate-chambers.

2. A star: defined as ≥ 2 tracks (or showers) originating from the original one. Here we require ≥ 3 sparks per track.

e) In order to be able to forecast the number of interactions expected in our sample, we have to know the interaction length for the interactions as defined above. Since we do observe only part of all interactions, we cannot take total cross sections from the literature, but have to perform a calibration. This was done by shooting π's, p's and K's of known momentum into assemblies of our chambers, identical to the ones used in the actual experiment, measuring the track length l_i of every test particle, and counting the number of interactions observed (by definition either 0 or 1):

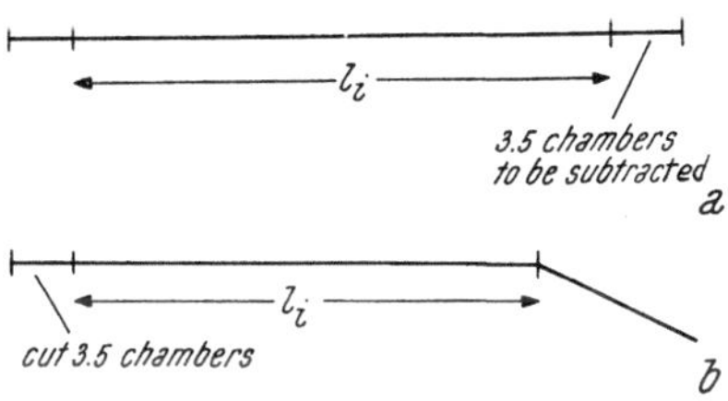

Fig. 36. Collection of track length in the interaction analysis for: a) non-interaction, b) interacting particles.

In case there is no interaction (Fig. 36a) there is no doubt how to count track length; one cuts on both ends the $3^1/_2$ chambers required for the recognition of a track, and takes what is left in the middle.

If there is an interaction (Fig. 36b), we cut again at the beginning — and then count up to the point of interaction.

The effective interaction-length is by definition: total track length divided by the total number of interactions observed, N:

$$\Lambda = \frac{\Sigma\, l_i}{N}. \qquad (14)$$

One easily convinces oneself that eq. (14), with our convention to count track-length, does give *exactly* what everybody would call a mean-free path, or interaction length, of an exponential absorption law.

The result of the calibration with pions and protons of different momenta is given in Fig. 37. For K's of 1 GeV/c we have only an upper limit for Λ of 350 gBE, since we had troubles with the muon contamination of the beam. We hope to square this out at the end of this week*.

* This has been done. Also the proton and pion calibrations have been considerably extended. Fig. 37 shows the state of affairs as of June 64. The conclusions as of February were not changed so far by the new calibrations. A small change in the number of predicted interactions (90 became 96) did not exceed the stated systematic uncertainties. But note that the measurements of Fig. 37 refer to Al or Al-brass mixture. There was no difference between the two, and we thought pure brass would differ from the curves shown by at most 10%. We better check this!

f) Collecting track length in the experiment.

The prescription, how to collect track length on single tracks, is absolutely clear. If we had chosen the same initial cut-off on both tracks there would be no difficulty in the experiment either. But since we have a different cut-off on long and short tracks respectively, we better are a bit careful. Let us look at the four patterns which we observe (Fig. 38):

1) If there is an interaction on the short track (Fig. 38a), the case is easy. We now know that the short track was due to a strongly interacting particle. It's track length, we are entitled to take, is between the initial cut-off and the interaction point. The initial cut-off is 50 gBE for a short track, because of the 80 gBE minimum required, we spend 30 gBE

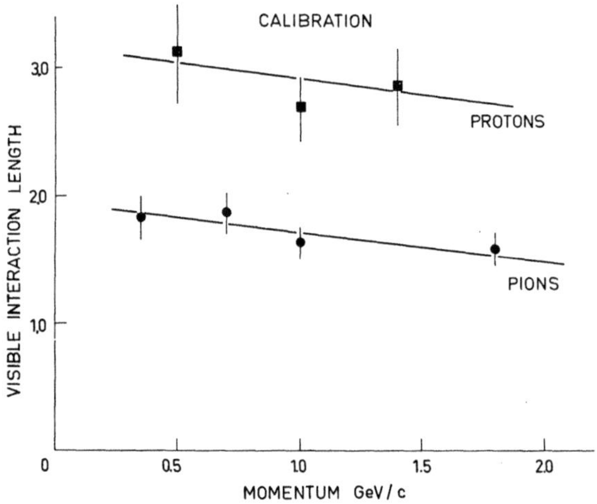

Fig. 37. Apparent interaction length Λ (in units of the geometrical interaction length $\Lambda_0 = 80\,g/cm^2$ Al or $105\,g/cm^2$ brass) for pions and protons. Preliminary results for the low-density region: Al or Al-brass mixture. Note added in proof: The final values differ only slightly from the ones shown. *But the π curve in pure brass lies appreciably above the curve shown* [33].

at the end and 50 gBE at the beginning. (This is our visibility criterion of 4 sparks in 7 gaps plus the desire to get away from the apex.)

2) In the complementary case, when the long track shows an interaction (Fig. 38b) we have to be careful. Let's assume first, the total track length of the short track exceeds 150 gBE. In this case we are entitled to cut only 50 gBE on the beginning of the long track. Why? — Well, we *know* it is a strongly interacting particle; and if it *had* interacted between 50 and 120 gBE we would have included it in the sample under category 1). We would have lost it, however, if $l_s < 150$ gBE, because then the event would not have fulfilled our acceptance criterion. Therefore, in this case (and only in this case) we have to start counting track length starting from 120 gBE, the initial cut-off on the long track.

3) If there are interactions on both tracks (Fig. 38c), we can do what we want. We chose to count on the short track, again between 50 gBE and interaction point. Cases like that are rare (only 8 out of 200). They are either neutron stars, or inelastic events where the muon was shorter than 50 gBE.

4) If there is no interaction at all (Fig. 38d) we got again to be careful: In this case we do not know which one was the strongly interacting

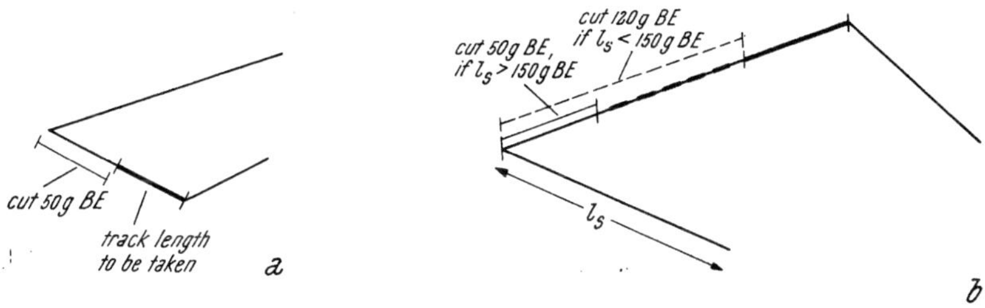

particle. We have therefore to try on both tracks; we subtract 150 (= 120 + 30) gBE from the total long track length, and 80 (= 50 + 30) gBE from the total short track length — and take the smaller of both. The number of interactions defined by the sum of these track lengths divided by the calibrated visible interaction length

$$N_{\text{exp}} = \frac{\Sigma l_i}{\Lambda} \qquad (14')$$

clearly gives a lower limit for the expected number of interactions in case our Sokratian assumption was correct. We made this counting in such a way that we either treat the track correctly, then this is not a limit but the right number, or we take less track length than what we had been entitled to take.

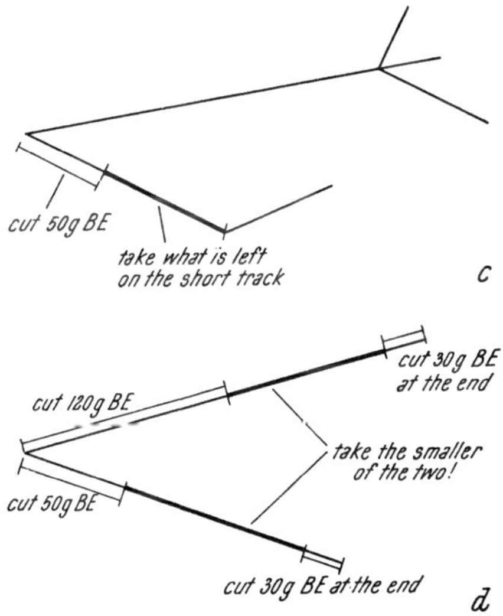

Fig. 38. Collection of track length for the events selected in the neutrino experiment.

g) Fraction of events due to protons or pions.

The bubble chamber gives the energy spectrum of protons from elastic reactions at low momentum transfers with reasonable accuracy. Let us remember that the four-momentum transfer squared q^2 is proportional to the proton kinetic energy:

$$-q^2 \approx 2M\,T_p \text{ [for (muon mass)}^2 = 0]. \qquad (15)$$

Fig. 28 gives therefore, apart from a scale factor, just the proton spectrum, we are interested in. For an extrapolation to higher proton energies, we take $F_A = F_V$ which fitted reasonably well at low q^2. With this distribution in hand we can calculate how much track length in a certain range interval of the shorter track is due to protons from elastic reactions. Let us assume, for the time being, that the rest of the track length is due to pions. (The 6% of π-track length leading to μ from decay-in-flight has been subtracted.)

h) We compare N_{exp} (lower limit) to N_{obs} as a function of the range of the shorter track. The result is seen in Fig. 39.

i) Integrated number: We predict $N_{\mathrm{exp}} = 96$ as the minimum number of interactions expected. But we predict also the variance expected

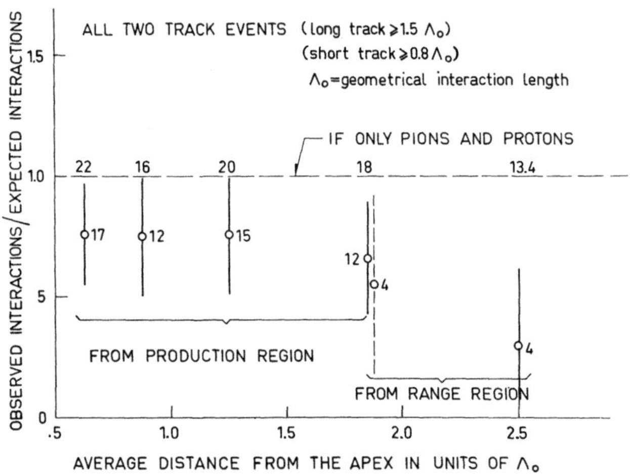

Fig. 39. Ratio of expected to observed number of interactions, as a function of the length of the analysed track. The expected number was derived from the calibration curves of Fig. 37; a possible contribution from K mesons was neglected. Note added in proof: *When one uses the new calibration curves for the brass part of the set-up, the deviations from* 1 *are no longer significant* [33].

on this number. (At numbers as large as that we can safely assume a Gaussian shape.) The variance comes mainly from the expected statistical fluctuation $\sqrt{N_{\mathrm{exp}}}$. But we have to include also the 5% uncertainty in calibration. The observed number of interactions is 60. According to statistical methods we get for the probability for this deviation 0.5%. Therefore the chance that this is a true observation is more than 99%. Thus *statistically* the effect is established.

k) Let us discuss systematic effects which can give us an effect like that. The K contamination is not known, but only K's >1 GeV would be dangerous. But to get $N_{\mathrm{obs}} = N_{\mathrm{exp}}$ one must assume that there are only protons and K's and no π's in the sample. Now a good fraction of the track length is in fact due to pions ($\gtrsim 40\%$), as we know from the bubble chamber data. Therefore our conclusion is, barring absolutely phantastic assumption about the ratio of K's and protons to pions, we have to consider the effect as real. To be sure: Even though we believe that there is no systematic flaw in our methods and calibrations, an interpretation of the observed effect in terms of muon-pairs does require a very detailed knowledge of the totality of neutrino interactions. For instance, we might have underestimated the amount of proton track length in our sample by not having considered inelastic processes at all. The argument, one may raise is, that a good part of the N^*-production processes may give us a high energy proton and a slow pion (which would have not made a track >50 gBE). Since protons interact much less visibly in the spark chamber than pions do, such an effect would have faked a discrepancy, in the direction observed, simply because we had more proton track length than we thought. In the publication, we are currently preparing, we shall discuss in more detail the effect, these different assumptions on the expected number of interactions has. Nevertheless, it would require extremely improbable assumptions to bring the observed deviation below 2.5 standard deviations.

l) What is the rate of μ-pairs if the observation is real? We make $N_{obs} = N_{exp}$ by mixing a certain amount of μ-track in (under the assumption that only pions and protons contribute otherwise). Then we conclude that our sample should contain 40 muon-pairs which is equal to 1% of all the neutrino events we observed.

5.3 *The search for muon-electron pairs*

a) We start again by selecting our sample. First we have to define what an electron is. I have shown examples of an electron (Fig. 26) and of a π° (Fig. 33). We make very simple criteria, which to a large extent sort out electrons from the π°'s. We require first that there is precisely one shower. The shower energy E_0 should be larger than 350 MeV in case the event was born in an aluminium spark chamber (A), and it should be larger than about 450 MeV in case it originated in one with a mixture of Al and brass (AB), or brass (B) alone. The reason for this differences is that because of the thickness of the brass plates one has there fewer sparks per MeV. In order to have a minimum number of sparks to play with we require the higher energy limit in case brass chambers are involved.

Secondly we required for a shower that $\gtrsim 90\%$ of all sparks lie within an opening angle of 15 degrees for (AB) or (B) and 9 degrees for (A). This serves to discriminate against showers made by π°'s, because the finite opening angle between the two decay gammas does usually give a shower pattern broader than that, even if the two showers merge together into one. Note that I did not take the arbitrary 90%-limit so dead serious; I never rejected a shower because it failed to reach the 90% by one or two sparks.

One may wonder why I did not require the obvious criterion that the shower starts right at the apex and is not converted at a distance.

The reason is simple: Most of the candidates are initiated in brass or the Al-brass mixture. There the conversion length is so short that most of the gammas convert anyway close to the apex. Even worse — also genuine electrons do show empty gaps close to the apex because of the robbing effect (in particular in the beginning of the run where the multiple particle efficiency was not yet at it's optimum). As a matter of fact, it was checked before the Siena Conference [5] that the distribution of empty gaps in the beginning of our selected e-type showers was identical to the one observed on *tracks* in 2- and 3-track events. We concluded from there that we would win nothing from rejecting showers with missing sparks; we might have thrown out some background, but would have certainly lost good events.

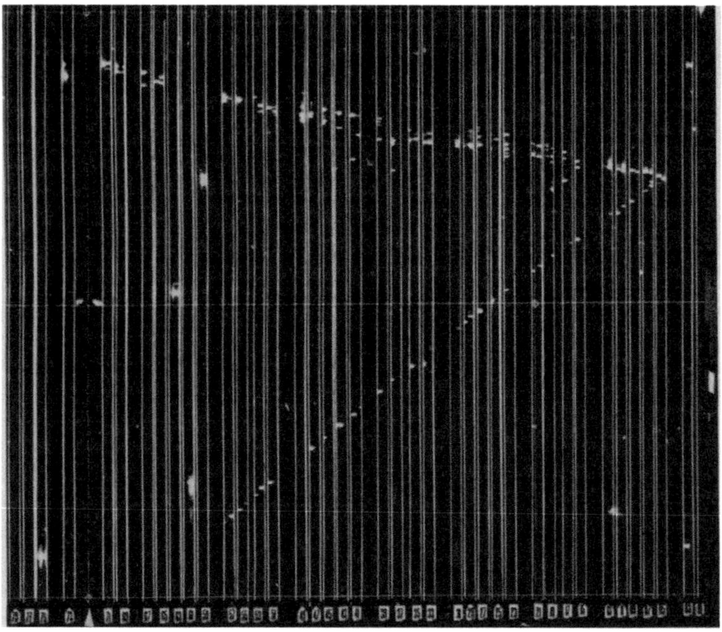

Fig. 40. A (μ e) candidate in the spark chamber.

After having defined what we call an electron, we require that it is accompanied by precisely one track, (defined by ≥ 4 sparks!). There are 41 events of this type in the sample of 3750 events. 8 of them have a (true) range <100 gBE, and are in fact compatible with being elastic ν_e-events. Since they give anyhow practically no useful track length, we keep only the 33 events with a (true) range >100 gBE (see Fig. 40).

b) The background associated with genuine electrons, is again composed of three classes. We list them in order of their importance:

$$e\pi^+, ep, eK^+.$$

They are all due to ν_e-processes, and this is a very attractive feature of the (μe) case. Here you have only the background of electron-type neutrinos (of the order of one or two percent) in contrast to the enormous background due to ν_μ for the $(\mu\mu)$ case. Therefore we start under much better conditions from the very beginning.

c) Now we make the same interaction analysis (according to SOKRATES) as described before. We have to subtract first the contribution from muons together with π^0's. Let's consider the case where both gammas of the π^0 are converted: We know the width of a shower initiated by a γ-ray of a given energy from calibration runs. You can calculate as a function of the π°-energy the fraction of the number of one shower to the total number of showers

$$f(E_p) = \frac{N_{1S}}{N_{1S} + N_{2S}}.$$

This is a computation which involves nothing but geometry and the widths of the shower which we take from calibration runs. This computation was done by FRANZINETTI [26]. Then we take the π^0-spectrum which we know from the bubble chamber data (the shape is the same as for the π^+), multiply the total number of π° present with this fraction and integrate over the whole spectrum. As a result we have in our sample after employing our selection criteria $\leq 10\%$ contribution from this sort of background. Because our chamber is short as compared to the potential ranges, the actual range of a particle is always determined by geometry. Therefore we have to substract $\approx 10\%$ of the track length.

This is not the only sort of background we have to worry about. There is a very unpleasant one of the type: muon together with *one* γ converting close to the apex. This 1γ-background may come from the following sources:

1. *A* π^0 with one γ escaping.
2. *A* π^0 with asymmetric decay.
3. Genuine 1γ-events due to internal Bremsstrahlung in $\nu_\mu \rightarrow \mu$ [34].

The analysis of these types of background is not yet finished. Preliminary estimates indicate that they may amount to another $(5-10)\%$ of the events.

Let me, with all due reserve, give the figures, I have at hand: The total track length would correspond to 29 interactions, if there were only pions involved. This should be approximately true, since the high energy proton background from elastic ν_e-reactions turns out to be very small. And as in the $(\mu\mu)$-case, it is very hard to see, from where a large contribution from high energy K-mesons should arise. Just to give you an impression, how large our effect is, let me take the 29 and substract the 10% estimated for $\mu\pi^0$. This gives 26 expected interactions, and if I take away another 10% because of the anticipated $(\mu\gamma)$-background I end up with 23. The observed number, however, is 6. — I cannot but calling this an effect. If this is true, the rate of (μe)-pairs, relative to all events, turns out to be about 0.8% (considering the selection biases involved). This would be in good agreement with the $(\mu\mu)$-pairs-rate estimated above.

Note added in writing down: In the mean-time, the analysis has been completed. The rate of inner bremsstrahlung was taken from theory [34]. Since it was obvious from the quoted numbers that the majority of the tracks show no interactions, doubt arose about the electron origin of the showers. Consequently the selection criteria imposed upon the showers were made more restrictive. As a result only half of the events survived. The new numbers are: 13 expected interactions (K's neglected) versus 3 seen. Note that total track length and number of observed interactions have been reduced by the same factor. Since an interacting track has to be associated with a true electron, this proves, at least to my mind, that the sharper acceptance criteria rejected good e-showers and γ-or π^0-showers with comparable efficiency. Therefore, I do not think we made the sample cleaner. But we have appeased our conscience.

Although we have still a 2.5 standard deviation effect, let me emphasize that again the magnitude of this effect does depend on our knowledge of the π°-spectrum and the like. There is no doubt that we have to get more data, notably in the aluminium chambers, where the distinction between e and γ is easy and clean.

5.4 What can we say about the boson mass?

a) From the rates. If one takes the measured rate of about 1% of the total events as real, one may compare it with what one would have expected believing the theoretical production cross section [31]. If we fold the neutrino spectrum into the rapidly raising cross section, which one can compute for any assumed boson mass, one finds that the rate has a sort of peak in the region between 6 and 7 GeV ν-energy. Based on a now obsolete neutrino spectrum [9] there was the suggestion [1, 35] that the mass should be $\gtrsim 1.3$ GeV. This implied of course assuming something about the relative decay-rate into lepton pairs and other channels (I think at least 20% leptonic decays). In fact the mass should be rather like $\gtrsim 1.4$ GeV according to new cross sections and better spectra.

b) From the distribution of transverse momenta of e^+ or μ^+. The advantage of the (μe)-case is that we know which particle came from the decay and which from production. The more characteristic one is the one from the decay and so we are looking for the e.

The boson, if it is so heavy, has to go into the forward direction. So the distribution of tranverse momenta is quite peaked, if we neglect Fermi motion of the target proton, and can never exceed $M_W/2$. (The neutrino has the same mass, for our purposes, as the muon or electron, namely zero.) Now there are two effects to disturb it: first of all, the boson is polarized and secondly because of the Fermi motion. As a result one gets a smeared-out distribution which extends to quite high transverse momenta. If one looks at the distribution of transverse momenta of e's from our (μe)-sample one sees two peaks. The first peak is at 200 MeV/c, where the maximum of ordinary events is, and has certainly nothing to do with the boson. The bump at about 0.6 GeV/c has some resemblance to what you would have expected theoretically. One would infer from this transverse momentum distribution that the boson mass is around 1.33 GeV (with large errors). This would be not in disagreement with what we concluded using the rates. But note that the distribution is terribly biased because of our selection criteria! The same is true for the μ^+-candidates which have been observed in the magnet.

c) "Directly" from kinematics.

The whole kinematics is fixed if we observe μ^-, recoil, and the decay-lepton. For the neutrino you must only know the direction but not the energy. The trouble with this argument is that we have very little information about the recoiling nucleus. If the production were coherent, we could neglect the recoil as compared to the other momenta. But the boson, if it is heavy, has to be produced incoherently and so a good deal of information is contained in the nuclear recoil which we do not measure. As a matter of fact, the observed candidates do *not* fit coherent boson production. I confess that we tried to get something out about the mass, under the assumption of an incoherent production. This required some wild assumptions — but we found that about 2/3 of the (μe)- and measured $(\mu\mu)$-candidates would be compatible with this hypothesis assuming a boson mass of (1.4 ± 0.2) GeV. How significant this is, I really don't know.

5.5 Bubble chamber information about the boson

Being stuck with the spark chamber data at this stage, one might wonder what the bubble chamber has to say. As outlined above, we cannot expect anything about the $(\mu\mu)$-case, but an e^+ together with a μ^--candidate would not have been overlooked. If we believe the tentative spark chamber rate, we would have expected (1.5 ± 1.3) events to show up in the 1963-experiment. As a matter of fact, one candidate has been seen (Fig. 41). The trouble with the event is that the shower development starts so early (at point b). It is impossible to reject the hypotheses that the shower was actually due to a Dalitz-pair from a π^0, with the other gamma escaping observation. If you ask me, this is a pretty improbable explanation: The shower energy is 1.22 GeV, and the number of $\mu^- + \pi^0$, with $E_{\pi^0} \leq 1$ GeV is only about 1% of all events. Considering that the probability of a Dalitz-pair is $^1/_{80}$, and multiplying another factor of $^1/_4$ in for not having seen the other gamma, I come out with an a priori probability of $\approx 3 \times 10^{-5}$, i. e. to 6×10^{-3} expected events of this type. — But I agree with the more sceptically minded people that a single event of this type is not a *proof*.

Let us turn then to the other possibility the bubble chamber group have, namely seeing the boson via pionic decays. This they have tried very hard. The idea is the following: one takes all events with ≥ 2 pions. Then one marries together 2, 3, 4... k pions in each event, ($k \leq$ the total pion multiplicity n), but in such a way that the total charge of the k-pion system is $+1$, and evaluates their invariant mass. If there are W-boson decay products amongst the events, they must show up as a peak in the mass distribution. (Note that the argument is not affected by nuclear re-interactions, since the W would decay miles outside the nucleus.)

What do they see? — Well, nothing very exciting. But there is a

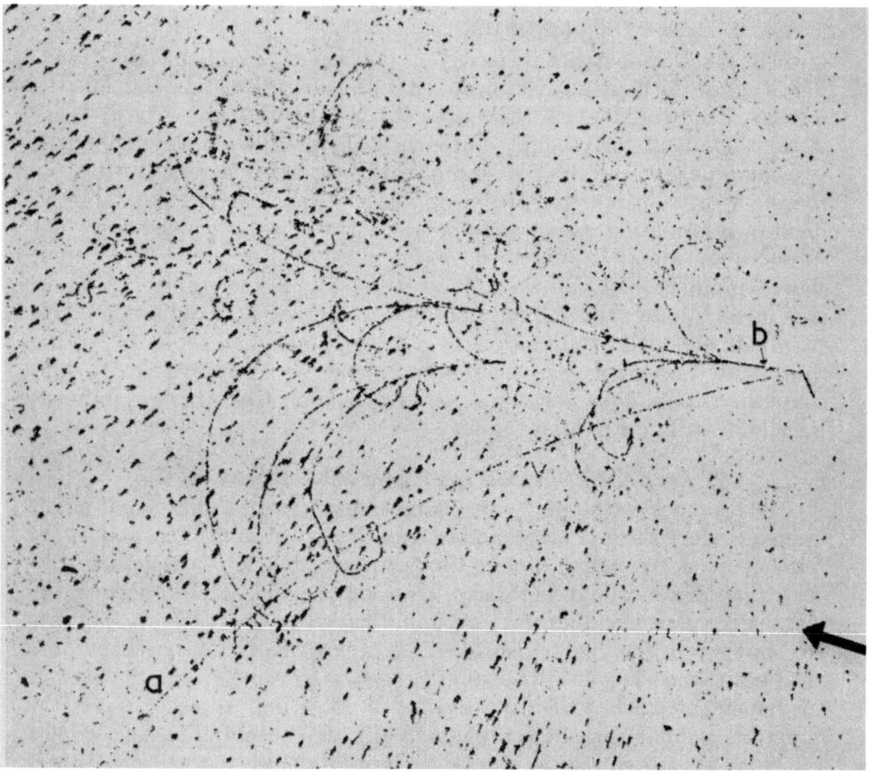

Fig. 41. A candidate for $(\mu^- e^+)$ in the bubble chamber. The most likely alternative interpretation is a π^- together with an e^+ from an inelastic $\bar{\nu}_e$ reaction; (the $\bar{\nu}_e$ flux is about 2×10^{-4} of the ν_μ flux!).

certain clustering of masses around 1.5 GeV. Perhaps one thing is significant: these masses around 1.5 GeV, seen in the charge $+1$ — state, seem to be less frequent in pion-combination with charge $+2$ or 0. Again, nobody is sure if this is an accident or an effect.

5.6 *Conclusions about the intermediate boson*

Let us try to summarize: As a result of the 1963-experiment the spark chamber has a small, but statistically significant effect, indicating the existence of lepton pairs $(\mu \bar{\mu})$ and $(\mu \bar{e})$, at a comparable rate of 0.5 to 1% each. No instrumental effect was found so far which could give this effect, but the interpretation in terms of lepton pairs does depend on our knowledge of the spectra of protons, π^0's, γ's and K's in ordinary neutrino interactions. The few kinematical tests, which could be made, are not inconsistent with part of the selected events being due to leptonic decays of a boson with a mass between 1.2 and 1.6 GeV. However, because of the systematic biases and uncertainties involved. they are far from being convincing.

The bubble chamber has one dubious $(\mu^- e^+)$-candidate. There are some suggestive features of many-pion events (none-the-least the rapid rise of their cross section with energy!), but there is no clear-cut evidence.

By the way, this is also true for 3 events observed in the spark chamber, which could be interpreted as pairs of electrons. There have already been speculations that they might represent boson production by electron neutrinos with subsequent electronical decay! But here the spark chamber is really at the end of it's capabilities.

The hardest fact is still the ≈ 2.5 standard deviation effect from the interaction analysis. As we always emphasized, it is indicative for lepton pairs — not necessarily for a weak boson. Could lepton pairs be produced without an intermediate boson? The answer is: Not with conventional processes at the rate under discussion. Within the framework of ordinary four-fermion-interaction the relatively most probable lepton-pair production process involves one four-fermion vertex and on top of that a virtual gamma ray, for instance

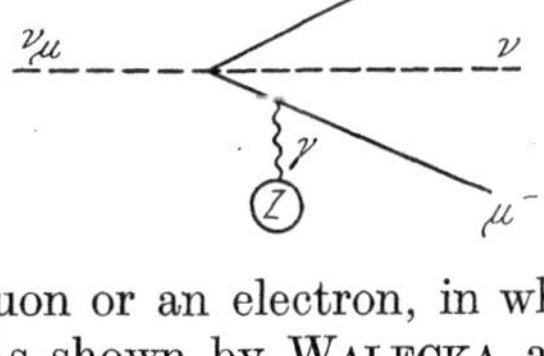

l^+ could be either a muon or an electron, in which case the process is inverse muon decay. As shown by WALECKA and CZYZ [36] the $(\mu \bar{e})$- and $(\mu \bar{\mu})$-rate expected from this process is orders of magnitudes down with respect to what we appear to see. One has to add principally new features to weak inter- actions, for instance a six-fermion-interaction [37] or other effectively many-fermion-interactions [38], in order to produce lepton pairs at the estimated rate. Of course, this is a possibility.

Note added in proof: The preliminary analysis of the 1964 experiment brought one decision: In the magnetized-iron spark chamber no μ^+ was found which had a momentum >1.2 GeV/c and was emitted together with a μ^--candidate with a momentum >500 MeV/c [33]. If one assumes that the boson decays at least to 25% into $\mu + \nu_\mu$, this places a lower limit to the boson mass of about 1.8 GeV. — No further $(\mu^- e^+)$-candidate was seen in the bubble chamber. And even worse: new calibration data obtained in brass spark chambers have weakened the case for lepton pairs in the 1963 experiment considerably. If some effect will persist — which then would have nothing to do with the boson — is not clear at present.

6. Some recent results at low energies in strangeness-conserving weak interactions

6.1 μ-capture in H_2

The last results were reported at the CERN conference in 1962. There were three independent experiments, one performed at Columbia employing counter techniques in liquid hydrogen [39] and two using bubble chambers performed at Chicago [40] and at CERN respectively [41]. The bubble chamber results are in good agreement with each other and in reasonable agreement with the predictions of the *V-A* theory.

The numbers of the capture rates are:
Counter $(515 \pm 85)\ \mathrm{sec}^{-1}$
B. C. $(425 \pm 60)\ \mathrm{sec}^{-1}$.

The theory predicts for capture in a state $\mu p \mu$ a rate of 560 and for a mixture of molecular and atomic capture 583. The reason of the remaining discrepancy is not clear. One tends to believe it is molecular physics.

6.2 μ-capture in He^3

The theoretical rate is $1450\ \mathrm{sec}^{-1}$ (WOLFENSTEIN [42]). The observed value [43], obtained by a technique which uses He^3 gas as target for stopping and as counter, is $(1520 \pm 20)\ \mathrm{sec}^{-1}$. This is in very good agreement theory. In particluar it is a test for the with value of the induced pseudoscalar coupling constant. The data can only be understood if you assume $g_P \approx 8\, g_A$. Incidentically a newly derived theoretical value (PETERMANN [44]) comes out to be a mere function of the masses of the particles involved (m_π, m_μ, M_N) and is 7.5.

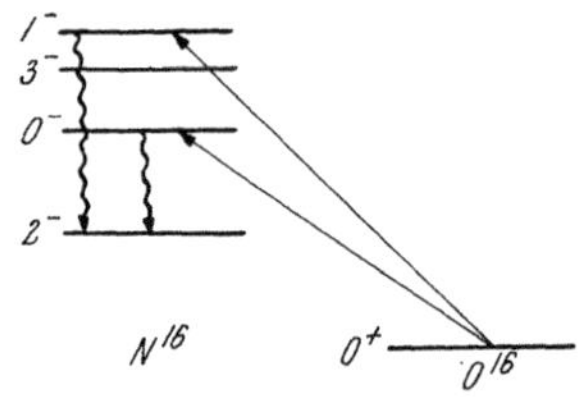

Fig. 42. Level scheme of 0^{16} and N^{16}.

6.3 μ-capture in O^{16}

This experiment is meant to determine the induced pseudoscalar coupling constant g_P in a very direct way. The μ-capture in one discrete level (from 0^+ to 0^-, see Fig. 42) must necessarily be pseudoscalar and so the differential capture rate is enormously sensitive to g_P. The experiment was carried out at Columbia [45] and also at Berkeley [46] and they measure the ratio of the emitted γ-rays from $0^- \to 2^-$ and $1^- \to 2^-$. The results for g_P/g_A are, unfortunately, a little bit confusing: Columbia reported first 16 and now 8 [47], Berkeley reports 2. The reasons for the discrepancy are unclear.

6.4 Dipole excitation

If you make a μ-capture you excite a sort of direct process, namely dipole oscillations in the nucleus quite similar to the one excited in the giant resonance with electromagnetic interaction. The process has been experimentally observed at CERN [48] and theoretically discussed [49].

6.5 Evidence that the conserved current is really conserved

The basic experiment was first done by comparison of the spectra of B^{12} and N^{12} which should show small deviations due to weak magnetism ($\approx$ a few permille, in contrast to the ν-experiment, where this effect is 20% of the total rate). The result was that the two spectra are different and deviate from each other by an amount compatible with GELL-MANN's prediction [50, 51].

A very precise measurement was done at CERN by HEINTZE, RUBBIA, SOERGEL and their coworkers [52] about the very rare decay mode of π^+

$$\pi^+ \to \pi^0 + e^+ + \nu_e$$

which is of the order 10^{-8} (this is just weak magnetism). The predicted value is by now almost precisely confirmed by the experiment.

6.6 Cabibbo [28]

explained in a very convincing way the low decay rate of the strange particles. The $\cos^2 \Theta$ gives, as you know, a correction to the comparison between the lifetime of the muon and the *ft*-value of 0^{14}. This lifetime of the muon has been remeasured with very great precision and agrees with the prediction of CABIBBO up to the third significant figure.

6.7 Heintze, Rubbia, Soergel et al. [53]

have also measured as a by-product the radiative decay of π^+

$$\pi^+ \to e^+ + \nu_e + \gamma.$$

The decay without γ is suppressed because of helicity reasons (a consequente of *V*-*A*), therefore it is down by a factor 10^{-4} with respect to the μ-mode. The graph for this process looks like

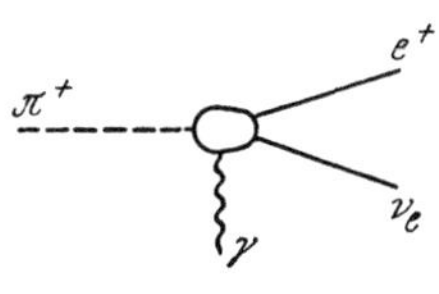

(13)

This is sort of an internal bremsstrahlung, and you get a very low rate if you calculate the process without any special assumptions about the structure in between the π and the e, ν. If you now assume a very special structure, i. e. insert a boson, you get a very substantical enhancement of this rate (there are factors of 10 or 15). There are, or course, other structures possible. Nevertheless HEINTZE, RUBBIA and SOERGEL extracted the boson mass from their observed rate

$$M_W = (1.4 \pm 0.3) \text{ GeV}$$

and this, I think, may be more than an accident.

Acknowledgements

Many people at CERN helped me to get this report together. I am particularly indebted to my friends in the neutrino bubble chamber group, M. BLOCK, C. FRANZINETTI and K. SCHULTZE, who supplied me with data. The subtle business of interaction analysis in the spark chamber I learned from my colleagues J. M. GAILLARD and B. HAHN. In understanding the problems involved the critical mind of G. v. DARDEL was of great value. About μ-capture I was instructed by N. LIPMAN and H. SENS. To all of them, and to the many others I have no space to mention, I extend my warmest thanks. Above all, however, I wish to thank Prof. P. URBAN for his kind invitation to Schladming. To him, to his co-workers, and to the publishers I want to express my gratitude for their painstaking work with this manuscript, and for their virtually unbounded patience.

References

1. RAMM, C. A., Proc. Internat. Conf. on the Fundamental Aspects of Weak Interactions, Brookhaven (1963), p. 91.
   VOSS, R. P. G., ibid, p. 114.
   FAISSNER, H., ibid, p. 137.
2. FAISSNER, H., Physikertagung Hamburg, Ed. E. BRÜCHE, (Physik-Verlag, Mosbach/Baden, 1963), p. 125.
3. PLASS, G., G. PLUYM, H. WACHSMUTH, and J. P. ZANASCO, Proc. Internat. Conf. on Elementary Particles, Sienna, Vol. I (1963), p. 523.
   GIESCH, M., S. VAN DER MEER, G. PLUYM, and K. M. VAHLBRUCH, ibid, p. 536.
4. BINGHAM, H. H., H. BURMEISTER, D. CUNDY, P. G. INNOCENTI, A. LECOURTOIS, R. MØLLERUD, G. MYATT, M. PATY, D. PERKINS, C. A. RAMM, K. SCHULTZE, H. SLETTEN, K. SOOP, R. P. G. VOSS and H. YOSHIKI, Proc. of the Internat. Conf. on Elementary Particles, Sienna, Vol. I (1963), p. 555.
5. BERNARDINI, G., G. VON DARDEL, P. EGLI, H. FAISSNER, F. FERRERO, C. FRANZINETTI, S. FUKUI, J.-M. GAILLARD, H. J. GERBER, B. HAHN, R. R. HILLIER, V. KAFTANOV, F. KRIENEN, M. REINHARZ, and R. A. SALMERON, ibid, p. 571.
   For technical details see H. FAISSNER, F. FERRERO, S. FUKUI, J.-M. GAILLARD, H. J. GERBER, B. HAHN, F. KRIENEN, G. MURATORI, M. REINHARZ, and R. A. SALMERON, ibid, p. 546.
6. The 1963 NPA Seminars: The Neutrino Experiment, Ed. C. FRANZINETTI, CERN-Report 63—37 (1963).
7. DANBY, G., J.-M. GAILLARD, K. GOULIANOS, L. M. LEDERMAN, N. MISTRY, M. SCHWARTZ, and J. STEINBERGER, Phys. Rev. Lett. 9 36 (1962), see also

DANBY, G., J.-M. GAILLARD, K. GOULIANOS, L. M. LEDERMAN, T. D. LEE, N. MISTRY, M. SCHWARTZ, and J. STEINBERGER, Phys. Rev. Lett. **10**, 260 (1963).
High energy neutrino experiments were initiated by B. PONTECORVO, Zh. Eksp. i. Teor. Fiz. **37**, 1751 (1959); translation: Soviet Phys., JETP **10**, 306 (1960).
SCHWARTZ, M., Phys. Rev. Lett. **4**, 306 (1960).
But there were several others thinking about them; see M. A. HARKOV "The Neutrino" Dubna Report D. 1269 (1963), p. 62 ff.
REINES, P., Ann. Rev. Nucl. Sci. **10** (1060), p. 16 ff. and ref. [19].

8. KUIPER, B., and G. PLASS, CERN-Report 59—30 (1959), see also ref. [6], p. 83.
9. VAN DER MEER, S., CERN-Report 61—7 (1961), see also ref. [6], p. 97.
10. BURMEISTER, H., G. VON DARDEL, and K. SCHULTZE, Proc. Internat. Conf. on Elementary Particles, Sienna, Vol. I 552 (1963).
11. KRIENEN, F., R. A. SALMERON, and J. STEINBERGER, intern. CERN-Report PS/int. EA 60—10 (1960).
12. FAISSNER, H., F. FERRERO, A. GHANI, E. HEER, F. KRIENEN, G. MURATORI, T. B. NOVEY, M. REINHARZ, and R. A. SALMERON, Nucl. Instr. & Meth. **20**, 213 (1963), see also
FAISSNER, H., in ref. [6], p. 43.
13. ALLEYN, E., H. FAISSNER, N. J. GERBER, E. HEER, and M. REINHARZ, CERN-Report 64—4 (1964).
14. FAISSNER, H., F. FERRERO, A. GHANI, and M. REINHARZ, Nucl. Instr. & Meth. **20**, 289 (1963).
15. CUCANČIĆ, L., and H. FAISSNER, CERN-Report (in preparation).
16. FEINBERG, G., F. GÜRSEY, and A. PAIS, Phys. Rev. Lett. **7**, 208 (1961).
BLUDMAN, S. A., UCRL 9667 (unpublished).
17. LIPMANOV, E. M., Soviet Phys. JETP **10**, 750 (1960).
BLUDMAN, S. A., Nuovo Cim. **27**, 751 (1963).
KABIR, P. K., Nuovo Cim. **28**, 165 (1963), see also
FRIEDBERG, R., Phys. Rev. **129**, 2298 (1962).
Lepton conservation was introduced by E. J. KONOPINSKI and H. M. MAHMOUD, Phys. Rev. **92**, 1045 (1953).
Remarkably enough they assigned already opposite lepton numbers to μ and e, getting thereby a μ-e selection rule.
18. FAISSNER, H., J. KJELLMANN, A. STAUDE, and T. ALVÄGER, Nuovo Cim. **32**, 782 (1964).
19. NISHIJIMA, K., Phys. Rev. **108**, 907 (1957).
20. LØVSETH, J., Phys. Lett. **5**, 199 (1963), ref. [6], p. 203, and private communication. The cross section for a free nucleon has been calculated by
LEE, T. D., and C. N. YANG, Phys. Rev. Lett. **4**, 307 (1960)
CABIBBO, N., and R. GATTO, Nuovo Cim. **15**, 304 (1960)
YAMAGUCHI, Y., Progr. Theoret. Phys. **23**, 1117 (1960) and CERN-Report 61—2 (1961).
21. BLOCK, M. M., H. BURMEISTER, D. C. CUNDY, B. EIBEN, G. FRANZINETTI, J. KEREN, R. MØLLERUD, G. MYATT, M. NIKOLIC, A. ORKIN-LECOUTOIS, M. PATY, D. PERKINS, C. A. RAMM, K. SCHULTZE, H. SLETTEN, K. SOOP, R. STUMP, W. VENNS, and H. YOSHIKI, Phys. Lett. **12**, 281 (1964),
22. BIENLEIN, J. K., A. BÖHM, G. VON DARDEL, H. FAISSNER, F. FERRERO, J. H. GAILLARD, H. J. GERBER, B. HAHN, V. KAFTANOV, F. KRIENEN.

M. Reinharz, R. A. Salmeron, R. G. Seiler, A. Staude, J. Stein, and H. J. Sterner, Phys. Lett. **13,** 80 (1964).

23. Dekkers, D., J. A. Geibel, R. Mermod, G. Weber, T. R. Willitts, K. Winter, B. Jordan, M. Vivargent, N. M. King, and E. J. N. Wilson, Internal CERN-Report NP/Int. 64—5 (1964) and Phys. Rev. to be published.
24. Block, M. M., Phys. Lett. (1964) to be published, and private communication.
25. Faissner, H., F. Ferrero, A. Ghani, F. Krienen, T. B. Novey, and M. Reinharz, Nucl. Instr. and Meth. **20,** 161 (1963).
26. Franzinetti, C., and H. J. Gerber, Internal CERN-Report NPA/Int. 64—10 (1964) unpublished.
27. Hegel, F. G. W., Sämtliche Werke, Ed. Georg Lasson, Philosophische Bibl. Felix Meiner, (Leipzig, 1905—1925), in particular Vol. 171a—171d.
28. Cabibbo, N., Phys. Rev. Lett. **10,** 531 (1963).
29. Block, M. M., Phys. Rev. Lett. **12,** 262 (1964).
30. Schwinger, J., Ann. of Physics (N. Y.) **2,** 407 (1957).
Lee, T. D., and C. N. Yang, Phys. Rev. **119,** 1410 (1960), and several others.
31. Lee, T. D., P. Markstein, and C. N. Yang, Phys. Rev. Lett. **7,** 429 (1961);
Bell, J. S., and M. Veltmann, Phys. Lett. **5,** 94 and 151 (1963);
Veltmann, M., Physica **29,** 161 (1963), also ref. [6], p. 21 and 179, and private communication;
Wu, A. C. T., C. P. Yang, K. Fuchel, and S. Heller, Phys. Rev. Lett. **12,** 57 (1964), and many others.
32. Sokrates in several Platonian Dialogues (see Platon, Collected Papers).
33. Bernardini, G., J. K. Bienlein, G. von Dardel, H. Faissner, F. Ferrero, J.-M. Gaillard, H. J. Gerber, B. Hahn, V. Kaftanov, F. Krienen, C. Manfredotti, M. Reinharz, and R. A. Salmeron, Phys. Lett. **13,** 86 (1964).
34. Løvseth, J., and J. D. Walecka, CERN-preprint (1964).
35. Veltman, M., Proc. Internat. Conf. on the Fundamental Aspects of Weak Interactions, Brookhaven (1963), p. 160, and private communication.
36. Czyz, W., and J. D. Walecka, Phys. Lett. **8,** 77 (1964).
Czyz, W., G. C. Sheppey, and J. D. Walecka, Nuovo Cim. **34,** 404 (1964).
37. Ericson, T., and S. L. Glashow, unpublished.
38. Kummer, W., CERN-preprint (1964).
39. Bleser, E., L. Lederman, J. Rosen, J. Rothberg, and E. Zavattini, Phys. Rev. Lett. **8,** 288 (1962).
40. Hildebrand, R. H., Phys. Rev. Lett. **8,** 34 (1962).
Hildebrand, R. H., and J. H. Doede, Proc. 1962 Internat. Conf. High-Energy Physics CERN (Genève 1962), p. 418.
41. Bertolini, E., A. Citron, G. Gialanella, S. Focardi, A. Mukhin, C. Rubbia, and S. Saporetti, ibid, p. 421.
42. Wolfenstein, L., ibid, p. 821, see also
Yano, A. F., Phys. Rev. Lett. **12,** 110 (1964).

43. AUERBACH, L. B., R. J. ESTERLING, R. E. HILL, D. A. JENKINS, J. T. LACH, and N. H. LIPMAN, Phys. Rev. Lett. **11,** 23 (1963); see also FALOMKIN, I. V., A. I. FILIPPOV, M. M. KULYUKIN, B. PONTECORVO, YU. A. SCHERBAKOV, R. M. SULYAEV, V. M. TSUPO-SITNIKOV, and O. A. ZAIMIDOROGA, Phys. Lett. **3,** 229 (1963); (Both groups measured the reaction $\mu^- + {}^3\mathrm{He} \to {}^3\mathrm{H} + \nu$.) The total capture rate was measured by
ZAIMIDOROGA, O. A., M. M. KULYUKIN, B. PONTECORVO, R. M. SULYAEV, I. V. FALOMKIN, A. I. FILLIPOV, V. M. TSUPKO-SITNIKOV, and YU. A. SCHERBAKOV, Phys. Lett. **6,** 100 (1963).
44. PETERMANN, A., to be published.
45. COHEN, R. H., S. DEVONS, and A. D. KANARIS, Phys. Rev. Lett. **11,** 134 (1963).
46. A. Berkeley group, unpublished rumors.
47. Private communication from Dr. J. C. SENS.
48. BARLOW, J., J. C. SENS, P. J. DUKE, and M. A. R. KEMP, Phys. Rev. Lett. **9,** 84 (1964), Erratum ibid, p. 318.
49. FOLDY, L., and J. D. WALECKA, CERN preprint.
50. MAYER-KUCKUK, TH., and F. C. MICHEL, Phys. Rev. Lett. **7,** 167 (1961).
51. LEE, Y. K., L. W. MO, and C. S. WU, Phys. Rev. Lett. **10,** 253 (1963).
52. DEPOMMIER, P., J. HEINTZE, A. MUHKIN, C. RUBBIA, V. SOERGEL, and K. WINTER, Phys. Lett. **2,** 23 (1962).
DEPOMMIER, P., J. HEINTZE, C. RUBBIA, and V. SOERGEL, Phys. Lett. **5,** 61 (1963).
53. DEPOMMIER, P., J. HEINTZE, C. RUBBIA, and V. SOERGEL, Phys. Lett. **7,** 285 (1963).

GPSR Compliance

The European Union's (EU) General Product Safety Regulation (GPSR) is a set of rules that requires consumer products to be safe and our obligations to ensure this.

If you have any concerns about our products, you can contact us on ProductSafety@springernature.com

In case Publisher is established outside the EU, the EU authorized representative is:

Springer Nature Customer Service Center GmbH
Europaplatz 3
69115 Heidelberg, Germany

Batch number: 10577896

Printed by Printforce, the Netherlands